Java
高并发编程详解
多线程与架构设计

Java Concurrency Programming
Multithreading and Architecture

汪文君 编著

图书在版编目（CIP）数据

Java 高并发编程详解：多线程与架构设计 / 汪文君著．—北京：机械工业出版社，2018.5
（2023.1 重印）
（Java 核心技术系列）

ISBN 978-7-111-59993-7

I. J⋯　II. 汪⋯　III. JAVA 语言 - 程序设计　IV. TP312.8

中国版本图书馆 CIP 数据核字（2018）第 095749 号

Java 高并发编程详解：多线程与架构设计

出版发行：机械工业出版社（北京市西城区百万庄大街 22 号　邮政编码：100037）
责任编辑：陈佳媛　　　　　　　　　　　责任校对：殷　虹
印　　刷：北京建宏印刷有限公司
版　　次：2023 年 1 月第 1 版第 9 次印刷
开　　本：186mm×240mm　1/16
印　　张：25
书　　号：ISBN 978-7-111-59993-7
定　　价：89.00 元

客服电话：(010) 88361066　68326294

版权所有・侵权必究
封底无防伪标均为盗版

Foreword 推荐序一

首先恭喜汪文君老师终于出书了，可喜可贺！汪文君老师一直是我敬佩和学习的楷模。十年之前，我在公司认识了新来的长发少年汪文君同学，至今依然记得文君他对人热情，对于工作、生活中接触的各种事物都充满了好奇心，总是在热情高涨地学习新技术，他每天晚上都会拿出时间学习，很多节假日也都抽出专门的时间来学习和编程。后来听说汪文君老师在电信、医疗、金融等多个行业从事架构设计、技术指导、编程等工作，经验非常丰富。其间还进行过创业，至今依然对架构设计、编程充满了热情，持续学习，持续成长，不仅仅自己学习实践，还录制了众多的视频传播技术与经验，根据自己的经历与心得进一步积累成书，是技术从业者中的佼佼者。

编程技术作为 IT 领域的关键技术，正在加速影响着越来越多行业的信息革命。IT 技术不仅仅引发了众多世界性的产品创新和技术革命，同时也引发了众多行业的变革，随着"互联网+"等的驱动，传统行业也正在加速技术革命带来的技术升级与产业升级。IT 技术正在加速改变我们的生活方式、沟通方式、学习方式、思维模式，涉及我们工作、学习、生活的方方面面，技术革新的力量成为了驱动经济变革与增长的最有效的引擎之一。

Java 技术自 1990 年由 James Gosling、Mike Sheridan、Sun 首席科学家 Bill Joy 等创建以来，在企业级应用、互联网应用、移动互联网应用等开发领域成为独一无二的霸主级语言，长盛不衰，形成了巨无霸的生态系统；其不仅仅是从业者的工具，也是学习深究的对象，而且一直都在不断地演进和重生。Java 创建时的宣言"一次编译、到处执行"（write once, run anywhere）将 Java 带给了所有的编程者。下面就来简单回顾一下 Java 的重大历程。

1994 年，"Java"之名正式诞生，Java 1.0a 版本开始提供下载。

1996 年 1 月，第一个 JDK——JDK1.0 诞生。

1997 年 2 月 18 日，JDK1.1 发布。

1998年12月8日，Java 2企业平台J2EE发布。

1999年6月，SUN公司发布Java的三个版本：标准版（J2SE）、企业版（J2EE）和微型版（J2ME）。

2004年9月30日18:00时，J2SE1.5发布，成为Java语言发展史上的又一里程碑。为了表示该版本的重要性，J2SE1.5更名为Java SE 5.0。

2005年6月，JavaOne大会召开，Sun公司公开Java SE 6。此时，Java的各种版本已经更名，以取消其中的数字"2"：J2EE更名为Java EE，J2SE更名为Java SE，J2ME更名为Java ME。

2006年12月，Sun公司发布JRE6.0。

2009年4月20日，Oracle公司以每股9.50美元，74亿美元的总额收购Sun公司。

2009年12月，Sun公司发布Java EE 6。

2011年7月28日，Oracle公司发布J2SE 7.0。

2014年3月18日，Oracle公司发布Java 8发行版（OTN）。

2017年9月21日，Oracle公司正式发布Java 9。

本书围绕Java编程中多线程编程的基础与应用设计分为四个部分来展开讲解，包括多线程技术知识、Java ClassLoader、深入理解volatile关键字、多线程设计架构模式。Java编程语言是工业级的编程语言，在诸多应用、诸多场景下被广泛使用，多线程技术作为Java语言和应用的基础能力，对其的学习、理解和掌握，不仅仅能够提升我们的技能，更能作为我们更好地理解面向对象编程、并发编程、高性能编程、分布式编程的入口，进一步还会涉及操作系统线程模型、JVM线程模型、应用场景优化。通过学习本书，我们能够更全面地拓展自己的编程能力，更进一步地充实编程设计和架构设计的系统性思维。

为了更好地运用Java编程，让我们从掌握多线程技术的知识点开始出发！

东软集团移动互联网事业部首席技术官　徐景辉

Foreword 推荐序二

汪文君是我们软件开发团队里特别有朝气的一员，平时不管是工作还是业余时间，其都会对软件开发遇到的难题、专题进行不折不挠的攻关研究，他是团队所有人心目中的技术大牛。我作为部门主管，从三个月前得知他开始写这本书时，就特别期待，之前他利用业余时间录制网上视频供大家学习与交流，现在他把这些知识再加以整理出版成书，相信他的所作所为能让许多人受益。

Java应该是目前使用者最多、应用场景最广的软件开发技术之一，这与Java技术本身的一个重要优点紧密相关。Java通过虚拟机技术隐藏了底层的复杂性以及机器和操作系统的差异性，使得程序员无须过多地关注底层技术，而是把精力集中在如何实现业务逻辑之上。在许多情况下，一般程序员只要了解了必要的Java API和语法，就已经能够满足日常开发的需要了。

然而，在企业级的软件开发场景里，要求远远不止如此简单。比方说，我们所在的金融领域，日常维护着上千个节点PB数据量级别的物理集群，对程序的并发性、稳定性和安全性都有极高的要求。在这样的情况下，高性能的物理硬件固然是基础，然而如果开发人员不了解与Java高并发相关的技术原理，就无法写出最优化的代码。

市面上Java相关的书籍，大多比较适合初学者，只涵盖基础内容，并不多见那种深入某个高级主题并富有思想性的专题书籍。虽然本书对读者的Java基础有一定的要求，但本书胜在内容丰富，讲解深入浅出，相信对于这个专题有兴趣的读者一定不会感到乏味和艰涩。作者在今日头条发布的《心蓝说Java》视频，也一定能让读者更容易深入了解这个课题。

Head of Data Engineering at HSBC　　Winne Chen

推荐序三 Foreword

我与汪文君共事过几年，我们负责的是全球性交易系统，对系统性能的要求极高，例如在高并发情况下，如何快速响应用户请求的同时又要保证数据的完整性，这就要求团队的技术人员需要有极高的专业素养。汪文君是团队里的技术骨干，在我们共事的日子里，汪文君解决了很多技术难题，更难能可贵的是，汪文君在设计阶段就敏捷地洞察到系统可能会出现的瓶颈并且提供解决方案，而且他还将每一个点子、每一个难题的解决之道激情地分享给团队的所有成员。

汪文君告诉我他要出版一本Java多线程方面的书，并且给我发来了一些章节，我阅读后觉得非常棒。这是一个逻辑性很强的技术牛人在有逻辑、有层次地展开Java多线程的话题，他能将每一个Java多线程相关的话题都讲解得很有深度。在整本书的构想方面，各个章节的内聚性都很强，章节与章节之间又是层层推进的关系，再加以精准的图示辅助理解，让读者阅读起来感觉非常舒服。

看汪文君的书，就如同听一个久经沙场的战士讲解如何玩转枪械，本书中所讲解的都是一个个的实战运用，对提升Java内功必然大有裨益。

Development manager at HSBC Global Banking and Markets　Bonson Zheng

Foreword 推荐序四

  Alex has spent years passionately promoting, mentoring and collaborating with the technology community. Whether this has been answering questions or creating tutorials for new and exciting technologies he always brings his keen intellect to bear and makes complex subjects both easy to understand and enjoyable to learn. In this book Alex distills all these years of passion and knowledge into a comprehensive book that covers everything you need to know about Java taking you from the basics all the way through to the most complex parts of the language. The result is a book that everyone working with Java should add to their bookshelf and will be a future classic text.（Alex 花了很多年的时间积极热情地推动着技术社区的发展，并为其做出贡献，无论是回答初学者的问题还是为最新的技术编著入门教程，他总是带着敏锐和智慧试图让一切复杂晦涩的技术内容变得通俗易懂，系统又有条理。在 Alex 的这本书中，他将这些年付诸的激情与知识提炼成册，涵盖了你所需要了解的有关 Java 多线程的大部分知识，从基础知识到最复杂的内容，他都做到了通俗易懂地娓娓道来，每一位从事 Java 开发的从业者都应该将这本书添加到他们的书架上，这本书将成为比较经典的文本资料。）

Chief Technology Officer at Octagon Strategy Ltd Andrew Davidson

前 言 Preface

为什么写这本书

从大学毕业到现在已有 11 年的时间，在这 11 年中，我一直保持着一个习惯，将工作中用到的技术梳理成系统性的文档。在项目的开发过程中，由于时间的紧迫性，我们对某个技术领域知识的掌握往往都是比较碎片化的，系统化地串联知识碎片，不仅能加深对该技术的掌握，还能方便日后快速查阅，帮助记忆。截至目前，我已经在互联网上发布了大约 12 本电子书，主要是围绕着项目构建工具、Java 开发的相关技术、NoSQL、J2EE 等内容展开的。

2017 年年初，很多人看过我写的《Java 8 之 Stream》电子书之后，给我发邮件，希望我写一本能够涵盖 Java 8 所有新特性的电子书。最开始一两个人这样提议的时候，我并没有在意，后来越来越多的朋友都有类似的需求，由于写电子书需要花费很长的时间，于是我尝试着将 Java 8 新特性录制成视频教程，大概花了一个月的业余时间我录制了 40 集《汪文君 Java8 实战视频》，视频一经推出收获了非常多的好评，所幸大家都比较喜欢我的授课风格，在过去的 2017 年，我一口气录制了 11 套视频教程，超过 400 集（每集都在 30 分钟左右），当然也包括 Java 高并发相关的内容。

在我的计划中，关于 Java 高并发的内容将会发布 4 套视频教程，分别是：第一阶段（Java 多线程基础），第二阶段（Java 内存模型，高并发设计模式），第三阶段（Java 并发包 JUC），第四阶段（Java 并发包源码剖析 AQS）。其中三个阶段都已经发布了，在今日头条《心蓝说 Java》中累计播放时长超过 20 万分钟，百度云盘下载量也超过了 30 万余次。由于内容太多，本书只涵盖了前两个阶段的内容，经过了数以万计读者对视频教程问题的指正，本书的写作相对比较顺利，本书内容不仅修复了视频讲解中无法修复的缺陷，而且还加入了我对 Java 高并发更深一层的领悟和体会。

本书是我第一本正式出版的书稿，关于本书的写作可以说是一次偶然的机缘巧合，在 2017 年 9 月初，机械工业出版社的策划编辑 Lisa 找到了我，她觉得我的视频内容比较系统，非常适合以书稿的形式发表，我们简单交流之后就快速敲定了本书内容的主体结构，围绕着高并发视频教程的前两个阶段展开，在今年我也会努力将高并发后两个阶段的内容编著成书，使之尽快与读者见面。

读者对象

- 计算机相关专业的在校学生。
- 从事 Java 语言的开发者。
- 从事 Java 系统架构的架构师。
- 使用 Java 作为开发语言的公司与集体。
- 开设 Java 课程的专业院校。
- 开设 Java 课程的培训机构。

如何阅读本书

本书主要分为四部分，其中，第一部分主要阐述 Thread 的基础知识，详细介绍线程的 API 使用、线程安全、线程间数据通信以及如何保护共享资源等内容，它是深入学习多线程内容的基础。

在第二部分中之所以引入 ClassLoader，是因为 ClassLoader 与线程不无关系，我们可以通过 synchronized 关键字，或者 Lock 等显式锁的方式在代码的编写阶段对共享资源进行数据一致性保护，那么一个 Class 在完成初始化的整个过程到最后在方法区（JDK8 以后在元数据空间）其数据结构是怎样确保数据一致性的呢？这就需要对 ClassLoader 有一个比较全面的认识和了解。

在本书的第三部分中，我用了三章的篇幅来详细、深入地介绍 volatile 关键字的语义，volatile 关键字在 Java 中非常重要，可以说它奠定了 Java 核心并发包的高效运行，在这一部分中，我们通过实例展示了如何使用 volatile 关键字并非常详细地介绍了 Java 内存模型等知识。

本书的最后一部分也就是第四部分，站在程序架构设计的角度讲解如何设计高效灵活的多线程应用程序，第四部分应该是内容最多的一部分，总共包含了 15 章。

勘误和支持

由于作者的水平有限，编写的时间也很仓促，书中难免会出现一些错误或者不准确的地方，恳请读者批评指正。如果在阅读的过程中发现任何问题都欢迎将您宝贵的意见发送到我的个人邮箱 532500648@qq.com，我会专门在我的今日头条《心蓝说 Java》开设专栏，用于修订书中出现的错误和不妥的地方，我真挚地期待着您的建议和反馈。

致谢

首先要感谢我的父亲，在我很小的时候，他就教育我做任何事情都要脚踏实地，一步一个脚印，做人不能浮躁，任何事情都不是一蹴而就的，这也致使我在遇到发展瓶颈的时候总能够耐得住性子寻求突破。

在本书最后一部分编写的过程中，我的妻子经历了十月怀胎为我生下了一对龙凤胎汪子敬、汪子兮兄妹，他俩的到来让我感觉到了初为人父的激动与喜悦，更加体会到了为人父母的不容易，感谢我的妻子，多谢你的支持和理解，本书的出版应该有一半你的功劳。

我还要感谢在我一路成长过程中带给我很多帮助的同事及朋友——徐景辉、Andrew Davidson、Bonson、Winne、Wilson、龙含等，在本书还是草稿阶段的时候，你们就给了我很多建设性的意见和建议。

当然也不能忘了感谢本书的策划编辑 Lisa 老师，是你直接促成了本书的诞生，在过去的半年多里，你反复不断地帮我审稿，修改错别字，调整不通顺的语句，你的专业水准和敬业精神帮助我最终顺利完稿。

最后一定要感谢我所在的研发团队——GBDS 的 Jack、Eachur、Jenny、Sebastian、Yuki、Kiki、Dillon、Gavin、Wendy、Josson、Echo、Ivy、Lik、Leo、Allen、Adrian、Kevin、Ken、Terrence，以及 VADM 的 Jeffrey、Robert、Amy、Randy 等，多谢你们在工作中对我的帮助。

谨以此书，献给我最亲爱的家人，以及众多热爱 Java 开发的朋友们。

汪文君（Alex Wang）

中国，广州，2018 年 3 月

Contents 目 录

推荐序一
推荐序二
推荐序三
推荐序四
前言

第一部分　多线程基础

第1章　快速认识线程 3
1.1 　线程的介绍 3
1.2 　快速创建并启动一个线程 3
　　1.2.1 　尝试并行运行 4
　　1.2.2 　并发运行交替输出 5
　　1.2.3 　使用 Jconsole 观察线程 6
1.3 　线程的生命周期详解 7
　　1.3.1 　线程的 NEW 状态 8
　　1.3.2 　线程的 RUNNABLE 状态 8
　　1.3.3 　线程的 RUNNING 状态 8
　　1.3.4 　线程的 BLOCKED 状态 8
　　1.3.5 　线程的 TERMINATED 状态 9
1.4 　线程的 start 方法剖析：模板设计模式在 Thread 中的应用 9

　　1.4.1 　Thread start 方法源码分析以及注意事项 9
　　1.4.2 　模板设计模式在 Thread 中的应用 11
　　1.4.3 　Thread 模拟营业大厅叫号机程序 13
1.5 　Runnable 接口的引入以及策略模式在 Thread 中的使用 16
　　1.5.1 　Runnable 的职责 16
　　1.5.2 　策略模式在 Thread 中的应用 16
　　1.5.3 　模拟营业大厅叫号机程序 18
1.6 　本章总结 19

第2章　深入理解Thread构造函数 20
2.1 　线程的命名 20
　　2.1.1 　线程的默认命名 21
　　2.1.2 　命名线程 21
　　2.1.3 　修改线程的名字 22
2.2 　线程的父子关系 22
2.3 　Thread 与 ThreadGroup 23
2.4 　Thread 与 Runnable 24

2.5	Thread 与 JVM 虚拟机栈	25		3.8.2	join 方法结合实战	50
	2.5.1 Thread 与 Stacksize	25	3.9	如何关闭一个线程		53
	2.5.2 JVM 内存结构	27		3.9.1	正常关闭	54
	2.5.3 Thread 与虚拟机栈	30		3.9.2	异常退出	56
2.6	守护线程	33		3.9.3	进程假死	56
	2.6.1 什么是守护线程	33	3.10	本章总结		58
	2.6.2 守护线程的作用	34				

第3章 Thread API的详细介绍 …… 35

第4章 线程安全与数据同步 …… 59

- 2.7 本章总结 …… 34
- 3.1 线程 sleep …… 35
  - 3.1.1 sleep 方法介绍 …… 35
  - 3.1.2 使用 TimeUnit 替代 Thread.sleep …… 36
- 3.2 线程 yield …… 37
  - 3.2.1 yield 方法介绍 …… 37
  - 3.2.2 yield 和 sleep …… 37
- 3.3 设置线程的优先级 …… 38
  - 3.3.1 线程优先级介绍 …… 38
  - 3.3.2 线程优先级源码分析 …… 39
  - 3.3.3 关于优先级的一些总结 …… 40
- 3.4 获取线程 ID …… 40
- 3.5 获取当前线程 …… 41
- 3.6 设置线程上下文类加载器 …… 41
- 3.7 线程 interrupt …… 42
  - 3.7.1 interrupt …… 42
  - 3.7.2 isInterrupted …… 43
  - 3.7.3 interrupted …… 45
  - 3.7.4 interrupt 注意事项 …… 46
- 3.8 线程 join …… 47
  - 3.8.1 线程 join 方法详解 …… 48

- 4.1 数据同步 …… 59
  - 4.1.1 数据不一致问题的引入 …… 59
  - 4.1.2 数据不一致问题原因分析 …… 61
- 4.2 初识 synchronized 关键字 …… 62
  - 4.2.1 什么是 synchronized …… 63
  - 4.2.2 synchronized 关键字的用法 …… 63
- 4.3 深入 synchronized 关键字 …… 65
  - 4.3.1 线程堆栈分析 …… 65
  - 4.3.2 JVM 指令分析 …… 67
  - 4.3.3 使用 synchronized 需要注意的问题 …… 70
- 4.4 This Monitor 和 Class Monitor 的详细介绍 …… 72
  - 4.4.1 this monitor …… 72
  - 4.4.2 class monitor …… 74
- 4.5 程序死锁的原因以及如何诊断 …… 77
  - 4.5.1 程序死锁 …… 77
  - 4.5.2 程序死锁举例 …… 77
  - 4.5.3 死锁诊断 …… 80
- 4.6 本章总结 …… 81

第5章 线程间通信 …… 82

- 5.1 同步阻塞与异步非阻塞 …… 82

	5.1.1 同步阻塞消息处理 …………82
	5.1.2 异步非阻塞消息处理 ………83
5.2	单线程间通信 ……………………84
	5.2.1 初识 wait 和 notify …………84
	5.2.2 wait 和 notify 方法详解………87
	5.2.3 关于 wait 和 notify 的注意事项 ……………………………89
	5.2.4 wait 和 sleep ………………90
5.3	多线程间通信 ……………………90
	5.3.1 生产者消费者 ………………90
	5.3.2 线程休息室 wait set …………93
5.4	自定义显式锁 BooleanLock ………94
	5.4.1 synchronized 关键字的缺陷 ……94
	5.4.2 显式锁 BooleanLock …………95
5.5	本章总结 …………………………104

第6章 ThreadGroup详细讲解 ………105

6.1	ThreadGroup 与 Thread …………105
6.2	创建 ThreadGroup ………………105
6.3	复制 Thread 数组和 ThreadGroup 数组 ……………………………106
	6.3.1 复制 Thread 数组 …………106
	6.3.2 复制 ThreadGroup 数组 ……109
6.4	ThreadGroup 操作 ………………109
	6.4.1 ThreadGroup 的基本操作 ……110
	6.4.2 ThreadGroup 的 interrupt ……113
	6.4.3 ThreadGroup 的 destroy ……114
	6.4.4 守护 ThreadGroup …………115
6.5	本章总结 …………………………116

第7章 Hook线程以及捕获线程执行异常 …………………………117

7.1	获取线程运行时异常 ……………117
	7.1.1 UncaughtExceptionHandler 的介绍 ……………………………117
	7.1.2 UncaughtExceptionHandler 实例 ……………………………118
	7.1.3 UncaughtExceptionHandler 源码分析 …………………119
7.2	注入钩子线程 ……………………121
	7.2.1 Hook 线程介绍 ……………121
	7.2.2 Hook 线程实战 ……………122
	7.2.3 Hook 线程应用场景以及注意事项 ……………………124
7.3	本章总结 …………………………124

第8章 线程池原理以及自定义线程池 ………………………………125

8.1	线程池原理 ………………………125
8.2	线程池实现 ………………………126
	8.2.1 线程池接口定义 ……………127
	8.2.2 线程池详细实现 ……………131
8.3	线程池的应用 ……………………139
8.4	本章总结 …………………………142

第二部分 Java ClassLoader

第9章 类的加载过程 …………………144

9.1	类的加载过程简介 ………………144
9.2	类的主动使用和被动使用 ………145

9.3 类的加载过程详解……………148
　9.3.1 类的加载阶段…………148
　9.3.2 类的连接阶段…………149
　9.3.3 类的初始化阶段………154
9.4 本章总结……………………156

第10章　JVM类加载器……………158

10.1 JVM 内置三大类加载器………158
　10.1.1 根类加载器介绍………159
　10.1.2 扩展类加载器介绍……159
　10.1.3 系统类加载器介绍……160
10.2 自定义类加载器………………161
　10.2.1 自定义类加载器，问候世界………………………161
　10.2.2 双亲委托机制详细介绍……165
　10.2.3 破坏双亲委托机制……167
　10.2.4 类加载器命名空间、运行时包、类的卸载等………170
10.3 本章总结……………………175

第11章　线程上下文类加载器……177

11.1 为什么需要线程上下文类加载器………………………177
11.2 数据库驱动的初始化源码分析………………………178
11.3 本章总结……………………180

第三部分　深入理解 volatile 关键字

第12章　volatile关键字的介绍……182

12.1 初识 volatile 关键字…………182

12.2 机器硬件 CPU…………………184
　12.2.1 CPU Cache 模型………184
　12.2.2 CPU 缓存一致性问题…186
12.3 Java 内存模型…………………187
12.4 本章总结……………………188

第13章　深入volatile关键字………189

13.1 并发编程的三个重要特性……189
　13.1.1 原子性…………………189
　13.1.2 可见性…………………190
　13.1.3 有序性…………………190
13.2 JMM 如何保证三大特性………191
　13.2.1 JMM 与原子性…………192
　13.2.2 JMM 与可见性…………193
　13.2.3 JMM 与有序性…………194
13.3 volatile 关键字深入解析………195
　13.3.1 volatile 关键字的语义…195
　13.3.2 volatile 的原理和实现机制……………………197
　13.3.3 volatile 的使用场景……198
　13.3.4 volatile 和 synchronized…199
13.4 本章总结……………………200

第14章　7种单例设计模式的设计…201

14.1 饿汉式………………………201
14.2 懒汉式………………………202
14.3 懒汉式 + 同步方法……………203
14.4 Double-Check…………………204
14.5 Volatile+Double-Check………206
14.6 Holder 方式…………………206

14.7	枚举方式 …………………… 207		17.2.1	接口定义 …………………… 232
14.8	本章总结 …………………… 208		17.2.2	程序实现 …………………… 234
		17.3	读写锁的使用 …………………… 239	
		17.4	本章总结 …………………… 242	

第四部分　多线程设计架构模式

第15章　监控任务的生命周期 ……… 212

- 15.1 场景描述 …………………… 212
- 15.2 当观察者模式遇到 Thread …… 212
  - 15.2.1 接口定义 …………………… 212
  - 15.2.2 ObservableThread 实现 …… 215
- 15.3 本章总结 …………………… 217
  - 15.3.1 测试运行 …………………… 217
  - 15.3.2 关键点总结 …………………… 219

第16章　Single Thread Execution 设计模式 …………………… 220

- 16.1 机场过安检 …………………… 220
  - 16.1.1 非线程安全 …………………… 221
  - 16.1.2 问题分析 …………………… 223
  - 16.1.3 线程安全 …………………… 225
- 16.2 吃面问题 …………………… 225
  - 16.2.1 吃面引起的死锁 …………… 226
  - 16.2.2 解决吃面引起的死锁问题 …………………… 228
  - 16.2.3 哲学家吃面 …………………… 229
- 16.3 本章总结 …………………… 230

第17章　读写锁分离设计模式 ……… 231

- 17.1 场景描述 …………………… 231
- 17.2 读写分离程序设计 …………… 232

第18章　不可变对象设计模式 ……… 244

- 18.1 线程安全性 …………………… 244
- 18.2 不可变对象的设计 …………… 244
  - 18.2.1 非线程安全的累加器 …… 245
  - 18.2.2 方法同步增加线程安全性 …………………… 247
  - 18.2.3 不可变的累加器对象设计 …………………… 248
- 18.3 本章总结 …………………… 249

第19章　Future设计模式 …………… 251

- 19.1 先给你一张凭据 …………… 251
- 19.2 Future 设计模式实现 ……… 251
  - 19.2.1 接口定义 …………………… 252
  - 19.2.2 程序实现 …………………… 253
- 19.3 Future 的使用以及技巧总结 …… 256
- 19.4 增强 FutureService 使其支持回调 …………………… 257
- 19.5 本章总结 …………………… 258

第20章　Guarded Suspension设计模式 …………………… 259

- 20.1 什么是 Guarded Suspension 设计模式 …………………… 259
- 20.2 Guarded Suspension 的示例 …… 259
- 20.3 本章总结 …………………… 261

第21章　线程上下文设计模式 ……… 262

- 21.1　什么是上下文 ……………… 262
- 21.2　线程上下文设计 …………… 263
- 21.3　ThreadLocal 详解 ………… 264
  - 21.3.1　ThreadLocal 的使用场景及注意事项 ………………… 265
  - 21.3.2　ThreadLocal 的方法详解及源码分析 ………………… 265
  - 21.3.3　ThreadLocal 的内存泄漏问题分析 ………………… 270
- 21.4　使用 ThreadLocal 设计线程上下文 …………………… 274
- 21.5　本章总结 …………………… 276

第22章　Balking设计模式 ……………… 277

- 22.1　什么是 Balking 设计 ……… 277
- 22.2　Balking 模式之文档编辑 … 278
  - 22.2.1　Document ……………… 278
  - 22.2.2　AutoSaveThread ……… 280
  - 22.2.3　DocumentEditThread … 281
- 22.3　本章总结 …………………… 283

第23章　Latch设计模式 ……………… 284

- 23.1　什么是 Latch ……………… 284
- 23.2　CountDownLatch 程序实现 … 285
  - 23.2.1　无限等待的 Latch ……… 285
  - 23.2.2　有超时设置的 Latch …… 289
- 23.3　本章总结 …………………… 291

第24章　Thread-Per-Message设计模式 …………………… 293

- 24.1　什么是 Thread-Per-Message 模式 ……………………… 293
- 24.2　每个任务一个线程 ………… 293
- 24.3　多用户的网络聊天 ………… 296
  - 24.3.1　服务端程序 ……………… 296
  - 24.3.2　响应客户端连接的 Handler ……………… 297
  - 24.3.3　聊天程序测试 …………… 299
- 24.4　本章总结 …………………… 300

第25章　Two Phase Termination 设计模式 ……………… 301

- 25.1　什么是 Two Phase Termination 模式 ……………………… 301
- 25.2　Two Phase Termination 的示例 ……………………… 302
  - 25.2.1　线程停止的 Two Phase Termination ……………… 302
  - 25.2.2　进程关闭的 Two Phase Termination ……………… 303
- 25.3　知识扩展 …………………… 304
  - 25.3.1　Strong Reference 及 LRUCache ……………… 304
  - 25.3.2　Soft Reference 及 SoftLRUCache ……………… 308
  - 25.3.3　Weak Reference ……… 311
  - 25.3.4　Phantom Reference …… 312
- 25.4　本章总结 …………………… 314

第26章　Worker-Thread设计模式 … 315

- 26.1　什么是 Worker-Thread 模式 … 315
- 26.2　Worker-Thread 模式实现 … 315
  - 26.2.1　产品及组装说明书 ……… 316

- 26.2.2 流水线传送带 ······ 317
- 26.2.3 流水线工人 ······ 319
- 26.3 本章总结 ······ 320
  - 26.3.1 产品流水线测试 ······ 320
  - 26.3.2 Worker-Thread 和 Producer-Consumer ······ 321

第27章 Active Objects设计模式 ······ 323

- 27.1 接受异步消息的主动对象 ······ 323
- 27.2 标准 Active Objects 模式设计 ······ 323
  - 27.2.1 OrderService 接口设计 ······ 325
  - 27.2.2 OrderServiceImpl 详解 ······ 325
  - 27.2.3 OrderServiceProxy 详解 ······ 326
  - 27.2.4 MethodMessage ······ 328
  - 27.2.5 ActiveMessageQueue ······ 330
  - 27.2.6 OrderServiceFactory 及测试 ······ 332
- 27.3 通用 Active Objects 框架设计 ······ 333
  - 27.3.1 ActiveMessage 详解 ······ 334
  - 27.3.2 @ActiveMethod ······ 336
  - 27.3.3 ActiveServiceFactory 详解 ······ 337
  - 27.3.4 ActiveMessageQueue 及其他 ······ 339
- 27.4 本章总结 ······ 341

第28章 Event Bus设计模式 ······ 342

- 28.1 Event Bus 设计 ······ 343
  - 28.1.1 Bus 接口详解 ······ 343
  - 28.1.2 同步 EventBus 详解 ······ 345
  - 28.1.3 异步 EventBus 详解 ······ 347
  - 28.1.4 Subscriber 注册表 Registry 详解 ······ 348
  - 28.1.5 Event 广播 Dispatcher 详解 ······ 350
  - 28.1.6 其他类接口设计 ······ 353
  - 28.1.7 Event Bus 测试 ······ 355
- 28.2 Event Bus 实战——监控目录变化 ······ 357
  - 28.2.1 WatchService 遇到 EventBus ······ 357
  - 28.2.2 FileChangeEvent ······ 359
  - 28.2.3 监控目录变化 ······ 359
- 28.3 本章总结 ······ 360

第29章 Event Driven设计模式 ······ 361

- 29.1 Event-Driven Architecture 基础 ······ 361
  - 29.1.1 Events ······ 361
  - 29.1.2 Event Handlers ······ 362
  - 29.1.3 Event Loop ······ 363
- 29.2 开发一个 Event-Driven 框架 ······ 364
  - 29.2.1 同步 EDA 框架设计 ······ 364
  - 29.2.2 异步 EDA 框架设计 ······ 370
- 29.3 Event-Driven 的使用 ······ 375
  - 29.3.1 Chat Event ······ 375
  - 29.3.2 Chat Channel（Handler）······ 376
  - 29.3.3 Chat User 线程 ······ 377
- 29.4 本章总结 ······ 379

PART1 · 第一部分

多线程基础

第一部分的内容是整本书的基础，其围绕着 Thread API 及 ThreadGroup API 等进行讲解，在编写第一部分内容的过程中，笔者大量参考了虚拟机规范以及 JDK 官方文档，并深入源码分析每一个方法的详细信息。

第 1 章 "快速认识线程"：本章主要介绍线程的概念，以及线程在 Java 中的主要作用，并且详细讲解了线程的生命周期，以及生命周期每个状态之间的切换方法。

第 2 章 "深入理解 Thread 构造函数"：本章主要介绍了所有与 Thread 有关的构造函数，线程的父子关系（并非继承关系，而是一种包含关系），Thread 和 ThreadGroup 之间的关系，Thread 与虚拟机栈的关系（学习这部分内容需要读者有 JVM 的相关基础，尤其是对栈内存要有深入的理解），最后还介绍了守护线程的概念、特点和使用场景。

第 3 章 " Thread API 的详细介绍"：本章深入分析了 Thread 的所有 API，熟练掌握 Thread 的 API 是学好 Thread 的前提。

第 4 章 "线程安全与数据同步"：本章首先从一个简单的例子入手，讲解了数据同步的概念，以及会引发数据不一致性问题的情况，然后非常详细地介绍了 synchronized 关键字以及与其对应的 JVM 指令。本章的最后还分析了几种可能引起程序进入死锁的原因，以及如何使用工具进行诊断，线程安全与数据同步是线程中最重要也是最复杂的知识点之一，掌握好本章的内容可以使得程序在多线程的情况下既高效又安全的运行。

第 5 章 "线程间通信"：我们在开发多线程程序的时候，往往不会只存在一个独立的线程，相反大多数情况下是需要多个线程之间进行协同工作的，如何在多个线程之间进行通信，是本章学习的重点。另外，本章的最后部分将会分析 synchronized 关键字的缺陷，我们手动实现了一个显式锁（BooleanLock）可以解决 synchronized

所不具备的功能，其中也需要用到线程间通信的知识。

第 6 章 "ThreadGroup 详细讲解"：如果在创建线程时并未指定所属的 Group，线程会默认和父线程加入同一个 Group 之中，ThreadGroup 为线程提供了一定的结构组织能力。通过本章的学习，读者将会深入掌握 ThreadGroup 所有 API 的使用以及 Thread 和 ThreadGroup 之间的关系。

第 7 章 "Hook 线程以及捕获线程执行异常"：不管是 Runnable 接口的 run 方法还是 Thread 本身的 run 方法，都是不允许抛出 checked 异常的，这样启动线程就会无法捕获到异常信息，JDK 为我们提供了 UncaughtExceptionHandler 接口。本章将通过对 UncaughtExceptionHandler 的源码进行分析，使读者能够清晰地了解当 Thread 执行任务时出现异常应如何进行 Handler 的回调。本章的最后，我们还将学习如何向一个 JVM 进程注入 Hook 线程，当 JVM 进程收到中断信号时 Hook 线程将被触发执行。

第 8 章 "线程池原理以及自定义线程池"：本章首先从线程池的原理入手，详细讲解了一个功能完善的线程池应该具备哪些要素，其中包括任务队列、线程数量管理、拒绝策略、线程工厂等，后文中也会经常使用我们创建的线程池。

CHAPTER1 · 第 1 章

快速认识线程

在计算机的世界里,当我们探讨并行的时候,实际上是指,一系列的任务在计算机中同时运行,比如在浏览网页的时候还能打开音乐播放器,在撰写邮件的时候,收件箱还能接收新的邮件。在单 CPU 的计算机中,其实并没有真正的并行,它只不过是 CPU 时间钟快速轮转调度带给你的错觉,而这种错觉让你产生了它们真的在同一时刻同时运行。当然如果是多核 CPU,那么并行运行还是真实存在的。

1.1 线程的介绍

现在几乎百分之百的操作系统都支持多任务的执行,对计算机来说每一个任务就是一个进程(Process),在每一个进程内部至少要有一个线程(Thread)是在运行中,有时线程也称为轻量级的进程。

线程是程序执行的一个路径,每一个线程都有自己的局部变量表、程序计数器(指向正在执行的指令指针)以及各自的生命周期,现代操作系统中一般不止一个线程在运行,当启动了一个 Java 虚拟机(JVM)时,从操作系统开始就会创建一个新的进程(JVM 进程),JVM 进程中将会派生或者创建很多线程。

1.2 快速创建并启动一个线程

本节中,我们将快速认识 Thread Class 的用法,并将尝试说明如何创建并启动一个线程。

这里还是以我们日常生活中的例子进行举例,假设你想在浏览网页看新闻的同时听听音乐，下面我们就来尝试着用 Java 的代码来实现这一功能。

1.2.1 尝试并行运行

代码清单 1-1　TryConcurrency.java

```java
package com.wangwenjun.concurrent.chapter01;

import java.util.concurrent.TimeUnit;

public class TryConcurrency
{
    public static void main(String[] args)
    {
        browseNews();
        enjoyMusic();
    }

    /**
     * Browse the latest news.
     */
    private static void browseNews()
    {
        for (; ; )
        {
            System.out.println("Uh-huh, the good news.");
            sleep(1);
        }
    }

    /**
     * Listening and enjoy the music.
     */
    private static void enjoyMusic()
    {
        for (; ; )
        {
            System.out.println("Uh-huh, the nice music.");
            sleep(1);
        }
    }

    /**
     * Simulate the wait and ignore exception.
     * @param seconds
     */
    private static void sleep(int seconds)
    {
        try
        {
```

```
                TimeUnit.SECONDS.sleep(seconds);
            } catch (InterruptedException e)
            {
                e.printStackTrace();
            }
        }
    }
```

代码清单 1-1 试图让听音乐和看新闻两个任务同时执行（在控制台输出表现为交替输出），不过很可惜，程序的输出永远都是在看新闻，而听音乐的任务永远都得不到执行，TryConcurrency 代码的输出如下：

Uh-huh, the good news.
Uh-huh, the good news.
Uhhuh, the good news.
...

1.2.2 并发运行交替输出

如果想让听音乐和看新闻两个事件并发执行，也就是在 Console 中看到它们彼此交替输出，就必须借助 Java 提供的 Thread 这个 class（关于 Thread 的用法和详解，后文中会有详细的讲解）。

只需要将代码清单 1-1 的 main 方法中的任意一个方法交给 Thread 即可，下面是加入了 Thread 之后的代码：

```
public static void main(String[] args)
{
    // 通过匿名内部类的方式创建线程，并且重写其中的 run 方法
    new Thread(){              //①
        @Override
        public void run()
        {
            enjoyMusic();
        }
    }.start();                 //②
    browseNews();
}
```

输出如下：

Uh-huh, the good news.
Uh-huh, the nice music.
Uh-huh, the good news.
Uh-huh, the nice music.
....

代码修改之后会发现听音乐和看新闻两个任务可以并行运行，并且在控制台交替输出。

> **注意** 上面的代码中，线程启动必须在其中一个任务之前，否则线程将永远得不到启动，因为前一个任务永远不会结束。
> ① 创建一个线程，并且重写其 run 方法，将 enjoyMusic 交给它执行。
> ② 启动新的线程，只有调用了 Thread 的 start 方法，才代表派生了一个新的线程，否则 Thread 和其他普通的 Java 对象没有什么区别，start 方法是一个立即返回方法，并不会让程序陷入阻塞。

如果用 Java 8 Lambda 改造上面的代码，那么代码看起来将会更加简洁，如下所示：

```java
public static void main(String[] args)
{
    new Thread(TryConcurrency::enjoyMusic).start();
    browseNews();
}
```

1.2.3 使用 Jconsole 观察线程

1.2.2 节创建了一个 Thread 并且启动，那么此时 JVM 中有多少个线程呢？除了我们创建的线程以外，还有哪些线程可以借助 Jconsole 或者 Jstack 命令来查看，如图 1-1 所示，这两个 JVM 工具都是由 JDK 自身提供的。

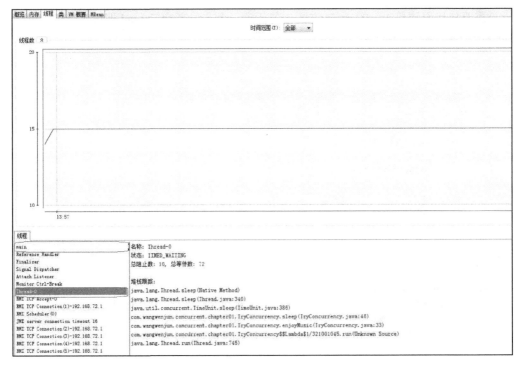

图 1-1　使用 Jconsole 观察 JVM 线程

在图 1-1 中，我用线框勾出了两个线程，其中一个是 main，另一个是 Thread-0，之前说过在操作系统启动一个 Java 虚拟机（JVM）的时候，其实是启动了一个进程，而在该进程里面启动了一个以上的线程，其中 Thread-0 这个线程就是 1.2.2 节中创建的，main 线程是由 JVM 启动时创建的，我们都知道 J2SE 程序的入口就是 main 函数，虽然我们在 1.2.2 节中显式地创建了一个线程，事实上还有一个 main 线程，当然还有一些其他的守护线程，比如垃圾回收线程、RMI 线程等。

1.3 线程的生命周期详解

前面提到过，每一个线程都有自己的局部变量表、程序计数器，以及生命周期等，本节就来分析一下线程的生命周期，如图 1-2 所示。

在开始解释线程的生命周期之前，请大家思考一个问题：执行了 Thread 的 start 方法就代表该线程已经开始执行了吗？

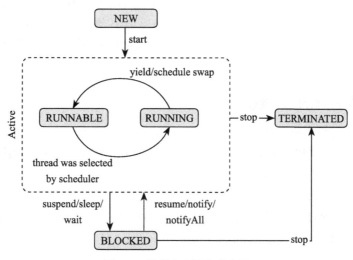

图 1-2　线程生命周期状态图

通过图 1-2 的展示可知，线程的生命周期大体可以分为如下 5 个主要的阶段。

❑ NEW

❑ RUNNABLE

❑ RUNNING

❑ BLOCKED

❑ TERMINATED

1.3.1 线程的 NEW 状态

当我们用关键字 new 创建一个 Thread 对象时，此时它并不处于执行状态，因为没有调用 start 方法启动该线程，那么线程的状态为 NEW 状态，准确地说，它只是 Thread 对象的状态，因为在没有 start 之前，该线程根本不存在，与你用关键字 new 创建一个普通的 Java 对象没什么区别。

NEW 状态通过 start 方法进入 RUNNABLE 状态。

1.3.2 线程的 RUNNABLE 状态

线程对象进入 RUNNABLE 状态必须调用 start 方法，那么此时才是真正地在 JVM 进程中创建了一个线程，线程一经启动就可以立即得到执行吗？答案是否定的，线程的运行与否和进程一样都要听令于 CPU 的调度，那么我们把这个中间状态称为可执行状态（RUNNABLE），也就是说它具备执行的资格，但是并没有真正地执行起来而是在等待 CPU 的调度。

由于存在 Running 状态，所以不会直接进入 BLOCKED 状态和 TERMINATED 状态，即使是在线程的执行逻辑中调用 wait、sleep 或者其他 block 的 IO 操作等，也必须先获得 CPU 的调度执行权才可以，严格来讲，RUNNABLE 的线程只能意外终止或者进入 RUNNING 状态。

1.3.3 线程的 RUNNING 状态

一旦 CPU 通过轮询或者其他方式从任务可执行队列中选中了线程，那么此时它才能真正地执行自己的逻辑代码，需要说明的一点是一个正在 RUNNING 状态的线程事实上也是 RUNNABLE 的，但是反过来则不成立。

在该状态中，线程的状态可以发生如下的状态转换。

- ❑ 直接进入 TERMINATED 状态，比如调用 JDK 已经不推荐使用的 stop 方法或者判断某个逻辑标识。
- ❑ 进入 BLOCKED 状态，比如调用了 sleep，或者 wait 方法而加入了 waitSet 中。
- ❑ 进行某个阻塞的 IO 操作，比如因网络数据的读写而进入了 BLOCKED 状态。
- ❑ 获取某个锁资源，从而加入到该锁的阻塞队列中而进入了 BLOCKED 状态。
- ❑ 由于 CPU 的调度器轮询使该线程放弃执行，进入 RUNNABLE 状态。
- ❑ 线程主动调用 yield 方法，放弃 CPU 执行权，进入 RUNNABLE 状态。

1.3.4 线程的 BLOCKED 状态

1.3.3 节中已经列举了线程进入 BLCOKED 状态的原因，此处就不再赘述了，线程在 BLOCKED 状态中可以切换至如下几个状态。

- 直接进入 TERMINATED 状态，比如调用 JDK 已经不推荐使用的 stop 方法或者意外死亡（JVM Crash）。
- 线程阻塞的操作结束，比如读取了想要的数据字节进入到 RUNNABLE 状态。
- 线程完成了指定时间的休眠，进入到了 RUNNABLE 状态。
- Wait 中的线程被其他线程 notify/notifyall 唤醒，进入 RUNNABLE 状态。
- 线程获取到了某个锁资源，进入 RUNNABLE 状态。
- 线程在阻塞过程中被打断，比如其他线程调用了 interrupt 方法，进入 RUNNABLE 状态。

1.3.5 线程的 TERMINATED 状态

TERMINATED 是一个线程的最终状态，在该状态中线程将不会切换到其他任何状态，线程进入 TERMINATED 状态，意味着该线程的整个生命周期都结束了，下列这些情况将会使线程进入 TERMINATED 状态。

- 线程运行正常结束，结束生命周期。
- 线程运行出错意外结束。
- JVM Crash，导致所有的线程都结束。

通过本节关于线程生命周期的分析，相信读者已经能够独立回答我们开始时提的问题了，线程的生命周期非常关键，很多工作多年的程序员，由于不重视这部分的内容，在编写程序的时候也经常会出现一些错误。

1.4 线程的 start 方法剖析：模板设计模式在 Thread 中的应用

在本节中，我们将分析 Thread 的 start 方法，在调用了 start 方法之后到底进行了什么操作，通过 1.3 节的内容讲解，相信大家已经明白了，start 方法启动了一个线程，并且该线程进入了可执行状态（RUNNABLE），在"代码清单 1-1　TryConcurrency"中，我们重写了 Thread 的 run 方法，但却调用了 start 方法，那么 run 方法和 start 方法有什么关系呢？带着诸多的疑问，我们一起在本节中寻找答案吧！

1.4.1　Thread start 方法源码分析以及注意事项

先来看一下 Thread start 方法的源码，如下所示：

```
public synchronized void start() {
    if (threadStatus != 0)
        throw new IllegalThreadStateException();
    group.add(this);

    boolean started = false;
```

```
        try {
            start0();
            started = true;
        } finally {
            try {
                if (!started) {
                    group.threadStartFailed(this);
                }
            } catch (Throwable ignore) {
            }
        }
    }
```

start 方法的源码足够简单,其实最核心的部分是 start0 这个本地方法,也就是 JNI 方法:

```
private native void start0();
```

也就是说在 start 方法中会调用 start0 方法,那么重写的那个 run 方法何时被调用了呢?单从上面是看不出来任何端倪的,但是打开 JDK 的官方文档,在 start 方法中有如下的注释说明:

※ Causes this thread to begin execution; the Java Virtual Machine calls the <code>run</code> method of this thread.

上面这句话的意思是:在开始执行这个线程时,JVM 将会调用该线程的 run 方法,换言之,run 方法是被 JNI 方法 start0() 调用的,仔细阅读 start 的源码将会总结出如下几个知识要点。

- Thread 被构造后的 NEW 状态,事实上 threadStatus 这个内部属性为 0。
- 不能两次启动 Thread,否则就会出现 IllegalThreadStateException 异常。
- 线程启动后将会被加入到一个 ThreadGroup 中,后文中我们将详细介绍 ThreadGroup。
- 一个线程生命周期结束,也就是到了 TERMINATED 状态,再次调用 start 方法是不允许的,也就是说 TERMINATED 状态是没有办法回到 RUNNABLE/RUNNING 状态的。

```
Thread thread = new Thread()
{
    @Override
    public void run()
    {
        try
        {
            TimeUnit.SECONDS.sleep(10);
        } catch (InterruptedException e)
        {
            e.printStackTrace();
        }
    }
```

```
};
thread.start();// 启动线程

thread.start();// 再次启动
```

执行上面的代码将会抛出 IllegalThreadStateException 异常,而我们将代码稍作改动,模拟一个线程生命周期的结束,再次启动看看会发生什么:

```
Thread thread = new Thread()
{
    @Override
    public void run()
    {
        try
        {
            TimeUnit.SECONDS.sleep(1);
        } catch (InterruptedException e)
        {
            e.printStackTrace();
        }
    }
};
thread.start();

TimeUnit.SECONDS.sleep(2);// 休眠主要是确保thread结束生命周期

thread.start();// 企图重新激活该线程
```

> **注意** 程序同样会抛出 IllegalThreadStateException 异常,但是这两个异常的抛出却有本质上的区别,第一个是重复启动,只是第二次启动是不允许的,但是此时该线程是处于运行状态的,而第二次企图重新激活也抛出了非法状态的异常,但是此时没有线程,因为该线程的生命周期已经被终结。

1.4.2 模板设计模式在 Thread 中的应用

通过 1.4.1 节的分析,我们不难看出,线程的真正的执行逻辑是在 run 方法中,通常我们会把 run 方法称为线程的执行单元,这也就回答了我们最开始提出的疑问,重写 run 方法,用 start 方法启动线程。Thread 中 run 方法的代码如下,如果我们没有使用 Runnable 接口对其进行构造,则可以认为 Thread 的 run 方法本身就是一个空的实现:

```
@Override
public void run() {
    if (target != null) {// 我们并没有使用 runnable 构造 Thread
        target.run();
    }
}
```

其实 Thread 的 run 和 start 就是一个比较典型的模板设计模式,父类编写算法结构代码，子类实现逻辑细节，下面通过一个简单的例子来看一下模板设计模式，然后读者可以参考该模式在 Thread 中的使用，示例代码如清单 1-2 所示。

代码清单 1-2　TemplateMethod.java

```java
package com.wangwenjun.concurrent.chapter01;
public class TemplateMethod {

    public final void print(String message) {
        System.out.println("###############");
        wrapPrint(message);
        System.out.println("###############");
    }

    protected void wrapPrint(String message) {

    }

    public static void main(String[] args) {
        TemplateMethod t1 = new TemplateMethod(){
            @Override
            protected void wrapPrint(String message) {
                System.out.println("*"+message+"*");
            }
        };
        t1.print("Hello Thread");

        TemplateMethod t2 = new TemplateMethod(){
            @Override
            protected void wrapPrint(String message) {
                System.out.println("+"+message+"+");
            }
        };

        t2.print("Hello Thread");
    }
}
```

print 方法类似于 Thread 的 start 方法，而 wrapPrint 则类似于 run 方法，这样做的好处是，程序结构由父类控制，并且是 final 修饰的，不允许被重写，子类只需要实现想要的逻辑任务即可，输出如下：

```
###############
*Hello Thread*
###############
###############
+Hello Thread+
###############
```

1.4.3　Thread 模拟营业大厅叫号机程序

相信很多人都去过银行、医院、移动营业厅、公积金中心等，在这些机构的营业大厅都有排队等号的机制，这种机制的主要作用就是限流，减轻业务受理人员的压力。当你走进营业大厅后，需要先领取一张流水号纸票，然后拿着纸票坐在休息区等待你的号码显示在业务办理的橱窗显示器上面，如图 1-3 所示。

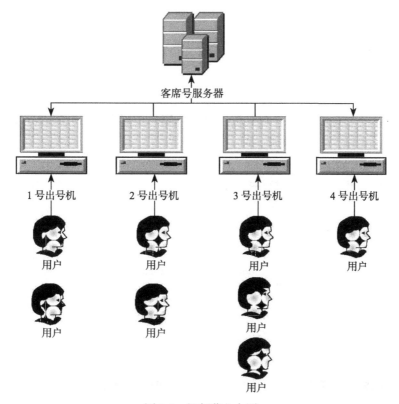

图 1-3　银行营业大厅

如图 1-3 所示，假设大厅共有四台出号机，这就意味着有四个线程在工作，下面我们用程序模拟一下叫号的过程，约定当天最多受理 50 笔业务，也就是说号码最多可以出到 50。

TicketWindow 代表大厅里的出号机器，代码如清单 1-3 所示。

代码清单 1-3　TicketWindow.java

```
package com.wangwenjun.concurrent.chapter01;
public class TicketWindow extends Thread {
    //柜台名称
    private final String name;

    //最多受理50笔业务
    private static final int MAX = 50;
```

```java
    private int index = 1;

    public TicketWindow(String name) {
        this.name = name;
    }

    @Override
    public void run() {
        while (index <= MAX) {

            System.out.println("柜台:" + name + "当前的号码是:" + (index++));
        }
    }
}
```

接下来,写一个 main 函数,对其进行测试,定义了四个 TicketWindow 线程,并且分别启动:

```java
public static void main(String[] args) {
    TicketWindow ticketWindow1 = new TicketWindow("一号出号机");
    ticketWindow1.start();

    TicketWindow ticketWindow2 = new TicketWindow("二号出号机");
    ticketWindow2.start();

    TicketWindow ticketWindow3 = new TicketWindow("三号出号机");
    ticketWindow3.start();

    TicketWindow ticketWindow4 = new TicketWindow("四号出号机");
    ticketWindow4.start();

}
```

运行之后的输出似乎令人大失所望,为何每一个 TickWindow 所出的号码都是从 1 到 50 呢?

```
柜台:一号柜台当前的号码是:1
柜台:三号柜台当前的号码是:1
柜台:二号柜台当前的号码是:1
柜台:三号柜台当前的号码是:2
柜台:三号柜台当前的号码是:3
柜台:三号柜台当前的号码是:4
柜台:三号柜台当前的号码是:5
柜台:三号柜台当前的号码是:6
...
```

之所以出现这个问题,根本原因是因为每一个线程的逻辑执行单元都不一样,我们新建了四个 Ticket Window 线程,它们的票号都是从 0 开始到 50 结束,四个线程并没有像图 1-3 所描述的那样均从客席号服务器进行交互,获取一个唯一的递增的号码,那么应该如何改进呢？无论 TicketWindow 被实例化多少次，只需要保证 index 是唯一的即可，我们会立即会想到使用 static 去修饰 index 以达到目的，改进后的代码如清单 1-4 所示：

代码清单 1-4　修改后的 TicketWindow.java

```java
public class TicketWindow extends Thread {

    private final String name;

    private static final int MAX = 50;

    private static int index = 1;

    public TicketWindow(String name) {
        this.name = name;
    }

    @Override
    public void run() {
        while (index <= MAX) {
            System.out.println("柜台：" + name + " 当前的号码是：" + (index++));
        }
    }
}
```

再次运行上面的 main 函数，会发现情况似乎有些改善，四个出号机交替着输出不同的号码，输出如下：

```
柜台：一号出号机当前的号码是 :1
柜台：一号出号机当前的号码是 :3
柜台：三号出号机当前的号码是 :2
柜台：一号出号机当前的号码是 :4
柜台：三号出号机当前的号码是 :5
柜台：一号出号机当前的号码是 :6
柜台：三号出号机当前的号码是 :7
柜台：一号出号机当前的号码是 :8
柜台：三号出号机当前的号码是 :9
柜台：一号出号机当前的号码是 :10
...
```

通过对 index 进行 static 修饰，做到了多线程下共享资源的唯一性，看起来似乎满足了我们的需求（事实上，如果将最大号码调整到 500、1000 等稍微大一些的数字就会出现线程安全的问题，关于这点将在后面的章节中详细介绍），但是只有一个 index 共享资源，如果共享资源很多呢？共享资源要经过一些比较复杂的计算呢？不可能都使用 static 修饰，而且 static 修饰的变量生命周期很长，所以 Java 提供了一个接口 Runnable 专门用于解决该问题，将线程的控制和业务逻辑的运行彻底分离开来。

1.5 Runnable 接口的引入以及策略模式在 Thread 中的使用

1.5.1 Runnable 的职责

Runnable 接口非常简单，只定义了一个无参数无返回值的 run 方法，具体如代码清单 1-5 所示。

代码清单 1-5　Runnable 接口

```
package java.lang;

public interface Runnable {
    void run();
}
```

在很多软文以及一些书籍中，经常会提到，创建线程有两种方式，第一种是构造一个 Thread，第二种是实现 Runnable 接口，这种说法是错误的，最起码是不严谨的，在 JDK 中代表线程的就只有 Thread 这个类，我们在前面分析过，线程的执行单元就是 run 方法，你可以通过继承 Thread 然后重写 run 方法实现自己的业务逻辑，也可以实现 Runnable 接口实现自己的业务逻辑，代码如下：

```
@Override
public void run() {
// 如果构造 Thread 时传递了 Runnable，则会执行 runnable 的 run 方法
    if (target != null) {
        target.run();
    }
// 否则需要重写 Thread 类的 run 方法
}
```

上面的代码段是 Thread run 方法的源码，我在其中加了两行注释更加清晰地说明了实现执行单元的两种方式，所以说创建线程有两种方式，一种是创建一个 Thread，一种是实现 Runnable 接口，这种说法是不严谨的。准确地讲，创建线程只有一种方式那就是构造 Thread 类，而实现线程的执行单元则有两种方式，第一种是重写 Thread 的 run 方法，第二种是实现 Runnable 接口的 run 方法，并且将 Runnable 实例用作构造 Thread 的参数。

1.5.2 策略模式在 Thread 中的应用

前面说过了，无论是 Runnable 的 run 方法，还是 Thread 类本身的 run 方法（事实上 Thread 类也是实现了 Runnable 接口）都是想将线程的控制本身和业务逻辑的运行分离开来，达到职责分明、功能单一的原则，这一点与 GoF 设计模式中的策略设计模式很相似，在本节中，我们一起来看看什么是策略模式，然后再来对比 Thread 和 Runnable 两者之间的关系。

相信很多人都做过关于 JDBC 的开发，下面我们在这里做一个简单的查询操作，只不过是把数据的封装部分抽取成一个策略接口，代码如清单 1-6 所示。

代码清单1-6　RowHandler.java

```java
package com.wangwenjun.concurrent.chapter01;

import java.sql.ResultSet;

public interface RowHandler<T>
{
    T handle(ResultSet rs);
}
```

RowHandler接口只负责对从数据库中查询出来的结果集进行操作，至于最终返回成什么样的数据结构，那就需要你自己去实现，类似于Runnable接口，示例代码如清单1-7所示。

代码清单1-7　RecordQuery.java

```java
package com.wangwenjun.concurrent.chapter01;

import java.sql.Connection;
import java.sql.PreparedStatement;
import java.sql.ResultSet;
import java.sql.SQLException;

public class RecordQuery
{
    private final Connection connection;

    public RecordQuery(Connection connection)
    {
        this.connection = connection;
    }

    public <T> T query(RowHandler<T> handler, String sql, Object... params)
            throws SQLException
    {
        try (PreparedStatement stmt = connection.prepareStatement(sql))
        {
            int index = 1;
            for (Object param : params)
            {
                stmt.setObject(index++, param);
            }

            ResultSet resultSet = stmt.executeQuery();
            return handler.handle(resultSet);// ①调用RowHandler
        }
    }
}
```

RecordQuery 中的 query 只负责将数据查询出来，然后调用 RowHandler 进行数据封装，至于将其封装成什么数据结构，那就得看你自己怎么处理了，下面我们来看看这样做有什么好处？

上面这段代码的好处是可以用 query 方法应对任何数据库的查询，返回结果的不同只会因为你传入 RowHandler 的不同而不同，同样 RecordQuery 只负责数据的获取，而 RowHandler 则负责数据的加工，职责分明，每个类均功能单一，相信通过这个简单的示例大家应该能够清楚 Thread 和 Runnable 之间的关系了。

> **注意** 重写 Thread 类的 run 方法和实现 Runnable 接口的 run 方法还有一个很重要的不同，那就是 Thread 类的 run 方法是不能共享的，也就是说 A 线程不能把 B 线程的 run 方法当作自己的执行单元，而使用 Runnable 接口则很容易就能实现这一点，使用同一个 Runnable 的实例构造不同的 Thread 实例。

1.5.3 模拟营业大厅叫号机程序

既然我们说使用 static 修饰 index 这个共享资源不是一种好的方式，那么我们在本节中使用 Runnable 接口来实现逻辑执行单元重构一下 1.4 节中的营业大厅叫号机程序。

首先我们将 Thread 的 run 方法抽取成一个 Runnable 接口的实现，代码如清单 1-8 所示。

代码清单 1-8　TicketWindowRunnable.java

```java
package com.wangwenjun.concurrent.chapter01;

public class TicketWindowRunnable implements Runnable {

    private int index = 1;// 不做 static 修饰

    private final static int MAX = 50;

    @Override
    public void run() {

        while (index <= MAX) {
            System.out.println(Thread.currentThread() + " 的号码是:" + (index++));
            try {
                Thread.sleep(100);
            } catch (InterruptedException e) {
                e.printStackTrace();
            }
        }
    }
}
```

可以看到上面的代码中并没有对 index 进行 static 的修饰，并且我们也将 Thread 中 run 的代码逻辑抽取到了 Runnable 的一个实现中，下面的代码构造了四个叫号机的线程，并且

开始工作:

```java
public static void main(String[] args) {

    final TicketWindowRunnable task = new TicketWindowRunnable();

    Thread windowThread1 = new Thread(task, "一号窗口");

    Thread windowThread2 = new Thread(task, "二号窗口");

    Thread windowThread3 = new Thread(task, "三号窗口");

    Thread windowThread4 = new Thread(task, "四号窗口");

    windowThread1.start();
    windowThread2.start();
    windowThread3.start();
    windowThread4.start();
}
```

程序的输出与 1.4.3 节改版之后的代码清单 1-4 的输出效果是一样的，四个叫号机线程，使用了同一个 Runnable 接口，这样它们的资源就是共享的，不会再出现每一个叫号机都从 1 打印到 50 这样的情况。

> **注意** 不管是如 1.4.3 节代码清单 1-4 中用 static 修饰 index 还是用实现 Runnable 接口的方式，这两个程序多运行几次或者 MAX 的值从 50 增加到 500、1000 或者更大都会出现一个号码出现两次的情况，也会出现某个号码根本不会出现的情况，更会出现超过最大值的情况，这是因为共享资源 index 存在线程安全的问题，我们在后面学习数据同步的时候会详细介绍。

1.6 本章总结

在本章中，我们学习了什么是线程，以及初步掌握了如何创建一个线程，并且通过重写 Thread 的 run 方法和实现 Runnable 接口的 run 方法进而实现线程的执行单元。

两个版本的叫号机程序，让读者对多线程的程序有了一个初步的认识，模板设计模式以及策略设计模式和 Thread 以及 Runnable 的结合使得大家能够更加清晰地掌握多线程的 API 是如何实现线程控制和业务执行解耦分离的。

当然本章中最为重要的内容就是线程的生命周期，在使用多线程的过程中，线程的生命周期将会贯穿始终，只有清晰地掌握生命周期各个阶段的切换，才能更好地理解线程的阻塞以及唤醒机制，同时也为掌握同步锁等概念打下一个良好的基础。

第 2 章 · CHAPTER2

深入理解 Thread 构造函数

Java 中的 Thread 为我们提供了比较丰富的构造函数，在本章中，我们将会逐一介绍每一个构造函数，以及分析其中一些可能并未引起你关注的细节。Thread 构造函数如图 2-1 所示。

Constructors

Constructor and Description
Thread()   Allocates a new Thread object.
Thread(**Runnable** target)   Allocates a new Thread object.
Thread(**Runnable** target, **String** name)   Allocates a new Thread object.
Thread(**String** name)   Allocates a new Thread object.
Thread(**ThreadGroup** group, **Runnable** target)   Allocates a new Thread object.
Thread(**ThreadGroup** group, **Runnable** target, **String** name)   Allocates a new Thread object so that it has target as its run object, has the specified name as its name, and belongs to the thread group referred to by group.
Thread(**ThreadGroup** group, **Runnable** target, **String** name, long stackSize)   Allocates a new Thread object so that it has target as its run object, has the specified name as its name, and belongs to the thread group referred to by group, and has the specified *stack size*.
Thread(**ThreadGroup** group, **String** name)   Allocates a new Thread object.

图 2-1 Thread 构造函数

2.1 线程的命名

在构造线程的时候可以为线程起一个有特殊意义的名字，这也是比较好的一种做法，尤其在一个线程比较多的程序中,为线程赋予一个包含特殊意义的名字有助于问题的排查和线程的跟踪,因此笔者强烈推荐在构造线程的时候赋予它一个名字。

2.1.1 线程的默认命名

下面的几个构造函数中,并没有提供为线程命名的参数,那么此时线程会有一个怎样的命名呢?

- Thread()
- Thread(Runnable target)
- Thread(ThreadGroup group, Runnable target)

打开 JDK 的源码会看到下面的代码:

```
public Thread(Runnable target) {
    init(null, target, "Thread-" + nextThreadNum(), 0);
}
/* For autonumbering anonymous threads. */
private static int threadInitNumber;
private static synchronized int nextThreadNum() {
    return threadInitNumber++;
}
```

如果没有为线程显式地指定一个名字,那么线程将会以"Thread-"作为前缀与一个自增数字进行组合,这个自增数字在整个 JVM 进程中将会不断自增:

```
public static void main(String[] args)
{
    IntStream.range(0, 5).boxed().map(i -> new Thread(
    () -> System.out.println(Thread.currentThread().getName())
    )).forEach(Thread::start);
}
```

执行上面的代码,这里使用无参的构造函数创建了 5 个线程,并且分别输出了各自的名字,会发现输出结果与我们对源码的分析是一致的,输出如下:

```
Thread-0
Thread-2
Thread-1
Thread-4
Thread-3
```

2.1.2 命名线程

在 2.1.1 节中,笔者强烈推荐在构造 Thread 的时候,为线程赋予一个特殊的名字是一种比较好的实战方式,Thread 同样也提供了这样的构造函数,具体如下。

- Thread(Runnable target, String name)

- Thread(String name)
- Thread(ThreadGroup group, Runnable target, String name)
- Thread(ThreadGroup group, Runnable target, String name, long stackSize)
- Thread(ThreadGroup group, String name)

示例代码如下：

```
private final static String PREFIX = "ALEX-";
public static void main(String[] args)
{
IntStream.range(0,5).mapToObj(ThreadConstruction::createThread)
          .forEach(Thread::start);
}
private static Thread createThread(final int intName)
{
    return new Thread(
        () -> System.out.println(Thread.currentThread().getName())
          , PREFIX + intName);
}
```

在上面的代码中，我们定义了一个新的前缀"ALEX-"，然后用 0 ~ 4 之间的数字作为后缀对线程进行了命名，代码执行输出的结果如下所示：

```
ALEX-0
ALEX-1
ALEX-2
ALEX-3
ALEX-4
```

2.1.3 修改线程的名字

不论你使用的是默认的函数命名规则，还是指定了一个特殊的名字，在线程启动之前还有一个机会可以对其进行修改，一旦线程启动，名字将不再被修改，下面是 Thread 的 setName 源码：

```
public final synchronized void setName(String name) {
    checkAccess();
    this.name = name.toCharArray();
    if (threadStatus != 0) { // 线程不是 NEW 状态，对其修改将不会生效
        setNativeName(name);
    }
}
```

2.2 线程的父子关系

Thread 的所有构造函数，最终都会去调用一个静态方法 init，我们截取片段代码对其进

行分析，不难发现新创建的任何一个线程都会有一个父线程：

```
private void init(ThreadGroup g, Runnable target, String name,
                  long stackSize, AccessControlContext acc) {
    if (name == null) {
        throw new NullPointerException("name cannot be null");
    }
    this.name = name.toCharArray();
    Thread parent = currentThread(); // 获取当前线程作为父线程
    SecurityManager security = System.getSecurityManager();
```

上面代码中的 currentThread() 是获取当前线程，在线程生命周期中，我们说过线程的最初状态为 NEW，没有执行 start 方法之前，它只能算是一个 Thread 的实例，并不意味着一个新的线程被创建，因此 currentThread() 代表的将会是创建它的那个线程，因此我们可以得出以下结论。

❑ 一个线程的创建肯定是由另一个线程完成的。
❑ 被创建线程的父线程是创建它的线程。

我们都知道 main 函数所在的线程是由 JVM 创建的，也就是 main 线程，那就意味着我们前面创建的所有线程，其父线程都是 main 线程。

2.3 Thread 与 ThreadGroup

在 Thread 的构造函数中，可以显式地指定线程的 Group，也就是 ThreadGroup（关于 ThreadGroup 会在后面的章节中做重点介绍）。

接着往下阅读 Thread init 方法的源码：

```
SecurityManager security = System.getSecurityManager();
if (g == null) {
    /* Determine if it's an applet or not */
    /* If there is a security manager, ask the security manager
       what to do. */
    if (security != null) {
        g = security.getThreadGroup();
    }
    /* If the security doesn't have a strong opinion of the matter
       use the parent thread group. */
    if (g == null) {
        g = parent.getThreadGroup();
    }
}
```

通过对源码进行分析，我们可以看出，如果在构造 Thread 的时候没有显示地指定一个 ThreadGroup，那么子线程将会被加入父线程所在的线程组，下面写一个简单的代码来测试一下，如代码清单 2-1 所示。

代码清单 2-1　ThreadConstruction.java

```java
package com.wangwenjun.concurrent.chapter02;

public class ThreadConstruction
{
    public static void main(String[] args)
    {
        //①
        Thread t1 = new Thread("t1");

        //②
        ThreadGroup group = new ThreadGroup("TestGroup");
        //③
        Thread t2 = new Thread(group, "t2");
        ThreadGroup mainThreadGroup = Thread.currentThread().getThreadGroup();
        System.out.println("Main thread belong group:" + mainThreadGroup.getName());
        System.out.println("t1 and main belong the same group:" + (mainThreadGroup
                    == t1.getThreadGroup()));
        System.out.println("t2 thread group not belong main group:" + (mainThreadGroup
                    == t2.getThreadGroup()));
        System.out.println("t2 thread group belong main TestGroup:" + (group ==
                    t2.getThreadGroup()));

    }
}
```

注释①创建了一个 Thread t1，注释②创建了一个 ThreadGroup，注释③创建了一个 Thread t2，并且将它加入到了 group 中，我们并没有给 t1 指定任何 Group，执行上面的代码，输出结果为：

```
Main thread belong group:main
t1 and main belong the same group:true
t2 thread group not belong main group:false
t2 thread group belong main TestGroup:true
```

通过对 Thread 源码的分析和我们自己的测试可以得出以下结论。
- main 线程所在的 ThreadGroup 称为 main。
- 构造一个线程的时候如果没有显式地指定 ThreadGroup，那么它将会和父线程同属于一个 ThreadGroup。

在默认设置中，当然除了子线程会和父线程同属于一个 Group 之外，它还会和父线程拥有同样的优先级，同样的 daemon，关于这点我们在后文中将会详细讲解。

2.4　Thread 与 Runnable

在本书的第 1 章中，详细介绍了为什么要有 Runnable，以及 Runnable 接口和 Thread

之间的关系，Thread 负责线程本身相关的职责和控制，而 Runnable 则负责逻辑执行单元的部分，这里就不再赘述了。

2.5 Thread 与 JVM 虚拟机栈

在 Thread 的构造函数中，可发现有一个特殊的参数 stackSize，这个参数的作用是什么呢？它的值对线程有什么影响呢？下面我们就来一起探讨这个问题。

2.5.1 Thread 与 Stacksize

打开 JDK 官方文档，将会发现 Thread 中对 stacksize 构造函数的文字说明，具体如下：

※ The stack size is the approximate number of bytes of address space that the virtual machine is to allocate for this thread's stack. The effect of the stackSize parameter, if any, is highly platform dependent.
On some platforms, specifying a higher value for the stackSize parameter may allow a thread to achieve greater recursion depth before throwing a StackOverflowError. Similarly, specifying a lower value may allow a greater number of threads to exist concurrently without throwing an OutOfMemoryError (or other internal error). The details of the relationship between the value of the stackSize parameter and the maximum recursion depth and concurrency level are platform-dependent. On some platforms, the value of the stackSize parameter may have no effect whatsoever.

一般情况下，创建线程的时候不会手动指定栈内存的地址空间字节数组，统一通过 xss 参数进行设置即可，通过上面这段官网文档的描述，我们不难发现 stacksize 越大则代表着正在线程内方法调用递归的深度就越深，stacksize 越小则代表着创建的线程数量越多，当然了这个参数对平台的依赖性比较高，比如不同的操作系统、不同的硬件。

在有些平台下，越高的 stack 设定，可以允许的递归深度越多；反之，越少的 stack 设定，则递归深度越浅。当然在某些平台下，该参数压根不会起到任何作用，如果将该参数设置为 0，也不会起到任何的作用，笔者在自己的 Windows 电脑上和 Ubuntu 虚拟机上分别作了测试，很明显地看到了 stacksize 给线程带来的影响。

代码清单 2-2 ThreadConstruction.java

```java
public class ThreadConstruction
{
    public static void main(String[] args)
    {
        if (args.length < 1)
        {
            System.out.println("Please enter the stack size.");
            System.exit(1);
        }

        ThreadGroup group = new ThreadGroup("TestGroup");
```

```java
        Runnable runnable = new Runnable()
        {
            final int MAX = Integer.MAX_VALUE;

            @Override
            public void run()
            {
                int i = 0;
                recurse(i);
            }

            private void recurse(int i)
            {
                System.out.println(i);
                if (i < MAX)
                {
                    recurse(i + 1);
                }
            }
        };
        Thread thread = new Thread(group, runnable, "Test", Integer.parseInt(args[0]));

        thread.start();
    }
}
```

在代码清单 2-2 中,我们在线程中,设定了一个简单的递归,就是不断调用自己,然后输出 int 值,为了使得效果更加明显,在运行上面的代码过程中请指定 JVM 内存参数为:

```
java -Xmx512m -Xms64m ThreadConstruction 1
```

由于不断地进行压栈弹栈操作,整个栈内存肯定会被压爆,也就是说最后都会抛出 StackOverflowError 异常,笔者分别在 Windows 7 和 Ubuntu 环境下进行了多次测试,发现在创建 Thread 时传入的 stacksize 对递归深度的影响具体如表 2-1 所示。

表 2-1 stacksize 与递归次数

platform \ stacksize	1	10	100	1 000	10 000	100 000	1 000 000	10 000 000
Windows 7	9 709	9 704	9 716	9 723	190	875	9 694	9 2066
Ubuntu 16.04	1 284	1 279	1 285	1 281	1 287	1 309	8 661	9 5207

Ubuntu 的测试数据还算合理,但是 Windows 的测试数据让笔者哭笑不得(10 000 和 100 000),图 2-2 是针对测试数据的统计对比图。

通过图 2-2,大家还是可以很清晰地看到,随着 stacksize 数量级的不断增加,递归的深度也变得越来越大,该参数一般情况下不会主动设置,采用系统默认的值就可以了,默认情况下会设置成 0。

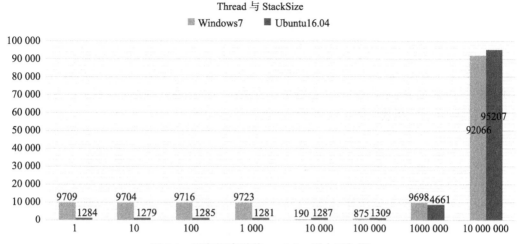

图 2-2　递归深度随着 stacksize 增大而加深

2.5.2　JVM 内存结构

虽然 stacksize 在构造的时候无须手动指定，但是我们通过 2.5.1 节的学习将会发现线程和栈内存的关系非常密切，想要了解他们之间到底有什么必然的联系，就需要了解 JVM 的内存分布机制，虽然本书不是讲解 JVM 原理和内存的书籍，但是笔者还是会向大家介绍一下 JVM 的内存分布，以及各个内存区域之间的关系，当然读者可以通过官方文档获取更权威的讲解：

http://www.oracle.com/technetwork/java/javase/memorymanagement-whitepaper-150215.pdf

JVM 在执行 Java 程序的时候会把对应的物理内存划分成不同的内存区域，每一个区域都存放着不同的数据，也有不同的创建与销毁时机，有些分区会在 JVM 启动的时候就创建，有些则是在运行时才创建，比如虚拟机栈，根据虚拟机规范，JVM 的内存结构如图 2-3 所示。

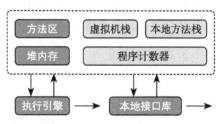

图 2-3　JVM 内存结构图

1. 程序计数器

无论任何语言，其实最终都是需要由操作系统通过控制总线向 CPU 发送机器指令，Java 也不例外，程序计数器在 JVM 中所起的作用就是用于存放当前线程接下来将要执行的字节码指令、分支、循环、跳转、异常处理等信息。在任何时候，一个处理器只执行其中一个线程中的指令，为了能够在 CPU 时间片轮转切换上下文之后顺利回到正确的执行位置，每条线程都需要具有一个独立的程序计数器，各个线程之间互相不影响，因此 JVM 将此块内存区域设计成了线程私有的。

2. Java 虚拟机栈

这里需要重点介绍虚拟机械内存区域，因为其与线程紧密关联，与程序计数器内存相类似，Java 虚拟机栈也是线程私有的，它的生命周期与线程相同，是在 JVM 运行时所创建的，在线程中，方法在执行的时候都会创建一个名为栈帧（stack frame）的数据结构，主要用于存放局部变量表、操作栈、动态链接、方法出口等信息，如图 2-4 所示，方法的调用也对应着栈帧在虚拟机栈中的压栈和弹栈过程。

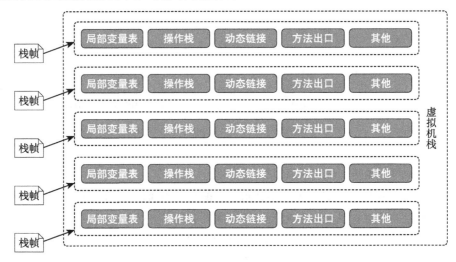

图 2-4 虚拟机栈结构图

每一个线程在创建的时候，JVM 都会为其创建对应的虚拟机栈，虚拟机栈的大小可以通过 -xss 来配置，方法的调用是栈帧被压入和弹出的过程，通过图 2-4 可以看出，同等的虚拟机栈如果局部变量表等占用内存越小则可被压入的栈帧就会越多，反之则可被压入的栈帧就会越少，一般将栈帧内存的大小称为宽度，而栈帧的数量则称为虚拟机栈的深度。

3. 本地方法栈

Java 中提供了调用本地方法的接口（Java Native Interface），也就是 C/C++ 程序，在线程的执行过程中，经常会碰到调用 JNI 方法的情况，比如网络通信、文件操作的底层，甚至是 String 的 intern 等都是 JNI 方法，JVM 为本地方法所划分的内存区域便是本地方法栈，这块内存区域其自由度非常高，完全靠不同的 JVM 厂商来实现，Java 虚拟机规范并未给出强制的规定，同样它也是线程私有的内存区域。

4. 堆内存

堆内存是 JVM 中最大的一块内存区域，被所有的线程所共享，Java 在运行期间创建的所有对象几乎都存放在该内存区域，该内存区域也是垃圾回收器重点照顾的区域，因此有些时候堆内存被称为"GC 堆"。

堆内存一般会被细分为新生代和老年代，更细致的划分为 Eden 区、From Survivor 区和 To Survivor 区，如图 2-5 所示。

5. 方法区

方法区也是被多个线程所共享的内存区域，他主要用于存储已经被虚拟机加载的类信息、常量、静态变量、即时编译器（JIT）编译后的代码等数据，虽然在 Java 虚拟机规范中，将方法区划分为堆内存

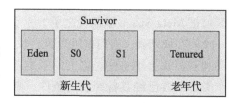

图 2-5　堆区的分代划分

的一个逻辑分区，但是它还是经常被称为"非堆"，有时候也被称为"持久代"，主要是站在垃圾回收器的角度进行划分，但是这种叫法比较欠妥，在 HotSpot JVM 中，方法区还会被细划分为持久代和代码缓存区，代码缓存区主要用于存储编译后的本地代码（和硬件相关）以及 JIT（Just In Time）编译器生成的代码，当然不同的 JVM 会有不同的实现。

6. Java 8 元空间

上述内容大致介绍了 JVM 的内存划分，在 JDK1.8 版本以前的内存大概都是这样划分的，但是自 JDK1.8 版本起，JVM 的内存区域发生了一些改变，实际上是持久代内存被彻底删除，取而代之的是元空间，图 2-6 与图 2-7 是使用分别使用不同版本的 jstat 命令对比 JVM 的 GC 内存分布。

JDK1.7 版本的 jstat

图 2-6　JDK1.8 以前没有元空间

JDK1.8 版本的 jstat

图 2-7　JDK1.8 加入了元空间

通过对比会发现在 JDK1.7 版本中存在持久代内存区域，而在 JDK1.8 版本中，该内存区域被 Meta Space 取而代之了，元空间同样是堆内存的一部分，JVM 为每个类加载器分配一块内存块列表，进行线性分配，块的大小取决于类加载器的类型，sun/ 反射 / 代理对应的类加载器块会小一些，之前的版本会单独卸载回收某个类，而现在则是 GC 过程中发现某个类加载器已经具备回收的条件，则会将整个类加载器相关的元空间全部回收，这样就可以减少内存碎片，节省 GC 扫描和压缩的时间。

2.5.3 Thread 与虚拟机栈

在 2.5.2 节中，我们简单地介绍了一下 JVM 的内存分布，其中程序计数器是比较小的一块内存，而且该部分内存是不会出现任何溢出异常的，与线程创建、运行、销毁等关系比较大的是虚拟机栈内存了，而且栈内存划分的大小将直接决定在一个 JVM 进程中可以创建多少个线程，请看下面的例子：

代码清单 2-3　ThreadCounter.java

```java
package com.wangwenjun.concurrent.chapter02;

import java.util.concurrent.TimeUnit;
import java.util.concurrent.atomic.AtomicInteger;

public class ThreadCounter extends Thread
{

    final static AtomicInteger counter = new AtomicInteger(0);

    public static void main(String[] args)
    {
        try
        {
            while (true)
            {
                new ThreadCounter().start();
            }
        } catch (Throwable e)
        {
            System.out.println("failed At=>" + counter.get());
        }
    }

    @Override
    public void run()
    {
        try
        {
            System.out.println("The " + counter.getAndIncrement() + " thread be
```

```
created.");
            TimeUnit.MINUTES.sleep(10);
        } catch (InterruptedException e)
        {
            e.printStackTrace();
        }
    }
}
```

在代码清单 2-3 中,我们不断地创建线程,直到 JVM 再也没有能力创建新的线程为止,我们通过设置栈内存的大小来分析栈内存大小对创建线程的影响,运行上面的程序很容易出现系统死机的情况,笔者为了这组测试数据重启了好几次电脑,简直不忍直视:

```
java -Xmx256m -Xms64m ThreadCounter
```

线程创建至某个数量后便不能再继续创建,我们通过不断地改变虚拟机栈内存的大小来查看可创建线程数量的变化,在开始测试之前,先要搞清楚 JVM 在默认情况下栈内存的大小,不同的版本大小不相同,可以通过下面的命令来查看(其中 ThreadStackSize 等价于 xss):

```
java -XX:+PrintFlagsFinal -version|grep ThreadStackSize
```

接下来我们固定堆内存的大小,不断增加栈内存的大小,以测试栈内存不断增大对 JVM 创建线程数量产生的影响。

通过表 2-2 中 5 组测试数据的对比我们不难看出,线程的创建数量是随着虚拟机栈内存的增多而减少的,也就是一种反比关系,如图 2-8 所示。

表 2-2 栈内存变化导致线程数量的变化

Xms	Xmx	Xss	Threads
512m	512m	1m	63 790
512m	512m	10m	60 803
512m	512m	64m	32 651
512m	512m	128m	20 678
512m	512m	512m	6 556

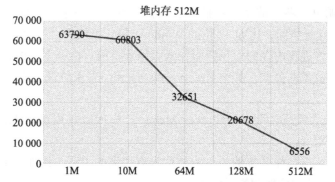

图 2-8 堆内存不变,栈内存越大,可创建的线程数量越小

其实上面的测试数据并不难理解,根据我们前面所学的知识可以得知,虚拟机栈内存是线程私有的,也就是说每一个线程都会占有指定的内存大小,我们粗略地认为一个 Java

进程的内存大小为：**堆内存 + 线程数量 * 栈内存**。

不管是 32 位操作系统还是 64 位操作系统，一个进程的最大内存是有限制的，比如 32 位的 Windows 操作系统所允许的最大进程内存为 2GB，因此根据上面的公式很容易得出，线程数量与栈内存的大小是反比关系，那么线程数量与堆内存的大小关系呢？当然也是反比关系，只不过堆内存是基数，而栈内存是系数而已，测试数据如表 2-3 所示。

表 2-3　堆内存变化导致线程数量变化

Xms	Xmx	Xss	Threads
64m	64m	1m	92 376
128m	128m	1m	90 948
256m	256m	1m	90 423
512m	512m	1m	87 931
1 024m	1 024m	1m	83 195

堆内存作为影响进程内存的基数，它的增大对线程数量的影响也是反比关系，但是并没有像栈内存那样明显，线程数量与堆内存大小的关系如图 2-9 所示。

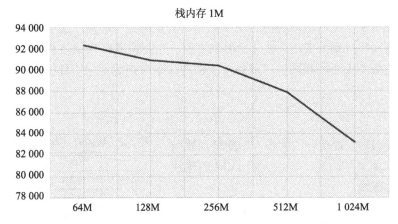

图 2-9　栈内存不变，堆内存越大，可创建的线程数量越小

在 JVM 中到底可以创建多少个线程，与堆内存、栈内存的大小有着直接的关系，只不过栈内存更加明显一些，前文中我们说过在操作系统中一个进程的内存大小是有限制的，这个限制称为地址空间，比如 32 位的 Windows 操作系统最大的地址空间约为 2G 多一点，操作系统则会将进程内存的大小控制在最大地址空间以内，下面的公式是一个相对比较精准的计算线程数量的公式，其中 ReservedOsMemory 是系统保留内存，一般在 136MB 左右：

线程数量 =（最大地址空间（MaxProcessMemory）- JVM 堆内存 - ReservedOsMemory）/ThreadStackSize（XSS）

当然线程数量还与操作系统的一些内核配置有很大的关系，比如在 Linux 下，下面三个内核配置信息也可以决定线程数量的大小。

- /proc/sys/kernel/threads-max
- /proc/sys/kernel/pid_max
- /proc/sys/vm/max_map_count

2.6 守护线程

守护线程是一类比较特殊的线程,一般用于处理一些后台的工作,比如 JDK 的垃圾回收线程,什么是守护线程?为什么要有守护线程,以及何时需要守护线程?本节我们就来一起探讨一下这个话题。

要回答关于守护线程的问题,就必须先搞清楚另外一个特别重要的问题:JVM 程序在什么情况下会退出?

※ `The Java Virtual Machine exits when the only threads running are all daemon threads.`

上面这句话来自于 JDK 的官方文档,当然这句话指的是正常退出的情况,而不是调用了 System.exit() 方法,通过这句话的描述,我们不难发现,在正常情况下,若 JVM 中没有一个非守护线程,则 JVM 的进程会退出。

2.6.1 什么是守护线程

我们先通过一个简单的程序,来认识一下守护线程和守护线程的特点:

```
package com.wangwenjun.concurrent.chapter02;

public class DaemonThread
{
    public static void main(String[] args) throws InterruptedException
    {
    //① main 线程开始
        Thread thread = new Thread(() ->
        {
            while (true)
            {
                try
                {
                    Thread.sleep(1);
                } catch (InterruptedException e)
                {
                    e.printStackTrace();
                }
            }
        });
        //thread.setDaemon(true); //②将 thread 设置为守护线程

        thread.start();   //③ 启动 thread 线程
        Thread.sleep(2_000L);
        System.out.println("Main thread finished lifecycle.");
        //④ main 线程结束
    }
}
```

上面的代码中存在两个线程,一个是由 JVM 启动的 main 线程,另外一个则是我们自已创建的线程 thread，运行上面的这段代码，你会发现 JVM 进程永远不会退出，即使 main 线程正常地结束了自己的生命周期（main 线程的生命周期是从注释①到注释④之间的那段代码），原因就是因为在 JVM 进程中还存在一个非守护线程在运行。

如果打开注释②，也就是通过 setDaemon 方法将 thread 设置为了守护线程，那么 main 进程结束生命周期后，JVM 也会随之退出运行，当然 thread 线程也会结束。

> **注意** 设置守护线程的方法很简单，调用 setDaemon 方法即可，true 代表守护线程，false 代表正常线程。

线程是否为守护线程和它的父线程有很大的关系，如果父线程是正常线程，则子线程也是正常线程，反之亦然，如果你想要修改它的特性则可以借助 setDaemon 方法。isDaemon 方法可以判断该线程是不是守护线程。

另外需要注意的就是，setDaemon 方法只在线程启动之前才能生效，如果一个线程已经死亡，那么再设置 setDaemon 则会抛出 IllegalThreadStateException 异常。

2.6.2 守护线程的作用

在了解了什么是守护线程，以及如何创建守护线程之后，我们来讨论一下为什么要有守护线程以及何时使用守护线程。

通过上面的分析，如果一个 JVM 进程中没有一个非守护线程，那么 JVM 会退出，也就是说守护线程具备自动结束生命周期的特性，而非守护线程则不具备这个特点，试想一下如果 JVM 进程的垃圾回收线程是非守护线程，如果 main 线程完成了工作，则 JVM 无法退出，因为垃圾回收线程还在正常的工作。再比如有一个简单的游戏程序，其中有一个线程正在与服务器不断地交互以获取玩家最新的金币、武器信息，若希望在退出游戏客户端的时候，这些数据同步的工作也能够立即结束，等等。

守护线程经常用作与执行一些后台任务，因此有时它也被称为后台线程，当你希望关闭某些线程的时候，或者退出 JVM 进程的时候，一些线程能够自动关闭，此时就可以考虑用守护线程为你完成这样的工作。

2.7 本章总结

在本章中，我们非常详细地讲解了 Thread 的构造函数，并且挖掘了其中的很多细节，尤其是 stacksize 对 Thread 的影响。

除此之外，本章也介绍了线程的父子关系，默认情况下子线程从父线程那里是否继承了守护线程、优先级、ThreadGroup 等特性。

最后，本章还分析了什么是守护线程，以及守护线程的特性以及其应该使用在何种场景之下。

CHAPTER3 · 第 3 章

Thread API 的详细介绍

第 1 章和第 2 章主要是从概念上去了解 Thread，而在本章中，我们将细致地学习 Thread 所有 API 的作用以及用法，同样，关键的地方我们也会去剖析源码。

3.1 线程 sleep

sleep 是一个静态方法，其有两个重载方法，其中一个需要传入毫秒数，另外一个既需要毫秒数也需要纳秒数。

3.1.1 sleep 方法介绍

- public static void sleep(long millis) throws InterruptedException
- public static void sleep(long millis, int nanos) throws InterruptedException

sleep 方法会使当前线程进入指定毫秒数的休眠，暂停执行，虽然给定了一个休眠的时间，但是最终要以系统的定时器和调度器的精度为准，休眠有一个非常重要的特性，那就是其不会放弃 monitor 锁的所有权（在后文中讲解线程同步和锁的时候会重点介绍 monitor），下面我们来看一个简单的例子：

```
package com.wangwenjun.concurrent.chapter03;

public class ThreadSleep
{
    public static void main(String[] args)
    {
        new Thread(() ->
        {
```

```
            long startTime = System.currentTimeMillis();
            sleep(2_000L);
            long endTime = System.currentTimeMillis();
            System.out.println(String.format("Total spend %d ms", (endTime - startTime)));
        }).start();

        long startTime = System.currentTimeMillis();
        sleep(3_000L);
        long endTime = System.currentTimeMillis();
        System.out.println(String.format("Main thread total spend %d ms", (endTime - startTime)));
    }

    private static void sleep(long ms)
    {
        try
        {
            Thread.sleep(ms);
        } catch (InterruptedException e)
        {
        }
    }
}
```

在上面的例子中，我们分别在自定义的线程和主线程中进行了休眠，每个线程的休眠互不影响，Thread.sleep 只会导致当前线程进入指定时间的休眠。

3.1.2 使用 TimeUnit 替代 Thread.sleep

在 JDK1.5 以后，JDK 引入了一个枚举 TimeUnit，其对 sleep 方法提供了很好的封装，使用它可以省去时间单位的换算步骤，比如线程想休眠 3 小时 24 分 17 秒 88 毫秒，使用 TimeUnit 来实现就非常的简便优雅了：

```
Thread.sleep(12257088L);
TimeUnit.HOURS.sleep(3);
TimeUnit.MINUTES.sleep(24);
TimeUnit.SECONDS.sleep(17);
TimeUnit.MILLISECONDS.sleep(88);
```

同样的时间表达，TimeUnit 显然清晰很多，笔者强烈建议，在使用 Thread.sleep 的地方，完全使用 TimeUnit 来代替，因为 sleep 能做的事，TimeUnit 全部都能完成，并且功能更加的强大，在本书后面的内容中，我将全部采用 TimeUnit 替代 sleep。

3.2 线程 yield

3.2.1 yield 方法介绍

yield 方法属于一种启发式的方法，其会提醒调度器我愿意放弃当前的 CPU 资源，如果 CPU 的资源不紧张，则会忽略这种提醒。

调用 yield 方法会使当前线程从 RUNNING 状态切换到 RUNNABLE 状态，一般这个方法不太常用：

```java
package com.wangwenjun.concurrent.chapter03;

import java.util.stream.IntStream;

public class ThreadYield
{
    public static void main(String[] args)
    {
        IntStream.range(0, 2).mapToObj(ThreadYield::create)
                .forEach(Thread::start);
    }

    private static Thread create(int index)
    {
        return new Thread(() ->
        {
            //①注释部分
            //if (index == 0)
            //    Thread.yield();
            System.out.println(index);
        });
    }
}
```

上面的程序运行很多次，你会发现输出的结果不一致，有时候是 0 最先打印出来，有时候是 1 最先打印出来，但是当你打开代码的注释部分，你会发现，顺序始终是 0，1。

因为第一个线程如果最先获得了 CPU 资源，它会比较谦虚，主动告诉 CPU 调度器释放了原本属于自己的资源，但是 yield 只是一个提示（hint），CPU 调度器并不会担保每次都能满足 yield 提示。

3.2.2 yield 和 sleep

看过前面的内容之后，会发现 yield 和 sleep 有一些混淆的地方，在 JDK1.5 以前的版本中 yield 的方法事实上是调用了 sleep(0)，但是它们之间存在着本质的区别，具体如下。

- sleep 会导致当前线程暂停指定的时间，没有 CPU 时间片的消耗。
- yield 只是对 CPU 调度器的一个提示，如果 CPU 调度器没有忽略这个提示，它会导致线程上下文的切换。
- sleep 会使线程短暂 block，会在给定的时间内释放 CPU 资源。
- yield 会使 RUNNING 状态的 Thread 进入 RUNNABLE 状态（如果 CPU 调度器没有忽略这个提示的话）。
- sleep 几乎百分之百地完成了给定时间的休眠，而 yield 的提示并不能一定担保。
- 一个线程 sleep 另一个线程调用 interrupt 会捕获到中断信号，而 yield 则不会。

3.3 设置线程的优先级

- public final void setPriority(int newPriority) 为线程设定优先级。
- public final int getPriority() 获取线程的优先级。

3.3.1 线程优先级介绍

进程有进程的优先级，线程同样也有优先级，理论上是优先级比较高的线程会获取优先被 CPU 调度的机会，但是事实上往往并不会如你所愿，设置线程的优先级同样也是一个 hint 操作，具体如下。

- 对于 root 用户，它会 hint 操作系统你想要设置的优先级别，否则它会被忽略。
- 如果 CPU 比较忙，设置优先级可能会获得更多的 CPU 时间片，但是闲时优先级的高低几乎不会有任何作用。

所以，不要在程序设计当中企图使用线程优先级绑定某些特定的业务，或者让业务严重依赖于线程优先级，这可能会让你大失所望。举个简单的例子，可能不同情况下的运行效果不会完全一样，但是我们只是想让优先级比较高的线程获得更多的信息输出机会，示例代码如下：

```
package com.wangwenjun.concurrent.chapter03;

public class ThreadPriority
{
    public static void main(String[] args)
    {
        Thread t1 = new Thread(() ->
        {
            while (true)
            {
                System.out.println("t1");
            }
        });
        t1.setPriority(3);
```

```
            Thread t2 = new Thread(() ->
            {
                while (true)
                {
                    System.out.println("t2");
                }
            });
            t2.setPriority(10);
            t1.start();
            t2.start();
    }
}
```

运行上面的程序，会发现 t2 出现的频率很明显要高一些，当然这也和笔者当前 CPU 的资源情况有关系，不同情况下的运行会有不一样的结果。

3.3.2 线程优先级源码分析

设置线程的优先级，只需要调用 setPriority 方法即可，下面我们打开 Thread 的源码，一起来分析一下：

```
public final void setPriority(int newPriority) {
    ThreadGroup g;
    checkAccess();
    if (newPriority > MAX_PRIORITY || newPriority < MIN_PRIORITY) {
        throw new IllegalArgumentException();
    }
    if((g = getThreadGroup()) != null) {
        if (newPriority > g.getMaxPriority()) {
            newPriority = g.getMaxPriority();
        }
        setPriority0(priority = newPriority);
    }
}
```

通过上面源码的分析，我们可以看出，线程的优先级不能小于 1 也不能大于 10，如果指定的线程优先级大于线程所在 group 的优先级，那么指定的优先级将会失效，取而代之的是 group 的最大优先级，下面我们通过一个例子来证明一下：

```
package com.wangwenjun.concurrent.chapter03;

public class ThreadPriority
{
    public static void main(String[] args)
    {
        //定义一个线程组
        ThreadGroup group = new ThreadGroup("test");
        //将线程组的优先级指定为 7
        group.setMaxPriority(7);
```

```
        //定义一个线程,将该线程加入到group中
        Thread thread = new Thread(group, "test-thread");
        //企图将线程的优先级设定为10
        thread.setPriority(10);
        //企图未遂
        System.out.println(thread.getPriority());
    }
}
```

上面的结果输出为 7,而不是 10,因为它超过了所在线程组的优先级别

3.3.3 关于优先级的一些总结

一般情况下,不会对线程设定优先级别,更不会让某些业务严重地依赖线程的优先级别,比如权重,借助优先级设定某个任务的权重,这种方式是不可取的,一般定义线程的时候使用默认的优先级就好了,那么线程默认的优先级是多少呢?

线程默认的优先级和它的父类保持一致,一般情况下都是 5,因为 main 线程的优先级就是 5,所以它派生出来的线程都是 5,示例代码如下:

```
package com.wangwenjun.concurrent.chapter03;

public class ThreadPriority
{
    public static void main(String[] args)
    {
        Thread t1 = new Thread();
        System.out.println("t1 priority " + t1.getPriority());
        Thread t2 = new Thread(() ->
        {
            Thread t3 = new Thread();
            System.out.println("t3 priority " + t3.getPriority());
        });

        t2.setPriority(6);
        t2.start();
        System.out.println("t2 priority " + t2.getPriority());
    }
}
```

上面程序的输出结果是 t1 的优先级为 5,因为 main 线程的优先级是 5;t2 的优先级是 6,因为显式地将其指定为 6;t3 的优先级为 6,没有显式地指定,因此其与父线程保持一致。

3.4 获取线程 ID

public long getId() 获取线程的唯一 ID,线程的 ID 在整个 JVM 进程中都会是唯一的,

并且是从 0 开始逐次递增。如果你在 main 线程（main 函数）中创建了一个唯一的线程，并且调用 getId() 后发现其并不等于 0，也许你会纳闷，不应该是从 0 开始的吗？之前已经说过了在一个 JVM 进程启动的时候，实际上是开辟了很多个线程，自增序列已经有了一定的消耗，因此我们自己创建的线程绝非第 0 号线程。

3.5 获取当前线程

public static Thread currentThread() 用于返回当前执行线程的引用，这个方法虽然很简单，但是使用非常广泛，我们在后面的内容中会大量的使用该方法，来看一段示例代码：

```
package com.wangwenjun.concurrent.chapter03;

public class CurrentThread
{
    public static void main(String[] args)
    {
        Thread thread = new Thread()
        {
            @Override
            public void run()
            {
                //always true
                System.out.println(Thread.currentThread() == this);
            }
        };
        thread.start();

        String name = Thread.currentThread().getName();
        System.out.println("main".equals(name));
    }
}
```

上面程序运行输出的两个结果都是 true。

3.6 设置线程上下文类加载器

- public ClassLoader getContextClassLoader() 获取线程上下文的类加载器，简单来说就是这个线程是由哪个类加载器加载的，如果是在没有修改线程上下文类加载器的情况下，则保持与父线程同样的类加载器。
- public void setContextClassLoader(ClassLoader cl) 设置该线程的类加载器，这个方法可以打破 JAVA 类加载器的父委托机制，有时候该方法也被称为 JAVA 类加载器的后门。

关于线程上下文类加载器的内容我们将在本书的第 11 章重点介绍，并且结合 jdbc 驱动包的源码分析 JDK 的开发者为什么要留有这样的后门。

3.7 线程 interrupt

线程 interrupt，是一个非常重要的 API，也是经常使用的方法，与线程中断相关的 API 有如下几个，在本节中我们也将 Thread 深入源码对其进行详细的剖析。

- public void interrupt()
- public static boolean interrupted()
- public boolean isInterrupted()

3.7.1 interrupt

如下方法的调用会使得当前线程进入阻塞状态，而调用当前线程的 interrupt 方法，就可以打断阻塞。

- Object 的 wait 方法。
- Object 的 wait(long) 方法。
- Object 的 wait(long,int) 方法。
- Thread 的 sleep(long) 方法。
- Thread 的 sleep(long,int) 方法。
- Thread 的 join 方法。
- Thread 的 join(long) 方法。
- Thread 的 join(long,int) 方法。
- InterruptibleChannel 的 io 操作。
- Selector 的 wakeup 方法。
- 其他方法。

上述若干方法都会使得当前线程进入阻塞状态，若另外的一个线程调用被阻塞线程的 interrupt 方法，则会打断这种阻塞，因此这种方法有时会被称为可中断方法，记住，打断一个线程并不等于该线程的生命周期结束，仅仅是打断了当前线程的阻塞状态。

一旦线程在阻塞的情况下被打断，都会抛出一个称为 InterruptedException 的异常，这个异常就像一个 signal（信号）一样通知当前线程被打断了，下面我们来看一个例子：

```
package com.wangwenjun.concurrent.chapter03;

import java.util.concurrent.TimeUnit;

public class ThreadInterrupt
{
```

```java
public static void main(String[] args) throws InterruptedException
{
    Thread thread = new Thread(() ->
    {
        try
        {
            TimeUnit.MINUTES.sleep(1);
        } catch (InterruptedException e)
        {
            System.out.println("Oh, i am be interrupted.");
        }
    });

    thread.start();

    //short block and make sure thread is started.
    TimeUnit.MILLISECONDS.sleep(2);
    thread.interrupt();
}
```

上面的代码创建了一个线程，并且企图休眠 1 分钟的时长，不过很可惜，大约在 2 毫秒之后就被主线程调用 interrupt 方法打断，程序的执行结果就是 "Oh, i am be interrupted."

interrupt 这个方法到底做了什么样的事情呢？在一个线程内部存在着名为 interrupt flag 的标识，如果一个线程被 interrupt，那么它的 flag 将被设置，但是如果当前线程正在执行可中断方法被阻塞时，调用 interrupt 方法将其中断，反而会导致 flag 被清除，关于这点我们在后面还会做详细的介绍。另外有一点需要注意的是，如果一个线程已经是死亡状态，那么尝试对其的 interrupt 会直接被忽略。

3.7.2　isInterrupted

isInterrupted 是 Thread 的一个成员方法，它主要判断当前线程是否被中断，该方法仅仅是对 interrupt 标识的一个判断，并不会影响标识发生任何改变，这个与我们即将学习到的 interrupted 是存在差别的，下面我们看一个简单的程序：

```java
package com.wangwenjun.concurrent.chapter03;

import java.util.concurrent.TimeUnit;

public class ThreadisInterrupted
{
    public static void main(String[] args) throws InterruptedException
    {
        Thread thread = new Thread()
```

```java
        {
            @Override
            public void run()
            {
                while (true)
                {
                    //do nothing, just empty loop.
                }
            }
        };

        thread.start();
        TimeUnit.MILLISECONDS.sleep(2);
        System.out.printf("Thread is interrupted ? %s\n", thread.isInterrupted());
        thread.interrupt();
        System.out.printf("Thread is interrupted ? %s\n", thread.isInterrupted());
    }
}
```

上面的代码中定义了一个线程，并且在线程的执行单元中（run 方法）写了一个空的死循环，为什么不写 sleep 呢？因为 sleep 是可中断方法，会捕获到中断信号，从而干扰我们程序的结果。下面是程序运行的结果，记得手动结束上面的程序运行，或者你也可以将上面定义的线程指定为守护线程，这样就会随着主线程的结束导致 JVM 中没有非守护线程而自动退出。

```
Thread is interrupted ? false
Thread is interrupted ? true
```

可中断方法捕获到了中断信号（signal）之后，也就是捕获了 InterruptedException 异常之后会擦除掉 interrupt 的标识，对上面的程序稍作修改，你会发现程序的结果又会出现很大的不同，示例代码如下：

```java
package com.wangwenjun.concurrent.chapter03;

import java.util.concurrent.TimeUnit;

public class ThreadisInterrupted
{
    public static void main(String[] args) throws InterruptedException
    {
        Thread thread = new Thread()
        {
            @Override
            public void run()
            {
                while (true)
                {
```

```
                try
                {
                    TimeUnit.MINUTES.sleep(1);
                } catch (InterruptedException e)
                {
                    //ignore the exception
                    //here the interrupt flag will be clear.
                    System.out.printf("I am be interrupted ? %s\n",
isInterrupted());
                }
            }
        }
    };

    thread.setDaemon(true);
    thread.start();
    TimeUnit.MILLISECONDS.sleep(2);
    System.out.printf("Thread is interrupted ? %s\n", thread.isInterrupted());
    thread.interrupt();
    TimeUnit.MILLISECONDS.sleep(2);
    System.out.printf("Thread is interrupted ? %s\n", thread.isInterrupted());
    }
}
```

由于在 run 方法中使用了 sleep 这个可中断方法，它会捕获到中断信号，并且会擦除 interrupt 标识，因此程序的执行结果都会是 false，程序输出如下：

```
Thread is interrupted ? false
I am be interrupted ? false
Thread is interrupted ? false
```

其实这也不难理解，可中断方法捕获到了中断信号之后，为了不影响线程中其他方法的执行，将线程的 interrupt 标识复位是一种很合理的设计。

3.7.3 interrupted

interrupted 是一个静态方法，虽然其也用于判断当前线程是否被中断，但是它和成员方法 isInterrupted 还是有很大的区别的，调用该方法会直接擦除掉线程的 interrupt 标识，需要注意的是，如果当前线程被打断了，那么第一次调用 interrupted 方法会返回 true，并且立即擦除了 interrupt 标识；第二次包括以后的调用永远都会返回 false，除非在此期间线程又一次地被打断，下面设计了一个简单的例子，来验证我们的说法：

```
package com.wangwenjun.concurrent.chapter03;

import java.util.concurrent.TimeUnit;

public class ThreadInterrupted
{
```

```java
public static void main(String[] args) throws InterruptedException
{
    Thread thread = new Thread()
    {
        @Override
        public void run()
        {
            while (true)
            {
                System.out.println(Thread.interrupted());
            }
        }
    };
    thread.setDaemon(true);
    thread.start();

    //shortly block make sure the thread is started.
    TimeUnit.MILLISECONDS.sleep(2);
    thread.interrupt();
}
```

同样由于不想要受到可中断方法如 sleep 的影响，在 Thread 的 run 方法中没有进行任何短暂的休眠，所以运行上面的程序会出现非常多的输出，但是我们通过对输出的检查会发现如下所示的内容，其足以作为对该方法的解释。

```
...
false
false
true
false
false
...
```

在很多的 false 包围中发现了一个 true，也就是 interrupted 方法判断到了其被中断，立即擦除了中断标识，并且只有这一次返回 true，后面的都将会是 false。

3.7.4　interrupt 注意事项

打开 Thread 的源码，不难发现，isInterrupted 方法和 interrupted 方法都调用了同一个本地方法：

```
private native boolean isInterrupted(boolean ClearInterrupted);
```

其中参数 ClearInterrupted 主要用来控制是否擦除线程 interrupt 的标识。
isInterrupted 方法的源码中该参数为 false，表示不想擦除：

```java
public boolean isInterrupted() {
    return isInterrupted(false);
}
```

而 interrupted 静态方法中该参数则为 true，表示想要擦除：

```java
public static boolean interrupted() {
    return currentThread().isInterrupted(true);
}
```

在比较详细地学习了 interrupt 方法之后，大家思考一个问题，如果一个线程在没有执行可中断方法之前就被打断，那么其接下来将执行可中断方法，比如 sleep 会发生什么样的情况呢？下面我们通过一个简单的实验来回答这个疑问：

```java
public static void main(String[] args)
{
    //① 判断当前线程是否被中断
    System.out.println("Main thread is interrupted? " + Thread.interrupted());

    //②中断当前线程
    Thread.currentThread().interrupt();

    //③判断当前线程是否已经被中断
    System.out.println("Main thread is interrupted? " + Thread.currentThread().isInterrupted());
    try
    {
        //④ 当前线程执行可中断方法
        TimeUnit.MINUTES.sleep(1);
    } catch (InterruptedException e)
    {
        //⑤捕获中断信号
        System.out.println("I will be interrupted still.");
    }
}
```

通过运行上面的程序，你会发现，如果一个线程设置了 interrupt 标识，那么接下来的可中断方法会立即中断，因此注释⑤的信号捕获部分代码会被执行，请大家注意注释①和注释③中判断线程中断方法的不同，也希望读者结合本节的内容思考为什么要这么做？

3.8 线程 join

Thread 的 join 方法同样是一个非常重要的方法，使用它的特性可以实现很多比较强大的功能，与 sleep 一样它也是一个可中断的方法，也就是说，如果有其他线程执行了对当前线程的 interrupt 操作，它也会捕获到中断信号，并且擦除线程的 interrupt 标识，Thread 的 API 为我们提供了三个不同的 join 方法，具体如下。

- public final void join() throws InterruptedException
- public final synchronized void join(long millis, int nanos)
  throws InterruptedException
- public final synchronized void join(long millis)
  throws InterruptedException

在本节中，笔者将会详细介绍 join 方法以及如何在实际应用中使用 join 方法。

3.8.1 线程 join 方法详解

join 某个线程 A，会使当前线程 B 进入等待，直到线程 A 结束生命周期，或者到达给定的时间，那么在此期间 B 线程是处于 BLOCKED 的，而不是 A 线程，下面就来通过一个简单的实例解释一下 join 方法的基本用法：

```java
package com.wangwenjun.concurrent.chapter03;

import java.util.List;
import java.util.concurrent.TimeUnit;
import java.util.stream.IntStream;

import static java.util.stream.Collectors.toList;

public class ThreadJoin
{
    public static void main(String[] args) throws InterruptedException
    {
        //① 定义两个线程，并保存在 threads 中
        List<Thread> threads = IntStream.range(1, 3)
                .mapToObj(ThreadJoin::create).collect(toList());

        //② 启动这两个线程
        threads.forEach(Thread::start);

        //③ 执行这两个线程的 join 方法
        for (Thread thread : threads)
        {
            thread.join();
        }

        //④ main 线程循环输出
        for (int i = 0; i < 10; i++)
        {
            System.out.println(Thread.currentThread().getName() + "#" + i);
            shortSleep();
        }
    }
```

```java
//构造一个简单的线程,每个线程只是简单的循环输出
private static Thread create(int seq)
{
    return new Thread(() ->
    {
        for (int i = 0; i < 10; i++)
        {
            System.out.println(Thread.currentThread().getName() + "#" + i);
            shortSleep();
        }
    }, String.valueOf(seq));
}

private static void shortSleep()
{
    try
    {
        TimeUnit.SECONDS.sleep(1);
    } catch (InterruptedException e)
    {
        e.printStackTrace();
    }
}
```

上面的代码结合 Java 8 的语法,创建了两个线程,分别启动,并且调用了每个线程的 join 方法(注意:join 方法是被主线程调用的,因此在第一个线程还没有结束生命周期的时候,第二个线程的 join 不会得到执行,但是此时,第二个线程也已经启动了),运行上面的程序,你会发现线程一和线程二会交替地输出直到它们结束生命周期,main 线程的循环才会开始运行,程序输出如下:

```
...
2#8
1#8
2#9
1#9
main#0
main#1
main#2
main#3
...
```

如果你将注释③下面的 join 全部注释掉,那么三个线程将会交替地输出,程序输出如下:

```
...
main#2
2#2
1#2
main#3
```

```
1#3
2#3
main#4
...
```

join 方法会使当前线程永远地等待下去,直到期间被另外的线程中断,或者 join 的线程执行结束,当然你也可以使用 join 的另外两个重载方法,指定毫秒数,在指定的时间到达之后,当前线程也会退出阻塞。同样思考一个问题,如果一个线程已经结束了生命周期,那么调用它的 join 方法的当前线程会被阻塞吗?

3.8.2 join 方法结合实战

本节我们将结合一个实际的案例,来看一下 join 方法的应用场景,假设你有一个 APP,主要用于查询航班信息,你的 APP 是没有这些实时数据的,当用户发起查询请求时,你需要到各大航空公司的接口获取信息,最后统一整理加工返回到 APP 客户端,如图 3-1 所示,当然 JDK 自带了很多高级工具,比如 CountDownLatch 和 CyclicBarrier 等都可以完成类似的功能,但是仅就我们目前所学的知识,使用 join 方法即可完成下面的功能。

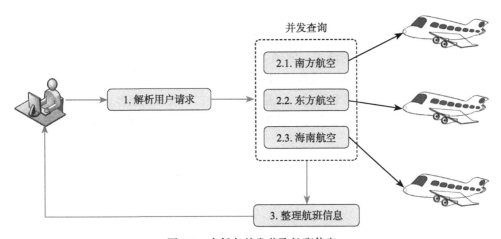

图 3-1 多任务并发获取航班信息

该例子是典型的串行任务局部并行化处理,用户在 APP 客户端输入出发地"北京"和目的地"上海",服务器接收到这个请求之后,先来验证用户的信息,然后到各大航空公司的接口查询信息,最后经过整理加工返回给客户端,每一个航空公司的接口不会都一样,获取的数据格式也不一样,查询的速度也存在着差异,如果再跟航空公司进行串行化交互(逐个地查询),很明显客户端需要等待很长的时间,这样的话,用户体验就会非常差。如果我们将每一个航空公司的查询都交给一个线程去工作,然后在它们结束工作之后统一对数据进行整理,这样就可以极大地节约时间,从而提高用户体验效果。

代码清单 3-1　查询接口 FightQuery

```
package com.wangwenjun.concurrent.chapter03;

import java.util.List;

public interface FightQuery
{
    List<String> get();
}
```

在代码清单 3-1 中，FightQuery 提供了一个返回方法，学到这里大家应该注意到了，不管是 Thread 的 run 方法，还是 Runnable 接口，都是 void 返回类型，如果你想通过某个线程的运行得到结果，就需要自己定义一个返回的接口。

查询 Fight 的 task，其实就是一个线程的子类，主要用于到各大航空公司获取数据，示例代码如下：

```
package com.wangwenjun.concurrent.chapter03;

import java.util.ArrayList;
import java.util.List;
import java.util.concurrent.ThreadLocalRandom;
import java.util.concurrent.TimeUnit;

public class FightQueryTask extends Thread
        implements FightQuery
{

    private final String origin;

    private final String destination;

    private final List<String> flightList = new ArrayList<>();

    public FightQueryTask(String airline, String origin, String destination)
    {
        super("[" + airline + "]");
        this.origin = origin;
        this.destination = destination;
    }

    @Override
    public void run()
    {
        System.out.printf("%s-query from %s to %s \n", getName(), origin, destination);
        int randomVal = ThreadLocalRandom.current().nextInt(10);
        try
        {
            TimeUnit.SECONDS.sleep(randomVal);
            this.flightList.add(getName() + "-" + randomVal);
            System.out.printf("The Fight:%s list query successful\n", getName());
```

```
            } catch (InterruptedException e)
            {
            }
        }

        @Override
        public List<String> get()
        {
            return this.flightList;
        }
    }
```

接口定义好了,查询航班数据的线程也有了,下面就来实现一下从SH(上海)到北京(BJ)的航班查询吧!示例代码如下:

```
package com.wangwenjun.concurrent.chapter03;

import java.util.ArrayList;
import java.util.Arrays;
import java.util.List;

import static java.util.stream.Collectors.toList;

public class FightQueryExample
{
    //①合作的各大航空公司
    private static List<String> fightCompany = Arrays.asList(
            "CSA", "CEA", "HNA"
    );

    public static void main(String[] args)
    {

        List<String> results = search("SH", "BJ");
        System.out.println("==========result==========");
        results.forEach(System.out::println);

    }

    private static List<String> search(String original, String dest)
    {
        final List<String> result = new ArrayList<>();

        //②创建查询航班信息的线程列表
        List<FightQueryTask> tasks = fightCompany.stream()
                .map(f -> createSearchTask(f, original, dest))
                .collect(toList());

        //③分别启动这几个线程
        tasks.forEach(Thread::start);
```

```java
    //④分别调用每一个线程的join方法,阻塞当前线程
    tasks.forEach(t ->
    {
        try
        {
            t.join();
        } catch (InterruptedException e)
        {
        }
    });

    //⑤在此之前,当前线程会阻塞住,获取每一个查询线程的结果,并且加入到result中
    tasks.stream().map(FightQuery::get).forEach(result::addAll);

    return result;
}

private static FightQueryTask createSearchTask(
        String fight,
        String original, String dest)
{
    return new FightQueryTask(fight, original, dest);
}
```

上面的代码,关键的地方已通过注释解释得非常清楚,主线程收到了 search 请求之后,交给了若干个查询线程分别进行工作,最后将每一个线程获取的航班数据进行统一的汇总。由于每个航空公司的查询时间可能不一样,所以用了一个随机值来反应不同的查询速度,返回给客户端(打印到控制台),程序的执行结果输出如下:

```
[CSA]-query from SH to BJ
[CEA]-query from SH to BJ
[HNA]-query from SH to BJ
The Fight:[HNA] list query successful
The Fight:[CSA] list query successful
The Fight:[CEA] list query successful
==========result==========
[CSA]-4
[CEA]-7
[HNA]-2
```

3.9 如何关闭一个线程

JDK 有一个 Deprecated 方法 stop,但是该方法存在一个问题,JDK 官方早已经不推荐使用,其在后面的版本中有可能会被移除,根据官网的描述,该方法在关闭线程时可能不

会释放掉 monitor 的锁，所以强烈建议不要使用该方法结束线程，本节将主要介绍几种关闭线程的方法。

3.9.1 正常关闭

1. 线程结束生命周期正常结束

线程运行结束，完成了自己的使命之后，就会正常退出，如果线程中的任务耗时比较短，或者时间可控，那么放任它正常结束就好了。

2. 捕获中断信号关闭线程

我们通过 new Thread 的方式创建线程，这种方式看似很简单，其实它的派生成本是比较高的，因此在一个线程中往往会循环地执行某个任务，比如心跳检查，不断地接收网络消息报文等，系统决定退出的时候，可以借助中断线程的方式使其退出，示例代码如下：

```java
package com.wangwenjun.concurrent.chapter03;

import java.util.concurrent.TimeUnit;

public class InterruptThreadExit
{
    public static void main(String[] args) throws InterruptedException
    {
        Thread t = new Thread()
        {
            @Override
            public void run()
            {
                System.out.println("I will start work");
                while (!isInterrupted())
                {
                    //working.
                }
                System.out.println("I will be exiting.");
            }
        };
        t.start();
        TimeUnit.MINUTES.sleep(1);
        System.out.println("System will be shutdown.");
        t.interrupt();
    }
}
```

上面的代码是通过检查线程 interrupt 的标识来决定是否退出的，如果在线程中执行某个可中断方法，则可以通过捕获中断信号来决定是否退出。

```
    @Override
    public void run()
    {
        System.out.println("I will start work");
        for (; ; )
        {
            //working.
            try
            {
                TimeUnit.MILLISECONDS.sleep(1);
            } catch (InterruptedException e)
            {
                break;
            }
        }
        System.out.println("I will be exiting.");
    }
```

上面的代码执行结果都会导致线程正常的结束，程序输出如下：

```
I will start work
System will be shutdown.
I will be exiting.
```

3. 使用 volatile 开关控制

由于线程的 interrupt 标识很有可能被擦除，或者逻辑单元中不会调用任何可中断方法，所以使用 volatile 修饰的开关 flag 关闭线程也是一种常用的做法，具体如下：

```
package com.wangwenjun.concurrent.chapter03;

import java.util.concurrent.TimeUnit;

public class FlagThreadExit
{

    static class MyTask extends Thread
    {

        private volatile boolean closed = false;

        @Override
        public void run()
        {
            System.out.println("I will start work");
            while (!closed && !isInterrupted())
            {
                // 正在运行
            }
            System.out.println("I will be exiting.");
        }
```

```java
    public void close()
    {
        this.closed = true;
        this.interrupt();
    }
}

public static void main(String[] args) throws InterruptedException
{
    MyTask t = new MyTask();
    t.start();
    TimeUnit.MINUTES.sleep(1);
    System.out.println("System will be shutdown.");
    t.close();
}
```

上面的例子中定义了一个 closed 开关变量，并且是使用 volatile 修饰（关于 volatile 关键字会在本书的第 3 部分中进行非常细致地讲解，volatile 关键字在 Java 中是一个革命性的关键字，非常重要，它是 Java 原子变量以及并发包的基础）运行上面的程序同样也可以关闭线程。

3.9.2 异常退出

在一个线程的执行单元中，是不允许抛出 checked 异常的，不论 Thread 中的 run 方法，还是 Runnable 中的 run 方法，如果线程在运行过程中需要捕获 checked 异常并且判断是否还有运行下去的必要，那么此时可以将 checked 异常封装成 unchecked 异常 (RuntimeException) 抛出进而结束线程的生命周期。

3.9.3 进程假死

相信很多程序员都会遇到进程假死的情况，所谓假死就是进程虽然存在，但没有日志输出，程序不进行任何的作业，看起来就像死了一样，但事实上它是没有死的，程序之所以出现这样的情况，绝大部分的原因就是某个线程阻塞了，或者线程出现了死锁的情况。

我们需要借助一些工具来帮助诊断，比如 jstack、jconsole、jvisualvm 等工具，在本节中，笔者将简单介绍一下 jvisualvm 这个可视化工具，在第 4 章中我们还会接触这些工具进行死锁的判断等操作。

笔者在写这本书的时候一直在用 IntelliJ IDEA 进行代码的编写，IntelliJ IDEA 其实也是一个 Java 进程，打开 jvisualvm，选择 IntelliJ IDEA 进程，如图 3-2 所示。

将右侧的 Tab 切换到【线程】，如图 3-3 所示。

第3章 Thread API的详细介绍 57

图 3-2　使用 jvisualvm 观察 JVM

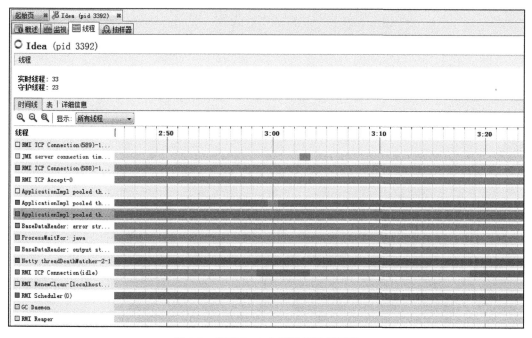

图 3-3　观察 JVM 内部线程运行情况

如果进程无法退出，则会出现假死的情况，可以打开 jvisualvm 查看有哪些活跃线程，它们的状态是什么,该线程在调用哪个方法而进入了阻塞。

3.10 本章总结

在本章中,我们比较详细地学习了 Thread 的大多数 API,其中有获取线程信息的方法,如 getId()、getName()、getPriority()、currentThread(),也有阻塞方法 sleep()、join() 方法等,并且结合若干个实战例子帮助大家更好地理解相关的 API,Thread 的 API 是掌握高并发编程的基础,因此非常有必要熟练掌握。

CHAPTER4 · 第 4 章

线程安全与数据同步

本章我们将学习多线程中最复杂也是最重要的内容之一，那就是数据同步、线程安全、锁等概念，在串行化的任务执行过程中，由于不存在资源的共享，线程安全的问题几乎不用考虑，但是串行化的程序，运行效率低下，不能最大化地利用 CPU 的计算能力，随着 CPU 核数的增加和计算速度的提升，串行化的任务执行显然是对资源的极大浪费，比如 B 客户提交了一个业务请求，只有等到 A 客户处理结束才能开始，这样的体验显然是用户无法忍受的。

无论是互联网系统，还是企业级系统，在追求稳定计算的同时也在追求更高的系统吞吐量，这也对系统的开发者提出了更高的要求，如何开发高效率的程序成了每个程序员必须掌握的技能，并发或者并行的程序并不意味着可以满足越多的 Thread，Thread 的多少对系统的性能来讲是一个抛物线，同时多线程的引入也带来了共享资源安全的隐患。在本章中，我们主要来探讨如何在安全的前提下高效地共享数据。

什么是共享资源？共享资源指的是多个线程同时对同一份资源进行访问（读写操作），被多个线程访问的资源就称为共享资源，如何保证多个线程访问到的数据是一致的，则被称为数据同步或者资源同步。

4.1 数据同步

4.1.1 数据不一致问题的引入

在第 1 章中，我们写了一个简单的营业大厅叫号机程序，当时我们设定的最大号码是 50（可能有些读者已经测试出了问题），现在我们对该程序稍加修改，就会出现数据不一致的情况，具体如下：

```java
package com.wangwenjun.concurrent.chapter04;

public class TicketWindowRunnable implements Runnable {

    private int index = 1;

    private final static int MAX = 500;

    @Override
    public void run() {

        while (index <= MAX) {
            System.out.println(Thread.currentThread() + " 的号码是:" + (index++));
        }
    }

    public static void main(String[] args)
    {
        final TicketWindowRunnable task = new TicketWindowRunnable();

        Thread windowThread1 = new Thread(task, "一号窗口");

        Thread windowThread2 = new Thread(task, "二号窗口");

        Thread windowThread3 = new Thread(task, "三号窗口");

        Thread windowThread4 = new Thread(task, "四号窗口");

        windowThread1.start();
        windowThread2.start();
        windowThread3.start();
        windowThread4.start();
    }
}
```

多次运行上述程序，每次都会有不一样的发现，但是总结起来主要有三个问题，具体如下。

❑ 第一，某个号码被略过没有出现。
❑ 第二，某个号码被多次显示。
❑ 第三，号码超过了最大值500。

笔者多次运行上面的程序，找出了数据不一致的几种情况，如图4-1所示。

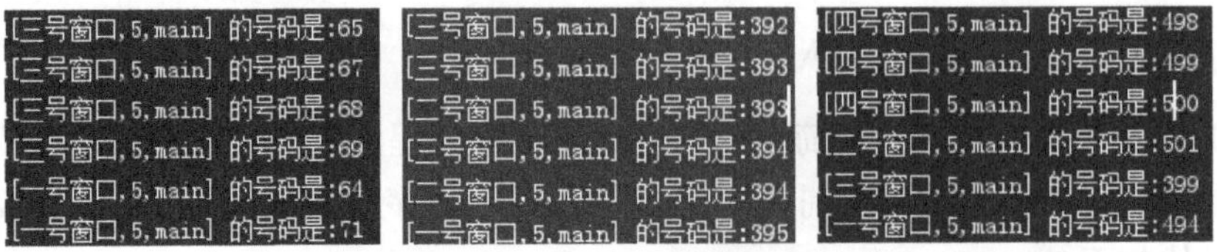

图4-1 数据不一致的运行结果

4.1.2 数据不一致问题原因分析

本节将针对上面出现的几种数据不一致问题分别进行分析。

1. 号码被略过

如图 4-2 所示，线程的执行是由 CPU 时间片轮询调度的，假设此时线程 1 和 2 都执行到了 index=65 的位置，其中线程 2 将 index 修改为 66 之后未输出之前，CPU 调度器将执行权利交给了线程 1，线程 1 直接将其累加到了 67，那么 66 就被忽略了。

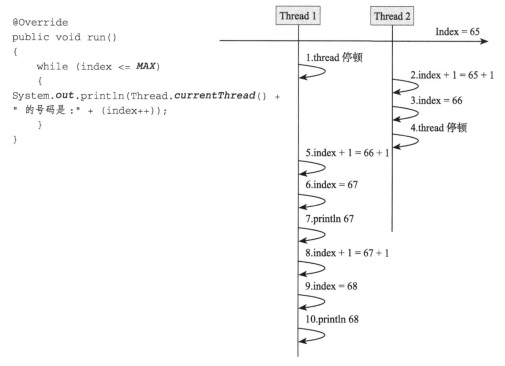

图 4-2 导致号码被略过的线程执行情况

2. 号码重复出现

线程 1 执行 index+1，然后 CPU 执行权落入线程 2 手里，由于线程 1 并没有给 index 赋予计算后的结果 393，因此线程 2 执行 index+1 的结果仍然是 393，所以会出现重复号码的情况。

3. 号码超过了最大值

下面来分析一下号码超过最大值的情况，当 index=499 的时候，线程 1 和线程 2 都看到条件满足，线程 2 短暂停顿，线程 1 将 index 增加到了 500，线程 2 恢复运行后又将 500 增加到了 501，此时就出现了超过最大值的情况。

我们虽然使用了时序图的方式对数据同步问题进行了分析，但是这样的解释还是不够严谨，本书的第三部分将会讲解 Java 的内存模型以及 CPU 缓存等知识，到时候会更加清晰和深入地讲解数据不一致的问题。

```
@Override
public void run()
{
    while (index <= MAX)
    {
        System.out.println(Thread.currentThread() +
" 的号码是:" + (index++));
    }
}
```

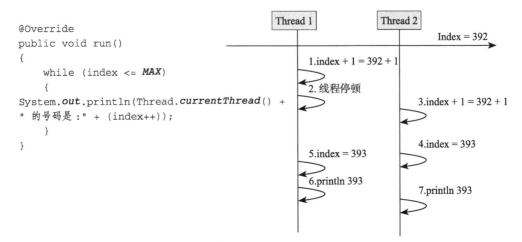

图 4-3　导致号码重复出现的线程执行情况

```
@Override
public void run()
{
    while (index <= MAX)
    {
        System.out.println(Thread.currentThread() +
" 的号码是:" + (index++));
    }
}
```

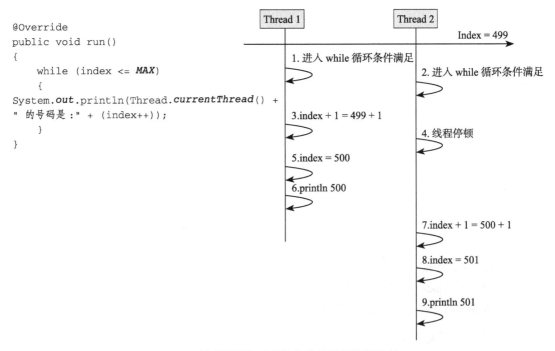

图 4-4　导致号码超过了最大值的线程执行情况

4.2　初识 synchronized 关键字

4.1.1 节出现的几个问题，究其原因就是因为多个线程对 index 变量（共享变量 / 资源）

同时操作引起的，在 JDK1.5 版本以前，要解决这个问题需要使用 synchronized 关键字，synchronized 提供了一种排他机制，也就是在同一时间只能有一个线程执行某些操作，在本章中，我们就来详细地探讨一下 synchronized 关键字的本质和用法。

4.2.1 什么是 synchronized

下面是一段来自于 JDK 官网对 synchronized 关键字比较权威的解释，如图 4-5 所示。

```
Synchronized keyword enable a simple strategy for preventing thread interference
and memory consistency errors: if an object is visible to more than one thread, all
reads or writes to that object's variables are done through synchronized methods.
```

图 4-5　synchronized 的官方解释

上述解释的意思是：synchronized 关键字可以实现一个简单的策略来防止线程干扰和内存一致性错误，如果一个对象对多个线程是可见的，那么对该对象的所有读或者写都将通过同步的方式来进行，具体表现如下。

- synchronized 关键字提供了一种锁的机制，能够确保共享变量的互斥访问，从而防止数据不一致问题的出现。
- synchronized 关键字包括 monitor enter 和 monitor exit 两个 JVM 指令，它能够保证在任何时候任何线程执行到 monitor enter 成功之前都必须从主内存中获取数据，而不是从缓存中，在 monitor exit 运行成功之后，共享变量被更新后的值必须刷入主内存（在本书的第三部分会重点介绍）。
- synchronized 的指令严格遵守 java happens-before 规则，一个 monitor exit 指令之前必定要有一个 monitor enter（在本书的第三部分会详细介绍）。

4.2.2　synchronized 关键字的用法

synchronized 可以用于对代码块或方法进行修饰，而不能够用于对 class 以及变量进行修饰。

1. 同步方法

同步方法的语法非常简单即：[default|public|private|protected] synchronized [static] type method()。示例代码如下：

```
public synchronized void sync()
{
    ...
    ...
}
public synchronized static void staticSync()
{
    ...
    ...
}
```

2. 同步代码块

同步代码块的语法示例如下：

```
private final Object MUTEX = new Object();
public void sync()
{
    synchronized (MUTEX)
    {
        ...
        ...
    }
}
```

介绍了什么是 synchronized 关键字以及它的基本用法之后，我们再次改写一下叫号程序：

```
package com.wangwenjun.concurrent.chapter04;

public class TicketWindowRunnable implements Runnable
{
    private int index = 1;

    private final static int MAX = 500;

    private final static Object MUTEX = new Object();

    @Override
    public void run()
    {
        synchronized (MUTEX)
        {
            while (index <= MAX)
            {
                System.out.println(Thread.currentThread() + " 的号码是:" + (index++));
            }
        }
    }

    public static void main(String[] args)
    {
```

```java
        final TicketWindowRunnable task = new TicketWindowRunnable();

        Thread windowThread1 = new Thread(task, "一号窗口");

        Thread windowThread2 = new Thread(task, "二号窗口");

        Thread windowThread3 = new Thread(task, "三号窗口");

        Thread windowThread4 = new Thread(task, "四号窗口");

        windowThread1.start();
        windowThread2.start();
        windowThread3.start();
        windowThread4.start();
    }
}
```

上面的程序无论运行多少次，都不会出现数据不一致的问题。

4.3 深入 synchronized 关键字

4.3.1 线程堆栈分析

synchronized 关键字提供了一种互斥机制，也就是说在同一时刻，只能有一个线程访问同步资源，很多资料、书籍将 synchronized（mutex）称为锁，其实这种说法是不严谨的，准确地讲应该是某线程获取了与 mutex 关联的 monitor 锁（当然写程序的时候知道它想要表达的语义即可），下面我们来看一个简单的例子对其进行说明：

```java
package com.wangwenjun.concurrent.chapter04;

import java.util.concurrent.TimeUnit;

public class Mutex
{
    private final static Object MUTEX = new Object();

    public void accessResource()
    {
        synchronized (MUTEX)
        {
            try
            {
                TimeUnit.MINUTES.sleep(10);
            } catch (InterruptedException e)
            {
                e.printStackTrace();
            }
        }
    }

    public static void main(String[] args)
```

```
    {
        final Mutex mutex = new Mutex();
        for (int i = 0; i < 5; i++)
        {
            new Thread(mutex::accessResource).start();
        }
    }
}
```

上面的代码中定义了一个方法 accessResource，并且使用同步代码块的方式对 accessResource 进行了同步，同时定义了 5 个线程调用 accessResource 方法，由于同步代码块的互斥性，只能有一个线程获取了 `mutex monitor` 的锁，其他线程只能进入阻塞状态，等待获取 `mutex monitor` 锁的线程对其进行释放，运行上面的程序然后打开 JConsole 工具监控，如图 4-6 所示。

选中要建立连接的本地进程，然后点击【连接】按钮进入 JConsole 控制台，将 tab 切换至【线程】，如图 4-7 所示。

随便选中程序中创建的某个线程，会发现只有一个线程在 `TIMED_WAITING(sleeping)` 状态，其他线程都进入了 `BLOCKED` 状态，如图 4-7 所示。

图 4-6　打开 JConsole，选择需要监控的 JVM 进程

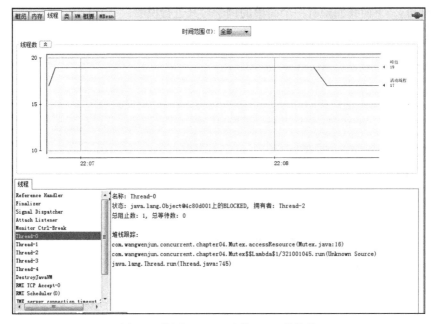

图 4-7　使用 jconsole 监控 JVM 的线程

使用 jstack 命令打印进程的线程堆栈信息，选取其中几处关键的地方对其进行分析。

Thread-2 持有 monitor <0x000000078b936ce0> 的锁并且处于休眠状态中，那么其他线程将会无法进入 accessResource 方法，如图 4-8 所示。

```
"Thread-2" #13 prio=5 os_prio=0 tid=0x000000000b371000 nid=0x1ffc waiting on condition [0x000000000c14e000]
   java.lang.Thread.State: TIMED_WAITING (sleeping)
    at java.lang.Thread.sleep(Native Method)
    at java.lang.Thread.sleep(Thread.java:340)
    at java.util.concurrent.TimeUnit.sleep(TimeUnit.java:386)
    at com.wangwenjun.concurrent.chapter04.Mutex.accessResource(Mutex.java:16)
    - locked <0x000000078b936ce0> (a java.lang.Object)
    at com.wangwenjun.concurrent.chapter04.Mutex$$Lambda$1/321001045.run(Unknown Source)
    at java.lang.Thread.run(Thread.java:745)
```

图 4-8　Thread-2 线程的堆栈信息

Thread-1 线程进入 BLOCKED 状态并且等待着获取 monitor <0x000000078b936ce0> 的锁，其他的几个线程同样也是 BLOCKED 状态，如图 4-9 所示。

```
"Thread-1" #12 prio=5 os_prio=0 tid=0x000000000b370000 nid=0x201c waiting for monitor entry [0x000000000c39f000]
   java.lang.Thread.State: BLOCKED (on object monitor)
    at com.wangwenjun.concurrent.chapter04.Mutex.accessResource(Mutex.java:16)
    - waiting to lock <0x000000078b936ce0> (a java.lang.Object)
    at com.wangwenjun.concurrent.chapter04.Mutex$$Lambda$1/321001045.run(Unknown Source)
    at java.lang.Thread.run(Thread.java:745)
```

图 4-9　Thread-1 线程的堆栈信息

4.3.2　JVM 指令分析

使用 JDK 命令 javap 对 Mutex class 进行反汇编，输出了大量的 JVM 指令，在这些指令中，你将发现 monitor enter 和 monitor exit 是成对出现的（有些时候会出现一个 monitor enter 多个 monitor exit，但是每一个 monitor exit 之前必有对应的 monitor enter，这是肯定的），运行下面的命令：

```
javap -c com.wangwenjun.concurrent.chapter04.Mutex

public class com.wangwenjun.concurrent.chapter04.Mutex {
  public com.wangwenjun.concurrent.chapter04.Mutex();
    Code:
       0: aload_0
       1: invokespecial #1         // Method java/lang/Object."<init>":()V
       4: return

  public void accessResource();
    Code:
       0: getstatic     #2         // Field MUTEX:Ljava/lang/Object;
       3: dup
```

```
               4: astore_1
               5: monitorenter
               6: getstatic      #3      // Field java/util/concurrent/TimeUnit.
MINUTES:Ljava/util/concurrent/TimeUnit;
               9: ldc2_w         #4      // long 101
              12: invokevirtual #6       // Method java/util/concurrent/TimeUnit.
sleep:(J)V
              15: goto           23
              18: astore_2
              19: aload_2
              20: invokevirtual #8       // Method java/lang/InterruptedException.
printStackTrace:()V
              23: aload_1
              24: monitorexit
              25: goto           33
              28: astore_3
              29: aload_1
              30: monitorexit
              31: aload_3
              32: athrow
              33: return
        Exception table:
           from    to  target type
              6    15      18 Class java/lang/InterruptedException
              6    25      28 any
             28    31      28 any

      public static void main(java.lang.String[]);
        Code:
               0: new            #9      // class com/wangwenjun/concurrent/
chapter04/Mutex
               3: dup
               4: invokespecial #10      // Method "<init>":()V
               7: astore_1
               8: iconst_0
               9: istore_2
              10: iload_2
              11: iconst_5
              12: if_icmpge      42
              15: new            #11     // class java/lang/Thread
              18: dup
              19: aload_1
              20: dup
              21: invokevirtual #12      // Method java/lang/Object.getClass:()
Ljava/lang/Class;
              24: pop
              25: invokedynamic #13,  0  // InvokeDynamic #0:run:(Lcom/wangwenjun/
concurrent/chapter04/Mutex;)Ljava/lang/Runnable;
              30: invokespecial #14      // Method java/lang/Thread."<init>" :
(Ljava/lang/Runnable;)V
```

```
            33: invokevirtual #15     // Method java/lang/Thread.start:()V
            36: iinc           2, 1
            39: goto           10
            42: return

    static {};
        Code:
            0: new             #16    // class java/lang/Object
            3: dup
            4: invokespecial   #1     // Method java/lang/Object."<init>":()V
            7: putstatic       #2     // Field MUTEX:Ljava/lang/Object;
           10: return
}
```

选取其中的片段,进行重点分析。①获取到 MUTEX 引用,然后执行② monitorenter JVM 指令,休眠结束之后 goto 至 ③ monitorexit 的位置(astore_<n> 存储引用至本地变量表;aload_<n> 从本地变量表加载引用;getstatic 从 class 中获得静态属性):

```
    public void accessResource();
        Code:
             0: getstatic          ①获取 MUTEX
             3: dup
             4: astore_1
             5: monitorenter       ②执行 monitorenter JVM 指令
             6: getstatic     #3        // Field java/util/concurrent/TimeUnit.MINUTES:Ljava/util/concurrent/TimeUnit;
             9: ldc2_w        #4        // long 101
            12: invokevirtual #6        // Method java/util/concurrent/TimeUnit.sleep:(J)V
            15: goto          23        ③跳转到 23 行
            18: astore_2
            19: aload_2
            20: invokevirtual #8        // Method java/lang/InterruptedException.printStackTrace:()V
            23: aload_1                 ④
            24: monitorexit             ⑤执行 monitor exit JVM 指令
            25: goto          33
            28: astore_3
            29: aload_1
            30: monitorexit
            31: aload_3
            32: athrow
            33: return
```

(1) Monitorenter

每个对象都与一个 monitor 相关联,一个 monitor 的 lock 的锁只能被一个线程在同一时间获得,在一个线程尝试获得与对象关联 monitor 的所有权时会发生如下的几件事情。

- 如果 monitor 的计数器为 0，则意味着该 monitor 的 lock 还没有被获得，某个线程获得之后将立即对该计数器加一，从此该线程就是这个 monitor 的所有者了。
- 如果一个已经拥有该 monitor 所有权的线程重入，则会导致 monitor 计数器再次累加。
- 如果 monitor 已经被其他线程所拥有，则其他线程尝试获取该 monitor 的所有权时，会被陷入阻塞状态直到 monitor 计数器变为 0，才能再次尝试获取对 monitor 的所有权。

（2）Monitorexit

释放对 monitor 的所有权，想要释放对某个对象关联的 monitor 的所有权的前提是，你曾经获得了所有权。释放 monitor 所有权的过程比较简单，就是将 monitor 的计数器减一，如果计数器的结果为 0，那就意味着该线程不再拥有对该 monitor 的所有权，通俗地讲就是解锁。与此同时被该 monitor block 的线程将再次尝试获得对该 monitor 的所有权。

4.3.3 使用 synchronized 需要注意的问题

在详细了解了 synchronized 关键字的用法和本质之后，笔者罗列了几个初学者容易出现的错误，以供读者参考。

1. 与 monitor 关联的对象不能为空

```
private final Object mutex = null;
public void syncMethod()
{
    synchronized (mutex)
    {
        //
    }
}
```

Mutex 为 null，很多人还是会犯这么简单的错误，每一个对象和一个 monitor 关联，对象都为 null 了，monitor 肯定无从谈起。

2. synchronized 作用域太大

由于 synchronized 关键字存在排他性，也就是说所有的线程必须串行地经过 synchronized 保护的共享区域，如果 synchronized 作用域越大，则代表着其效率越低，甚至还会丧失并发的优势，示例代码如下：

```
public static class Task implements Runnable
{
    @Override
    public synchronized void run()
    {
        //
    }
}
```

上面的代码对整个线程的执行逻辑单元都进行了 synchronized 同步,从而丧失了并发的能力,synchronized 关键字应该尽可能地只作用于共享资源(数据)的读写作用域。

3. 不同的 monitor 企图锁相同的方法

笔者利用业余时间从事互联网培训的一段时间里,发现很多人都容易犯这个错误,下面我们就来举个例子说明一下:

```java
public static class Task implements Runnable
{
    private final Object MUTEX = new Object();
    @Override
    public void run()
    {
        //...
        synchronized (MUTEX)
        {
            //...
        }
        //...
    }
}
public static void main(String[] args)
{
    for (int i = 0; i < 5; i++)
    {
        new Thread(Task::new).start();
    }
}
```

上面的代码构造了五个线程,同时也构造了五个 Runnable 实例,Runnable 作为线程逻辑执行单元传递给 Thread,然后你将发现,synchronized 根本互斥不了与之对应的作用域,线程之间进行 monitor lock 的争抢只能发生在与 monitor 关联的同一个引用上,上面的代码每一个线程争抢的 monitor 关联引用都是彼此独立的,因此不可能起到互斥的作用。

4. 多个锁的交叉导致死锁

多个锁的交叉很容易引起线程出现死锁的情况,程序并没有任何错误输出,但就是不工作,如下面的代码所示(本书的 4.5 节中会分析死锁的原因以及教大家如何对其进行诊断):

```java
private final Object MUTEX_READ = new Object();
private final Object MUTEX_WRITE = new Object();
public void read()
{
    synchronized (MUTEX_READ)
    {
        synchronized (MUTEX_WRITE)
        {
            //...
```

```java
            }
        }
    }
    public void write()
    {
        synchronized (MUTEX_WRITE)
        {
            synchronized (MUTEX_READ)
            {
                //...
            }
        }
    }
}
```

4.4 This Monitor 和 Class Monitor 的详细介绍

通过 4.2 节和 4.3 节的学习，想必大家一定非常清楚，多个线程争抢同一个 monitor 的 lock 会陷入阻塞进而达到数据同步、资源同步的目的，在本章中我们将通过实例认识两个比较特别的 monitor。

4.4.1 this monitor

在下面的代码 ThisMonitor 中，两个方法 method1 和 method2 都被 synchronized 关键字修饰，启动了两个线程分别访问 method1 和 method2，在开始运行之前请读者思考一个问题：synchronized 关键字修饰了同一个实例对象的两个不同方法，那么与之对应的 monitor 是什么？两个 monitor 是否一致呢？

```java
package com.wangwenjun.concurrent.chapter04;

import java.util.concurrent.TimeUnit;

import static java.lang.Thread.currentThread;

public class ThisMonitor
{
    public synchronized void method1()
    {
        System.out.println(currentThread().getName() + " enter to method1");
        try
        {
            TimeUnit.MINUTES.sleep(10);
        } catch (InterruptedException e)
        {
            e.printStackTrace();
        }
```

```
    }
    public synchronized void method2()
    {
        System.out.println(currentThread().getName() + " enter to method2");
        try
        {
            TimeUnit.MINUTES.sleep(10);
        } catch (InterruptedException e)
        {
            e.printStackTrace();
        }
    }

    public static void main(String[] args)
    {
        ThisMonitor thisMonitor = new ThisMonitor();
        new Thread(thisMonitor::method1, "T1").start();
        new Thread(thisMonitor::method2, "T2").start();
    }
}
```

带着上面的疑问，运行程序将会发现只有一个方法被调用，另外一个方法根本没有被调用，分析线程的堆栈信息，执行 jdk 自带的 jstack pid 命令，如图 4-10 所示。

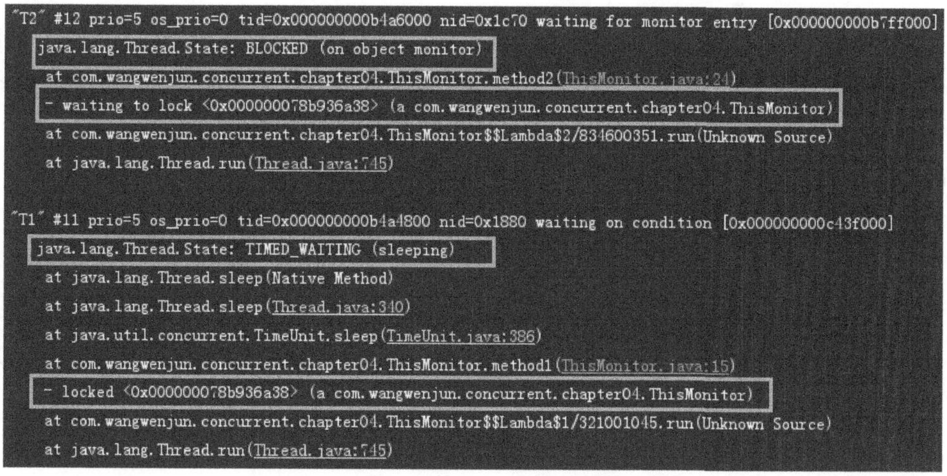

图 4-10　使用 this monitor 的线程堆栈信息

笔者将重点的地方用红色的框标识了出来，T1 线程获取了 <0x000000078b936a38> monitor 的 lock 并且处于休眠状态，而 T2 线程企图获取 <0x000000078b936a38> monitor 的 lock 时陷入了 BLOCKED 状态，可见使用 synchronized 关键字同步类的不同实例方法，争抢的是同一个 monitor 的 lock，而与之关联的引用则是 ThisMonitor 的实例引用，为了证实我们的推论，将上面的代码稍作修改，如下所示：

```java
public synchronized void method1()
{
    System.out.println(currentThread().getName() + " enter to method1");
    try
    {
        TimeUnit.MINUTES.sleep(10);
    } catch (InterruptedException e)
    {
        e.printStackTrace();
    }
}
public void method2()
{
    synchronized (this)
    {
        System.out.println(currentThread().getName() + " enter to method2");
        try
        {
            TimeUnit.MINUTES.sleep(10);
        } catch (InterruptedException e)
        {
            e.printStackTrace();
        }
    }
}
```

其中，method1 保持方法同步的方式，method2 则采用了同步代码块的方式，并且使用的是 this 的 monitor，运行修改后的代码将会发现效果完全一样，在 JDK 官方文档中也有这样的描述（见 https://docs.oracle.com/javase/tutorial/essential/concurrency/locksync.html）：

> *When a thread invokes a synchronized method, it automatically acquires the intrinsic lock for that method's object and releases it when the method returns. The lock release occurs even if the return was caused by an uncaught exception.*

4.4.2 class monitor

同样的方式，来看下面的例子，有两个类方法（静态方法）分别使用 synchronized 对其进行同步：

```java
package com.wangwenjun.concurrent.chapter04;

import java.util.concurrent.TimeUnit;

import static java.lang.Thread.currentThread;

public class ClassMonitor
{

    public static synchronized void method1()
```

```
    {
        System.out.println(currentThread().getName() + " enter to method1");
        try
        {
            TimeUnit.MINUTES.sleep(10);
        } catch (InterruptedException e)
        {
            e.printStackTrace();
        }
    }

    public static synchronized void method2()
    {
        System.out.println(currentThread().getName() + " enter to method2");
        try
        {
            TimeUnit.MINUTES.sleep(10);
        } catch (InterruptedException e)
        {
            e.printStackTrace();
        }
    }

    public static void main(String[] args)
    {
        new Thread(ClassMonitor::method1, "T1").start();
        new Thread(ClassMonitor::method2, "T2").start();
    }
}
```

运行上面的例子，在同一时刻只能有一个线程访问 ClassMonitor 的静态方法，我们仍旧使用 jstack 命令分析其线程堆栈信息，如图 4-11 所示。

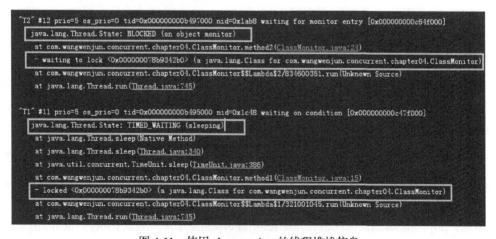

图 4-11　使用 class monitor 的线程堆栈信息

同样，笔者将关键的地方用红色方框标识了出来，T1 线程持有 <0x000000078b9342b0> monitor 的锁在正在休眠，而 T2 线程在试图获取 <0x000000078b9342b0> monitor 锁的时候陷入了 BLOCKED 状态，因此我们可以得出用 synchronized 同步某个类的不同静态方法争抢的也是同一个 monitor 的 lock，再仔细对比堆栈信息会发现与 4.4.1 节中关于 monitor 信息不一样的地方在于 (**a java.lang.Class** for com.wangwenjun.concurrent.chapter04.ClassMonitor)，由此可以推断与该 monitor 关联的引用是 ClassMonitord.class 实例。

对上面的代码稍作修改，然后运行会发现具有同样的效果，示例代码如下：

```java
public static synchronized void method1()
{
    System.out.println(currentThread().getName() + " enter to method1");
    try
    {
        TimeUnit.MINUTES.sleep(10);
    } catch (InterruptedException e)
    {
        e.printStackTrace();
    }
}
public static void method2()
{
    synchronized (ClassMonitor.class)
    {
        System.out.println(currentThread().getName() + " enter to method2");
        try
        {
            TimeUnit.MINUTES.sleep(10);
        } catch (InterruptedException e)
        {
            e.printStackTrace();
        }
    }
}
```

其中静态方法 method1 继续保持同步方法的方式，而 method2 则修改为同步代码块的方式，使用 ClassMonitor.class 的实例引用作为 monitor，同样在 JDK 官方文档中对 class monitor 也有比较权威的说明：

> since a static method is associated with a class, not an object. In this case, the thread acquires the intrinsic lock for the Class object associated with the class. Thus access to class's static fields is controlled by a lock that's distinct from the lock for any instance of the class.

详见：https://docs.oracle.com/javase/tutorial/essential/concurrency/locksync.html。

4.5 程序死锁的原因以及如何诊断

关于死锁，在 4.3.3 节中有过简单介绍，在本节中我们将详细介绍在什么情况下会发生死锁，以及死锁之后如何诊断。程序死锁可以类比于如图 4-12 所示的交通堵塞现象。

4.5.1 程序死锁

（1）交叉锁可导致程序出现死锁

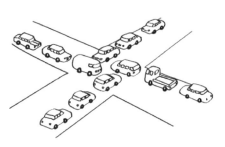

图 4-12　程序进入死锁

线程 A 持有 R1 的锁等待获取 R2 的锁，线程 B 持有 R2 的锁等待获取 R1 的锁（典型的哲学家吃面），这种情况最容易导致程序发生死锁的问题，我们在本节课程中会举例说明。

（2）内存不足

当并发请求系统可用内存时，如果此时系统内存不足，则可能会出现死锁的情况。举个例子，两个线程 T1 和 T2，执行某个任务，其中 T1 已经获取了 10MB 内存，T2 获取了 20MB 内存，如果每个线程的执行单元都需要 30MB 的内存，但是剩余可用的内存刚好为 20MB，那么两个线程有可能都在等待彼此能够释放内存资源。

（3）一问一答式的数据交换

服务端开启某个端口，等待客户端访问，客户端发送请求立即等待接收，由于某种原因服务端错过了客户端的请求，仍然在等待一问一答式的数据交换，此时服务端和客户端都在等待着双方发送数据（笔者在刚参加工作的时候就犯过这样的错误）。

（4）数据库锁

无论是数据库表级别的锁，还是行级别的锁，比如某个线程执行 for update 语句退出了事务，其他线程访问该数据库时都将陷入死锁。

（5）文件锁

同理，某线程获得了文件锁意外退出，其他读取该文件的线程也将会进入死锁直到系统释放文件句柄资源。

（6）死循环引起的死锁

程序由于代码原因或者对某些异常处理不得当，进入了死循环，虽然查看线程堆栈信息不会发现任何死锁的迹象，但是程序不工作，CPU 占有率又居高不下，这种死锁一般称为系统假死，是一种最为致命也是最难排查的死锁现象，由于重现困难，进程对系统资源的使用量又达到了极限，想要做出 dump 有时候也是非常困难的。

4.5.2 程序死锁举例

在本节中，我们将举例说明程序由于交叉锁引起的死锁的情况，交叉锁不仅是指自己写的代码出现了交叉的情况，如果使用某个框架或者开源库，由于对源码 API 的不熟悉，很有可能也会引起死锁，由于使用不当而出现后者死锁，排查的困难则要高于前者，所以在使用框架或者开源库的时候做到了如指掌还是很有必要的。示例代码如下：

```java
package com.wangwenjun.concurrent.chapter04;

import static java.lang.Thread.currentThread;

public class DeadLock
{
    private final Object MUTEX_READ = new Object();
    private final Object MUTEX_WRITE = new Object();

    public void read()
    {
        synchronized (MUTEX_READ)
        {
            System.out.println(currentThread().getName() + " get READ lock");
            synchronized (MUTEX_WRITE)
            {
                System.out.println(currentThread().getName() + " get WRITE lock");
            }
            System.out.println(currentThread().getName() + " release WRITE lock");
        }
        System.out.println(currentThread().getName() + " release READ lock");
    }


    public void write()
    {
        synchronized (MUTEX_WRITE)
        {
            System.out.println(currentThread().getName() + " get WRITE lock");
            synchronized (MUTEX_READ)
            {
                System.out.println(currentThread().getName() + " get READ lock");
            }
            System.out.println(currentThread().getName() + " release READ lock");
        }
        System.out.println(currentThread().getName() + " release WRITE lock");
    }

    public static void main(String[] args)
    {
        final DeadLock deadLock = new DeadLock();
        new Thread(() ->
        {
            while (true)
```

```
            {
                deadLock.read();
            }
        }, "READ-THREAD").start();

        new Thread(() ->
        {
            while (true)
            {
                deadLock.write();
            }
        }, "WRITE-THREAD").start();
    }
}
```

上面的程序再明显不过了，一眼就可以看出程序有死锁的风险，如果使用一些开源库，API 的调用层次比较深，那么看代码是不容易发现死锁风险的，比如 JDK 中的 HashMap，文档很明显地指出了该数据结构不是线程安全的类，如果在多线程同时写操作的情况下不对其进行同步化封装，则很容易出现死循环引起的死锁，程序运行一段时间后 CPU 等资源高居不下，各种诊断工具很难派上用场，因为死锁引起的进程往往会榨干 CPU 等几乎所有资源，诊断工具由于缺少资源一时间也很难启动，比如下面的例子（如果读者运行下面的代码，记得保存手头的工作，该代码有可能会导致你的电脑死机）：

```
package com.wangwenjun.concurrent.chapter04;

import java.util.HashMap;

public class HashMapDeadLock
{
    private final HashMap<String, String> map = new HashMap<>();

    public void add(String key, String value)
    {
        this.map.put(key, value);
    }

    public static void main(String[] args)
    {
        final HashMapDeadLock hmdl = new HashMapDeadLock();
        for (int x = 0; x < 2; x++)
            new Thread(() ->
            {
                for (int i = 1; i < Integer.MAX_VALUE; i++)
                {
                    hmdl.add(String.valueOf(i), String.valueOf(i));
                }
```

```
            }).start();
    }
}
```

HashMap 不具备线程安全的能力，如果想要使用线程安全的 map 结构请使用 ConcurrentHashMap 或者使用 Collections.synchronizedMap 来代替。

4.5.3 死锁诊断

大致知道了引起死锁的几种原因之后，也看了两个具体的实例，本节中，我们将借助诊断工具对其进行诊断。

（1）交叉锁引起的死锁

运行 DeadLock 代码，程序将陷入死锁，打开 jstack 工具或者 jconsole 工具，Jstack-l PID 会直接发现死锁的信息，示例代码如下：

```
Found one Java-level deadlock:
=============================
"WRITE-THREAD":
    waiting to lock monitor 0x0000000009d1b0f8 (object 0x000000078b936800, a java.lang.Object),
    which is held by "READ-THREAD"
"READ-THREAD":
    waiting to lock monitor 0x0000000009d1b048 (object 0x000000078b936810, a java.lang.Object),
    which is held by "WRITE-THREAD"

Java stack information for the threads listed above:
===================================================
"WRITE-THREAD":
        at com.wangwenjun.concurrent.chapter04.DeadLock.write(DeadLock.java:33)
        - waiting to lock <0x000000078b936800> (a java.lang.Object)
        - locked <0x000000078b936810> (a java.lang.Object)
        at com.wangwenjun.concurrent.chapter04.DeadLock.lambda$main$1(DeadLock.java:55)
        at com.wangwenjun.concurrent.chapter04.DeadLock$$Lambda$2/791452441.run(Unknown Source)
        at java.lang.Thread.run(Thread.java:745)
"READ-THREAD":
        at com.wangwenjun.concurrent.chapter04.DeadLock.read(DeadLock.java:18)
        - waiting to lock <0x000000078b936810> (a java.lang.Object)
        - locked <0x000000078b936800> (a java.lang.Object)
        at com.wangwenjun.concurrent.chapter04.DeadLock.lambda$main$0(DeadLock.java:47)
        at com.wangwenjun.concurrent.chapter04.DeadLock$$Lambda$1/424058530.run(Unknown Source)
        at java.lang.Thread.run(Thread.java:745)

Found 1 deadlock.
```

一般交叉锁引起的死锁线程都会进入 BLOCKED 状态，CPU 资源占用不高，很容易借助工具来发现。

（2）死循环引起的死锁（假死）

运行 HashMapDeadLock 程序，也可以使用 jstack、jconsole、jvisualvm 工具或者 jProfiler（收费的）工具进行诊断，但是不会给出很明显的提示，因为工作的线程并未 BLOCKED，而是始终处于 RUNNABLE 状态，CPU 使用率高居不下，甚至都不能够正常运行你的命令。

如图 4-13 所示的是使用 jprofile 工具抓出来的方法线程运行状态，可以发现某个线程在执行 hashmap 的 put 方法时陷入了死循环，而且 CPU 占用率非常高，一个很普通的方法调用导致了接近 100 毫秒的耗时很显然是不正常的。

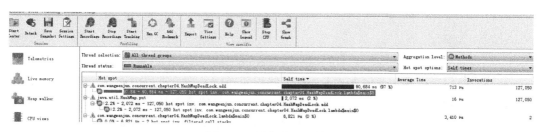

图 4-13　程序进入死循环

严格意义上来说死循环会导致程序假死，算不上真正的死锁，但是某个线程对 CPU 消耗过多，导致其他线程等待 CPU，内存等资源也会陷入死锁等待。

4.6　本章总结

本章继续使用营业厅叫号程序引入数据不一致的问题，并且重点分析了在多线程下出现数据不一致情况的各种可能性。

synchronized 关键字在 Java 中提供了同步语义，它可以保证在同一时间只允许一个线程访问共享数据资源，长期以来 synchronized 关键字的性能一直被诟病，但是随着 JDK 的不断发展，synchronized 关键字在同步过程中的性能也是逐步提升，使用得当性能不输 LockSupport，本章不仅详细介绍了 synchronized 关键字的作用和用法，而且更进一步深入地分析了 synchronized 关键字的内在原理。

在多线程访问共享资源的情况下，对线程驾驭不得当很容易引起死锁的情况发生，在本章中我们列举了多种引起死锁的可能性，并且用程序实现了一个死锁示例让读者能在开发中尽量规避死锁的发生。当然，如果程序出现死锁，那就必须要掌握如何对其进行诊断，在本章中我们也介绍了 jstack、jvisualvm 等工具。

第 5 章 CHAPTER5

线程间通信

与网络通信等进程间通信方式不一样,线程间通信又称为进程内通信,多个线程实现互斥访问共享资源时会互相发送信号或等待信号,比如线程等待数据到来的通知,线程收到变量改变的信号等。本章将通过对一些案例的分析来学习 Java 提供的原生通信 API,以及这些通信机制背后的内幕。

5.1 同步阻塞与异步非阻塞

本章将分析对比同步阻塞消息处理机制和异步非阻塞消息处理机制的优缺点。

5.1.1 同步阻塞消息处理

假如有这样一个系统功能,客户端提交 Event 至服务器,服务器接收到客户请求之后开辟线程处理客户请求,经过比较复杂的业务计算后将结果返回给客户端,如图 5-1 所示。

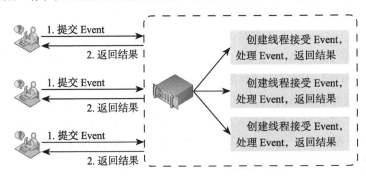

图 5-1 同步的方式提交业务请求

图 5-1 所示的设计存在几个显著的缺陷，具体如下。

- 同步 Event 提交，客户端等待时间过长（提交 Event 时长 + 接受 Event 创建 thread 时长 + 业务处理时长 + 返回结果时长）会陷入阻塞，导致二次提交 Event 耗时过长。
- 由于客户端提交的 Event 数量不多，导致系统同时受理业务数量有限，也就是系统整体的吞吐量不高。
- 这种一个线程处理一个 Event 的方式，会导致出现频繁的创建开启与销毁，从而增加系统额外开销。
- 在业务达到峰值的时候，大量的业务处理线程阻塞会导致频繁的 CPU 上下文切换，从而降低系统性能。

5.1.2 异步非阻塞消息处理

前面分析了同步阻塞消息的处理方式存在诸多缺陷，尤其是系统的吞吐量低，很难应对比较高的业务并发量。如果我们将同步阻塞的方式换成异步非阻塞的方式，则不仅可以提高系统的吞吐量，而且业务处理线程的数量也能够控制在一个固定的范围，以增加系统的稳定性，如图 5-2 所示。

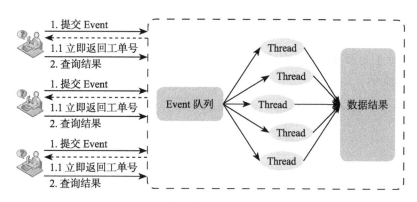

图 5-2 异步的方式提交业务请求

客户端提交 Event 后会得到一个相应的工单并且立即返回，Event 则会被放置在 Event 队列中。服务端有若干个工作线程，不断地从 Event 队列中获取任务并且进行异步处理，最后将处理结果保存至另外一个结果集中，如果客户端想要获得处理结果，则可凭借工单号再次查询。

两种方式相比较，你会发现异步非阻塞的优势非常明显，首先客户端不用等到结果处理结束之后才能返回，从而提高了系统的吞吐量和并发量；其次若服务端的线程数量在一个可控的范围之内是不会导致太多的 CPU 上下文切换从而带来的额外开销的；再次服务端线程可以重复利用，这样就减少了不断创建线程带来的资源浪费。但是异步处理的方式同样也存在缺陷，比如客户端想要得到结果还需要再次调用接口方法进行查询（在本书的第 4

部分会详细讲解如何利用异步回调接口的方式来解决这个问题)。

5.2 单线程间通信

在 4.6.2 节中，服务端有若干个线程会从队列中获取相应的 Event 进行异步处理，那么这些线程又是如何从队列中获取数据的呢？换句话说就是如何知道队列里此时是否有数据呢？比较笨的办法是不断地轮询：如果有数据则读取数据并处理，如果没有则等待若干时间再次轮询。还有一种比较好的方式就是通知机制：如果队列中有 Event，则通知工作的线程开始工作；没有 Event，则工作线程休息并等待通知。

5.2.1 初识 wait 和 notify

在本节中，我们将会学习线程之间如何进行通信，首先实现一个 EventQueue，该 Queue 有如下三种状态：

- 队列满——最多可容纳多少个 Event，好比一个系统最多同时能够受理多少业务一样；
- 队列空——当所有的 Event 都被处理并且没有新的 Event 被提交的时候，此时队列将是空的状态；
- 有 Event 但是没有满——有新的 Event 被提交，但是此时没有到达队列的上限。

示例代码如下：

```java
package com.wangwenjun.concurrent.chapter05;

import java.util.LinkedList;

import static java.lang.Thread.currentThread;

public class EventQueue
{
    private final int max;

    static class Event
    {
    }

    private final LinkedList<Event> eventQueue
            = new LinkedList<>();

    private final static int DEFAULT_MAX_EVENT = 10;

    public EventQueue()
    {
```

```java
        this(DEFAULT_MAX_EVENT);
    }

    public EventQueue(int max)
    {
        this.max = max;
    }

    public void offer(Event event)
    {
        synchronized (eventQueue)
        {
            if (eventQueue.size() >= max)
            {
                try
                {
                    console(" the queue is full.");
                    eventQueue.wait();
                } catch (InterruptedException e)
                {
                    e.printStackTrace();
                }
            }

            console(" the new event is submitted");
            eventQueue.addLast(event);
            eventQueue.notify();
        }
    }

    public Event take()
    {
        synchronized (eventQueue)
        {
            if (eventQueue.isEmpty())
            {
                try
                {
                    console(" the queue is empty.");
                    eventQueue.wait();
                } catch (InterruptedException e)
                {
                    e.printStackTrace();
                }
            }

            Event event = eventQueue.removeFirst();
            this.eventQueue.notify();
            console(" the event " + event + " is handled.");
```

```
            return event;
        }
    }

    private void console(String message)
    {
        System.out.printf("%s:%s\n", currentThread().getName(), message);
    }
}
```

上述代码中，在 EventQueue 中定义了一个队列，offer 方法会提交一个 Event 至队尾，如果此时队列已经满了，那么提交的线程将会被阻塞，这是调用了 wait 方法的结果（后文中会重点介绍 wait 方法）。同样 take 方法会从队头获取数据，如果队列中没有可用数据，那么工作线程就会被阻塞，这也是调用 wait 方法的直接结果。此外，还可以看到一个 notify 方法，该方法的作用是唤醒那些曾经执行 monitor 的 wait 方法而进入阻塞的线程。

EventQueue 的代码大概介绍完了，下面简单写两个线程对其进行测试，现在假设提交 Event 几乎没有任何延迟，而处理 Event 可能要花费比提交更多的时间，代码如下：

```
package com.wangwenjun.concurrent.chapter05;

import java.util.concurrent.TimeUnit;

public class EventClient
{
    public static void main(String[] args)
    {
        final EventQueue eventQueue = new EventQueue();
        new Thread(() ->
        {
            for (; ; )
            {
                eventQueue.offer(new EventQueue.Event());
            }
        }, "Producer").start();

        new Thread(() ->
        {
            for (; ; )
            {
                eventQueue.take();
                try
                {
                    TimeUnit.MILLISECONDS.sleep(10);
                } catch (InterruptedException e)
                {
                    e.printStackTrace();
                }
```

```
        }
    }, "Consumer").start();
}
```

其中，Producer 线程模拟提交 Event 的客户端几乎没有任何的延迟，而 Consumer 线程则用于模拟处理请求的工作线程（上面的 EventQueue 目前只支持一个线程的 Producer 和一个线程的 Consumer，也就是单线程间的通信，多个线程的生产者与消费者的会在 5.3 节中讲到）。运行上面的代码，分析一下输出的结果，具体如下：

```
Producer: the new event is submitted
Producer: the new event is submitted
Producer: the new event is submitted
Producer: the new event is submitted
Producer: the new event is submitted
Producer: the new event is submitted
Producer: the new event is submitted
Producer: the new event is submitted
Producer: the new event is submitted
Producer: the new event is submitted
Producer: the queue is full.
Consumer: the event com.wangwenjun.concurrent.chapter04.EventQueue$Event@530142c9 is handled.
Producer: the new event is submitted
Producer: the queue is full.
Consumer: the event com.wangwenjun.concurrent.chapter04.EventQueue$Event@5dc6b6d9 is handled.
Producer: the new event is submitted
Producer: the queue is full.
Consumer: the event com.wangwenjun.concurrent.chapter04.EventQueue$Event@59f929b9 is handled.
```

通过上述的输出日志可以看出，Producer 线程很快就提交了 10 个 Event 数据，此时队列已经满了，那么它将会执行 eventQueue 的 wait 方法进而进入阻塞状态，Consumer 线程由于要处理数据，所以会花费大概 10 毫秒的时间来处理其中的一条数据，然后通知 Producer 线程可以继续提交数据了，如此循环往复。

5.2.2 wait 和 notify 方法详解

看到这里想必读者对 wait 和 notify 方法有了一个大概的认识，wait 和 notify 方法并不是 Thread 特有的方法，而是 Object 中的方法，也就是说在 JDK 中的每一个类都拥有这两个方法，那么这两个方法到底有什么神奇之处可以使线程阻塞又可以唤醒线程呢？我们先来说说 wait 方法，下面是 wait 方法的三个重载方法。

```
public final void wait() throws InterruptedException
```

```
public final void wait(long timeout) throws InterruptedException
public final void wait(long timeout, int nanos) throws InterruptedException
```

- wait 方法的这三个重载方法都将调用 wait（long timeout）这个方法，前文使用的 wait() 方法等价于 wait(0)，0 代表着永不超时。
- Object 的 wait（long timeout）方法会导致当前线程进入阻塞，直到有其他线程调用了 Object 的 notify 或者 notifyAll 方法才能将其唤醒，或者阻塞时间到达了 timeout 时间而自动唤醒。
- wait 方法必须拥有该对象的 monitor，也就是 wait 方法必须在同步方法中使用。
- 当前线程执行了该对象的 wait 方法之后，将会放弃对该 monitor 的所有权并且进入与该对象关联的 wait set 中，也就是说一旦线程执行了某个 object 的 wait 方法之后，它就会释放对该对象 monitor 的所有权，其他线程也会有机会继续争抢该 monitor 的所有权。

对照上面所展示的 eventQueue 中的代码，可以看到，由于有多个线程操作 eventQueue，因此它自然而然就是共享资源，为了防止多线程对共享资源操作引起数据不一致的问题，我们需要对共享资源进行同步处理，这里使用 synchronized 关键字对其进行同步。如果当前队列 event 数量已经达到了上限，那么它会调用 eventQueue 的 wait 方法使当前线程进入 wait set 中并且释放 monitor 的锁。示例代码如下：

```
public void offer(Event event)
    {
        synchronized (eventQueue)
        {
            if (eventQueue.size() >= max)
            {
                try
                {
                    console(" the queue is full.");
                    eventQueue.wait();
                ...
                eventQueue.addLast(event);
                eventQueue.notify();
                ...
```

如果当前队列的数量还没有达到上限，则会将 event 插入队尾，并且企图唤醒已经加入 wait set 中的线程，由于我们使用的是单线程的插入和移除，因此在 offer 中执行的唤醒只能是执行 take 时被阻塞的线程，下面再来分析一下 notify 方法的作用

```
public final native void notify();
```

- 唤醒单个正在执行该对象 wait 方法的线程。
- 如果有某个线程由于执行该对象的 wait 方法而进入阻塞则会被唤醒，如果没有则会忽略。

- 被唤醒的线程需要重新获取对该对象所关联 monitor 的 lock 才能继续执行。

5.2.3　关于 wait 和 notify 的注意事项

- wait 方法是可中断方法，这也就意味着，当前线程一旦调用了 wait 方法进入阻塞状态，其他线程是可以使用 interrupt 方法将其打断的；根据 3.7 节的介绍，可中断方法被打断后会收到中断异常 InterruptedException，同时 interrupt 标识也会被擦除。
- 线程执行了某个对象的 wait 方法以后，会加入与之对应的 wait set 中，每一个对象的 monitor 都有一个与之关联的 wait set（5.3.2 节中会详细讲解 wait set 的知识）。
- 当线程进入 wait set 之后，notify 方法可以将其唤醒，也就是从 wait set 中弹出，同时中断 wait 中的线程也会将其唤醒。
- 必须在同步方法中使用 wait 和 notify 方法，因为执行 wait 和 notify 的前提条件是必须持有同步方法的 monitor 的所有权，运行下面任何一个方法都会抛出非法的 monitor 状态异常 IllegalMonitorStateException：

```
private void testWait()
{
    try
    {
        this.wait();
    } catch (InterruptedException e)
    {
        e.printStackTrace();
    }
}

private void testNotify()
{
    this.notify();
}
```

- 同步代码的 monitor 必须与执行 wait notify 方法的对象一致，简单地说就是用哪个对象的 monitor 进行同步，就只能用哪个对象进行 wait 和 notify 操作。运行下面代码中的任何一个方法，同样都会抛出 IllegalMonitorStateException 异常信息：

```
private final Object MUTEX = new Object();
private synchronized void testWait()
{
    try
    {
        MUTEX.wait();
    } catch (InterruptedException e)
    {
        e.printStackTrace();
    }
```

```
}
private synchronized void testNotify()
{
    MUTEX.notify();
}
```

上述同步方法中 monitor 引用的是 this，而 wait 和 notify 方法使用的却是 MUTEX 的方法，其实这并不难理解，虽然是在同步方法中执行 wait 和 notify 方法，但是 wait 和 notify 方法的执行并未以获取 MUTEX 的 monitor 为前提。

5.2.4　wait 和 sleep

从表面上看，wait 和 sleep 方法都可以使当前线程进入阻塞状态，但是两者之间存在着本质的区别，下面我们将总结两者的区别和相似之处。

- wait 和 sleep 方法都可以使线程进入阻塞状态。
- wait 和 sleep 方法均是可中断方法，被中断后都会收到中断异常。
- wait 是 Object 的方法，而 sleep 是 Thread 特有的方法。
- wait 方法的执行必须在同步方法中进行，而 sleep 则不需要。
- 线程在同步方法中执行 sleep 方法时，并不会释放 monitor 的锁，而 wait 方法则会释放 monitor 的锁。
- sleep 方法短暂休眠之后会主动退出阻塞，而 wait 方法（没有指定 wait 时间）则需要被其他线程中断后才能退出阻塞。

5.3　多线程间通信

在 5.2 节中，讲解了两个线程间的通信，只有一个线程对 EventQueue 进行 offer 操作，也只有一个线程对 EventQueue 进行 take 操作，如果多个线程同时进行 take 或者 offer，那么上面的程序就会出现问题。在本节中，我们将学习如何在多线程的情况下进行通信，并且还会讲解 wait set 线程休息室的相关内容。

5.3.1　生产者消费者

1. notifyAll 方法

多线程间通信需要用到 Object 的 notifyAll 方法，该方法与 notify 比较类似，都可以唤醒由于调用了 wait 方法而阻塞的线程，但是 notify 方法每次只能唤醒其中的一个线程，而 notifyAll 方法则可以同时唤醒全部的阻塞线程，同样被唤醒的线程仍需要继续争抢 monitor 的锁。

2. 生产者消费者

在 5.2.1 节中曾定义了一个 EventQueue，该队列在多个线程同时并发的情况下会出现数据不一致的情况，读者可以自行增加 EventClient 中的线程数量进行测试，在笔者的测试中出现了数据不一致的情况，大致可分为两类：其一是 LinkedList 中没有元素的时候仍旧调用了 removeFirst 方法，其二是当 LinkedList 中的元素超过 10 个的时候仍旧执行了 addLast 方法，下面通过图示的方法分别对其进行分析。

（1）LinkedList 为空时执行 removeFirst 方法

也许有读者会有疑问，EventQueue 中的方法都增加了 synchronized 数据同步，为何还会存在数据不一致的情况？假设 EventQueue 中的元素为空，两个线程在执行 take 方法时分别调用 wait 方法进入了阻塞之中，另外一个 offer 线程执行 addLast 方法之后唤醒了其中一个阻塞的 take 线程，该线程顺利消费了一个元素之后恰巧再次唤醒了一个 take 线程，这时就会导致执行空 LinkedList 的 removeFirst 方法，执行过程如图 5-3 所示。

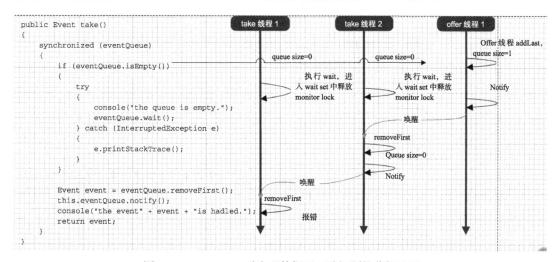

图 5-3　removeFirst 时出现数据不一致问题的分析过程

（2）LinkedList 元素为 10 时执行 addLast 方法

假设某个时刻 EventQueue 中存在 10 个 Event 数据，其中两个线程在执行 offer 方法的时候分别因为调用了 wait 方法而进入阻塞中，另外的一个线程执行 take 方法消费了 event 元素并且唤醒了一个 offer 线程，而该 offer 线程执行了 addLast 方法之后，queue 中的元素为 10，并且再次执行唤醒方法，恰巧另外一个 offer 线程也被唤醒，因此可以绕开阈值检查 eventQueue.size()>=max，致使 EventQueue 中的元素超过 10 个，执行过程如图 5-4 所示。

（3）改进

在分析完多线程情况下出现的问题之后，我们将对其进行改进。实际上在真实的开发中，绝大多数时候遇到的都是多线程间通信的情况，其中生产者消费者的例子就是最好的模型，示例如下：

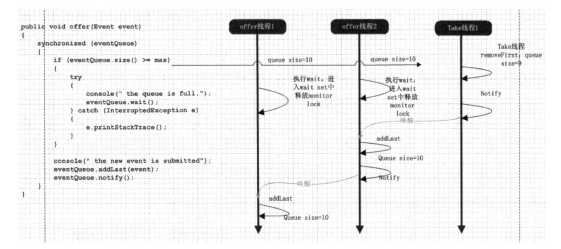

图 5-4　多线程执行 addLast 出现数据不一致问题的分析过程

```
public void offer(Event event)
{
    synchronized (eventQueue)
    {
        while (eventQueue.size() >= max)
        {
            try
            {
                console(" the queue is full.");
                eventQueue.wait();
            } catch (InterruptedException e)
            {
                e.printStackTrace();
            }
        }

        console(" the new event is submitted");
        eventQueue.addLast(event);
        eventQueue.notifyAll();
    }
}

public Event take()
{
    synchronized (eventQueue)
    {
        while (eventQueue.isEmpty())
        {
            try
```

```
            {
                console(" the queue is empty.");
                eventQueue.wait();
            } catch (InterruptedException e)
            {
                e.printStackTrace();
            }
        }

        Event event = eventQueue.removeFirst();
        this.eventQueue.notifyAll();
        console(" the event " + event + " is handled.");
        return event;
    }
}
```

只需要将临界值的判断 if 更改为 while，将 notify 更改为 notifyAll 即可。

5.3.2 线程休息室 wait set

在虚拟机规范中存在一个 wait set（wait set 又被称为线程休息室）的概念，至于该 wait set 是怎样的数据结构，JDK 官方并没有给出明确的定义，不同厂家的 JDK 有着不同的实现方式，甚至相同的 JDK 厂家不同的版本也存在着差异，但是不管怎样，线程调用了某个对象的 wait 方法之后都会被加入与该对象 monitor 关联的 wait set 中，并且释放 monitor 的所有权。

图 5-5 是若干个线程调用了 wait 方法之后被加入与 monitor 关联的 wait set 中，当另外一个线程调用该 monitor 的 notify 方法之后，其中一个线程会从 wait set 中弹出，至于是随机弹出还是以先进先出的方式弹出，虚拟机规范同样也没有给出强制的要求

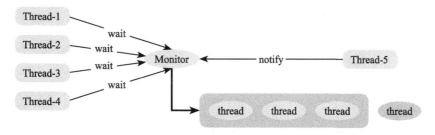

图 5-5　notify 方法只会唤醒在 wait set 中休息的一个线程

而执行 notifyAll 则不需要考虑哪个线程会被弹出，因为 wait set 中的所有 wait 线程都将被弹出，如图 5-6 所示。

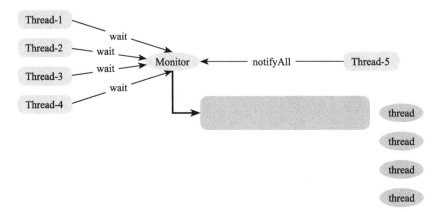

图 5-6　notifyAll 方法只会唤醒在 wait set 中休息的所有线程

5.4　自定义显式锁 BooleanLock

在本节中，我们将利用前面所掌握的知识，构建一个自定义的显式锁，类似于 Java utils 包下的 Lock 接口，并且分析 synchronized 关键字的缺陷。

5.4.1　synchronized 关键字的缺陷

synchronized 关键字提供了一种排他式的数据同步机制，某个线程在获取 monitor lock 的时候可能会被阻塞，而这种阻塞有两个很明显的缺陷：第一，无法控制阻塞时长。第二，阻塞不可被中断。下面通过示例来进行分析，如下：

```
package com.wangwenjun.concurrent.chapter05;

import java.util.concurrent.TimeUnit;

public class SynchronizedDefect
{
    public synchronized void syncMethod()
    {
        try
        {
            TimeUnit.HOURS.sleep(1);
        } catch (InterruptedException e)
        {
            e.printStackTrace();
        }
    }

    public static void main(String[] args) throws InterruptedException
```

```
    {
        SynchronizedDefect defect = new SynchronizedDefect();
        Thread t1 = new Thread(defect::syncMethod, "T1");
        //make sure the t1 started.
        t1.start();
        TimeUnit.MILLISECONDS.sleep(2);

        Thread t2 = new Thread(defect::syncMethod, "T2");
        t2.start();
    }
}
```

上面的代码中有一个同步方法 syncMethod，这里启动了两个线程分别调用该方法，在该方法中线程会休眠 1 小时的时间，为了确保 T1 线程能够最先进入同步方法，在 T1 线程启动后主线程休眠了 2 毫秒的时间。T2 线程启动执行 syncMethod 方法时会进入阻塞，T2 什么时候能够获得 syncMethod 的执行完全取决于 T1 何时对其释放，如果 T2 计划最多 1 分钟获得执行权，否则就放弃，很显然这种方式是做不到的，这也就是前面所说的阻塞时长无法控制。

第二个缺陷是 T2 若因争抢某个 monitor 的锁而进入阻塞状态，那么它是无法中断的，虽然可以设置 T2 线程的 interrupt 标识，但是 synchronized 阻塞不像 sleep 和 wait 方法一样能够捕获得到中断信号。下面将 main 方法的代码稍作修改，试图打断 T2 线程，来看看会怎样。

```
public static void main(String[] args) throws InterruptedException
{
    SynchronizedDefect defect = new SynchronizedDefect();
    Thread t1 = new Thread(defect::syncMethod, "T1");
    //make sure the t1 started.
    t1.start();
    TimeUnit.MILLISECONDS.sleep(2);
    Thread t2 = new Thread(defect::syncMethod, "T2");
    t2.start();
    //make sure the t2 started.
    TimeUnit.MILLISECONDS.sleep(2);
    t2.interrupt();
    System.out.println(t2.isInterrupted());
    System.out.println(t2.getState());
}
```

程序的输出一定会让你感到失望，但是同时也证明了被 synchronized 同步的线程不可被中断。

5.4.2 显式锁 BooleanLock

在本节中，我们将利用前面所学的知识，构造一个显式的 BooleanLock，使其在具备 synchronized 关键字所有功能的同时又具备可中断和 lock 超时的功能。

1. 定义 Lock 接口

示例代码如下：

```
package com.wangwenjun.concurrent.chapter05;

import java.util.List;

public interface Lock
{
    void lock() throws InterruptedException;

    void lock(long mills) throws InterruptedException, TimeoutException;

    void unlock();

    List<Thread> getBlockedThreads();
}
```

在上述代码中：

- lock() 方法永远阻塞，除非获取到了锁，这一点和 synchronized 非常类似，但是该方法是可以被中断的，中断时会抛出 InterruptedException 异常。
- lock（long mills）方法除了可以被中断以外，还增加了对应的超时功能。
- unlock() 方法可用来进行锁的释放。
- getBlockedThreads() 用于获取当前有哪些线程被阻塞。

2. 实现 BooleanLock

BooleanLock 是 Lock 的一个 Boolean 实现，通过控制一个 Boolean 变量的开关来决定是否允许当前的线程获取该锁，首先定义三个非常重要的成员变量，如下所示：

```
package com.wangwenjun.concurrent.chapter05;

import java.util.ArrayList;
import java.util.List;

public class BooleanLock implements Lock
{
    private Thread currentThread;

    private boolean locked = false;

    private final List<Thread> blockedList = new ArrayList<>();
```

其中 currentThread 代表当前拥有锁的线程，locked 是一个 boolean 开关，false 代表当前该锁没有被任何线程获得或者已经释放，true 代表该锁已经被某个线程获得，该线程就是 currentThread；blockedList 用来存储哪些线程在获取当前线程时进入了阻塞状态。下面继续实现 lock() 方法，代码如下：

```java
@Override
public void lock() throws InterruptedException
{
    synchronized (this)// ①
    {
        while (locked) // ②
        {
            blockedList.add(currentThread());
            this.wait();
        }
        blockedList.remove(currentThread());// ③
        this.locked = true;// ④
        this.currentThread = currentThread();// ⑤
    }
}
```

在上述代码中：

① Lock() 方法使用同步代码块的方式进行方法同步。

② 如果当前锁已经被某个线程获得，则该线程将加入阻塞队列，并且使当前线程 wait 释放对 this monitor 的所有权。

③ 如果当前锁没有被其他线程获得，则该线程将尝试从阻塞队列中删除自己（注意：如果当前线程从未进入过阻塞队列，删除方法不会有任何影响；如果当前线程是从 wait set 中被唤醒的，则需要从阻塞队列中将自己删除）。

④ locked 开关被指定为 true。

⑤ 记录获取锁的线程。

继续实现带有超时功能的 lock（long mills）方法，代码如下：

```java
@Override
public void lock(long mills) throws InterruptedException, TimeoutException
{
    synchronized (this)
    {
        if (mills <= 0)       ①
        {
            this.lock();
        } else
        {
            long remainingMills = mills;
            long endMills = currentTimeMillis() + remainingMills;
            while (locked)
            {
                if (remainingMills <= 0)  // ②
                    throw new TimeoutException("can not get the lock during " + mills
                if (!blockedList.contains(currentThread()))
```

```
                blockedList.add(currentThread());
                this.wait(remainingMills);    //③
                remainingMills = endMills - currentTimeMillis();  //④
            }
            blockedList.remove(currentThread());    //⑤
            this.locked = true;
            this.currentThread = currentThread();
        }
    }
}
```

相比较 lock() 方法，该方法稍微有些复杂，我们逐步解释。在上述代码中：

①如果 mills 不合法，则默认调用 lock() 方法，当然也可以抛出参数非法的异常，一般来说，抛出异常是一种比较好的做法。

②如果 remainingMills 小于等于 0，则意味着当前线程被其他线程唤醒或者在指定的 wait 时间到了之后还没有获得锁，这种情况下会抛出超时的异常。

③等待 remainingMills 的毫秒数，该值最开始是由其他线程传入的，但在多次 wait 的过程中会重新计算。

④重新计算 remainingMills 时间。

⑤获得该锁，并且从 block 列表中删除当前线程，将 locked 的状态修改为 true 并且指定获得锁的线程就是当前线程。

unlock() 方法需要做的仅仅是将 locked 状态修改为 false，并且唤醒 wait set 中的其他线程，再次争抢锁资源。但是需要注意的一点是，哪个线程加的锁只能由该线程来解锁：

```
@Override
public void unlock()
{
    synchronized (this)
    {
        if (currentThread == currentThread())//①
        {
            this.locked = false;     // ②
            this.notifyAll();         // ③
        }
    }
}
```

在上述代码中：

①判断当前线程是否为获取锁的那个线程，只有加了锁的线程才有资格进行解锁。

②将锁的 locked 状态修改为 false。

③通知其他在 wait set 中的线程，你们可以再次尝试抢锁了，这里使用 notify 和 notifyAll 都可以。

下面是 BooleanLock 的完整代码：

```java
package com.wangwenjun.concurrent.chapter05;

import java.util.ArrayList;
import java.util.Collections;
import java.util.List;
import java.util.concurrent.TimeoutException;

import static java.lang.System.currentTimeMillis;
import static java.lang.Thread.currentThread;

public class BooleanLock implements Lock
{

    private Thread currentThread;

    private boolean locked = false;

    private final List<Thread> blockedList = new ArrayList<>();

    @Override
    public void lock() throws InterruptedException
    {
        synchronized (this)
        {
            while (locked)
            {
                if (!blockedList.contains(currentThread()))
                    blockedList.add(currentThread());
                this.wait();
            }

            blockedList.remove(currentThread());
            this.locked = true;
            this.currentThread = currentThread();
        }
    }

    @Override
    public void lock(long mills) throws InterruptedException, TimeoutException
    {
        synchronized (this)
        {
            if (mills <= 0)
            {
                this.lock();
            } else
            {
                long remainingMills = mills;
                long endMills = currentTimeMillis() + remainingMills;
```

```java
                while (locked)
                {
                    if (remainingMills <= 0)
                        throw new TimeoutException("can not get the lock during "
                            + mills + " ms.");
                    if (!blockedList.contains(currentThread()))
                        blockedList.add(currentThread());
                    this.wait(remainingMills);
                    remainingMills = endMills - currentTimeMillis();
                }

                blockedList.remove(currentThread());
                this.locked = true;
                this.currentThread = currentThread();
            }
        }
    }

    @Override
    public void unlock()
    {
        synchronized (this)
        {
            if (currentThread == currentThread())
            {
                this.locked = false;
                Optional.of(currentThread().getName() + " release the lock.")
                    .ifPresent(System.out::println);
                this.notifyAll();
            }
        }
    }

    @Override
    public List<Thread> getBlockedThreads()
    {
        return Collections.unmodifiableList(blockedList);
    }
}
```

至此 BooleanLock 的基本功能已经完成，当然读者也可以对其进行扩充，比如增加 tryLock 的功能，也就是说能获得锁便获得，获得不了就退出，压根不会阻塞。

3. 使用 BooleanLock

5.2 节中充分利用了 wait、notify 方法的功能，实现了一个显式锁，可中断被阻塞的线程。在使用该显式锁的时候，务必借助于 try finally 语句来确保每次获取到锁之后都可以正常的释放，本节将列举一些简单的例子来展示如何使用 BooleanLock。

（1）多个线程通过 lock() 方法争抢锁

示例代码如下：

```java
package com.wangwenjun.concurrent.chapter05;

import java.util.concurrent.TimeUnit;
import java.util.stream.IntStream;

import static java.lang.Thread.currentThread;
import static java.util.concurrent.ThreadLocalRandom.current;

public class BooleanLockTest
{
    // 定义 BooleanLock
    private final Lock lock = new BooleanLock();

    // 使用 try..finally 语句块确保 lock 每次都能被正确释放
    public void syncMethod()
    {
        // 加锁
        lock.lock();
        try
        {
            int randomInt = current().nextInt(10);
            System.out.println(currentThread() + " get the lock.");
            TimeUnit.SECONDS.sleep(randomInt);
        } catch (InterruptedException e)
        {
            e.printStackTrace();
        } finally
        {
            // 释放锁
            lock.unlock();
        }
    }

    public static void main(String[] args)
    {
        BooleanLockTest blt = new BooleanLockTest();
        // 定义一个线程并且启动
        IntStream.range(0, 10)
                .mapToObj(i -> new Thread(blt::syncMethod))
                .forEach(Thread::start);
    }
}
```

上面的代码比较简单，代码注释已给出相应的说明，运行上面代码，查看执行结果，输出如下：

```
Thread[Thread-0,5,main] get the lock.
```

```
Thread-0 release the lock monitor.
Thread[Thread-9,5,main] get the lock.
Thread-9 release the lock monitor.
Thread[Thread-1,5,main] get the lock.
Thread-1 release the lock monitor.
Thread[Thread-6,5,main] get the lock.
Thread-6 release the lock monitor.
Thread[Thread-2,5,main] get the lock.
Thread-2 release the lock monitor.
Thread[Thread-5,5,main] get the lock.
Thread-5 release the lock monitor.
Thread[Thread-3,5,main] get the lock.
Thread-3 release the lock monitor.
Thread[Thread-7,5,main] get the lock.
Thread-7 release the lock monitor.
Thread[Thread-4,5,main] get the lock.
Thread-4 release the lock monitor.
Thread[Thread-8,5,main] get the lock.
Thread-8 release the lock monitor.
```

根据控制台输出可以看到，每次都会确保只有一个线程能够获得锁的执行权限，这一点已经与 synchronized 同步非常类似。接下来再讲解 BooleanLock 的可中断特性。

（2）可中断被阻塞的线程

将 main 函数中的代码稍作修改，启动 T1 线程以确保能够获得锁，紧接着启动另外一个 T2，main 线程将稍作休眠然后打断 T2 线程，代码如下：

```
BooleanLockTest blt = new BooleanLockTest();
new Thread(blt::syncMethod, "T1").start();
TimeUnit.MILLISECONDS.sleep(2);
Thread t2 = new Thread(blt::syncMethod, "T2");
t2.start();
TimeUnit.MILLISECONDS.sleep(10);
t2.interrupt();
```

运行上面的程序，在 T2 线程启动 10 毫秒以后，主动将其中断，T2 线程会收到中断信号，也就是 InterruptedException 异常，这样也就弥补了 Synchronized 同步方式不可被中断的缺陷。上述程序的运行结果如下：

```
Thread[T1,5,main] get the lock.
java.lang.InterruptedException
    at java.lang.Object.wait(Native Method)
    at java.lang.Object.wait(Object.java:502)
    at com.wangwenjun.concurrent.chapter05.BooleanLock.lock(BooleanLock.java:31)
    at com.wangwenjun.concurrent.chapter05.BooleanLockTest.syncMethod(BooleanLock-
Test.java:18)
    at com.wangwenjun.concurrent.chapter05.BooleanLockTest$$Lambda$2/834600351.run
(Unknown Source)
```

```
        at java.lang.Thread.run(Thread.java:745)
T1 release the lock monitor.
```

但是，BooleanLock 还存在一个问题，如果某个线程被中断，那么它将有可能还存在于 blockList 中，该问题的修复也非常简单，可以对 BooleanLock 的 lock 方法进行进一步的增强加以修复，另外一个 lock 重载方法的实现思路与之类似，如下代码所示：

```
...
    @Override
    public void lock() throws InterruptedException
    {
        synchronized (this)
        {
            while (locked)
            {
                // 暂存当前线程
                final Thread tempThread = currentThread();
                try
                {
                    if (!blockedList.contains(tempThread))
                        blockedList.add(tempThread);
                    this.wait();
                }
                catch(InterruptedException e)
                {
                    // 如果当前线程在 wait 时被中断，则从 blockedList 中将其删除，避免内存泄漏
                    blockedList.remove(tempThread);
                    // 继续抛出中断异常
                    throw e;
                }
            }
            blockedList.remove(currentThread());
            this.locked = true;
            this.currentThread = currentThread();
        }
    }
...
```

（3）阻塞的线程可超时

最后再写一个具有超时功能 lock 的使用示例，同样定义两个线程 T1 和 T2，确保 T1 先执行能够最先获得锁，T2 稍后启动，在 1000ms 以内未获得锁则会抛出超时异常，代码如下：

```
public void syncMethodTimeoutable()
{
    try
    {
        lock.lock(1000);
```

```
            System.out.println(currentThread() + " get the lock.");
            int randomInt = current().nextInt(10);
            TimeUnit.SECONDS.sleep(randomInt);
        } catch (InterruptedException | TimeoutException e)
        {
            e.printStackTrace();
        } finally
        {
            lock.unlock();
        }
    }
    public static void main(String[] args) throws InterruptedException
    {
        BooleanLockTest blt = new BooleanLockTest();
        new Thread(blt::syncMethod, "T1").start();
        TimeUnit.MILLISECONDS.sleep(2);
        Thread t2 = new Thread(blt::syncMethodTimeoutable, "T2");
        t2.start();
        TimeUnit.MILLISECONDS.sleep(10);
    }
```

这里的代码比较简单,不再赘述其实现细节,程序的输出结果如下所示:

```
Thread[T1,5,main] get the lock.
java.util.concurrent.TimeoutException: can not get the lock during 1000 ms.
    at com.wangwenjun.concurrent.chapter05.BooleanLock.lock(BooleanLock.java:55)
    at com.wangwenjun.concurrent.chapter05.BooleanLockTest.syncMethodTimeoutable
(BooleanLockTest.java:35)
    at com.wangwenjun.concurrent.chapter05.BooleanLockTest$$Lambda$2/834600351.run
(Unknown Source)
    at java.lang.Thread.run(Thread.java:745)
T1 release the lock monitor.
```

5.5 本章总结

5.1 节通过对比同步阻塞消息处理机制和异步非阻塞消息处理机制的案例,引入了线程间通信的需求;5.2 节中详细讲解了 wait 和 notify 方法的使用以及在开发中应该注意的事项;生产者与消费者模型是多线程开发中最常用的模型之一,也是多线程间通信的最好范例,因此在 5.3 节中进行了详细的讲解,并且重点分析了 wait set 线程休息室;由于 synchronized 同步方法具有不可中断性以及无法超时的缺陷,因此在争抢锁的过程中无法获取锁的线程将会无休止地陷入阻塞之中,因此我们结合之前的知识开发了一个显式的 BooleanLock,BooleanLock 除了具备 synchronized 关键字的互斥访问共享资源的语义之外,还增加了可中断以及可超时等特点。

CHAPTER 6 · 第 6 章

ThreadGroup 详细讲解

在 2.3 节中曾经介绍过，创建线程的时候如果没有显式地指定 ThreadGroup，那么新的线程会被加入与父线程相同的 ThreadGroup 中，在本章中，将详细讲解有关 ThreadGroup 的知识以及各个 API 的使用情况。

6.1 ThreadGroup 与 Thread

在 Java 程序中，默认情况下，新的线程都会被加入到 main 线程所在的 group 中，main 线程的 group 名字同线程名。如同线程存在父子关系一样，ThreadGroup 同样也存在父子关系。图 6-1 就很好地说明了父子 thread、父子 threadGroup 以及 thread 和 group 之间的层次关系。

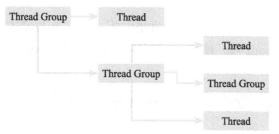

图 6-1　Thread 与 ThreadGroup 的关系图

无论如何，线程都会被加入某个 Thread Group 之中。

6.2 创建 ThreadGroup

创建 ThreadGroup 的语法如下：

```
public ThreadGroup(String name)
public ThreadGroup(ThreadGroup parent,String name)
```

创建 ThreadGroup 的语法非常简单，可通过上面某个构造函数来创建，第一个构造函数为 ThreadGroup 赋予了名字，但是该 ThreadGroup 的父 ThreadGroup 是创建它的线程所在的 ThreadGroup；第二个 ThreadGroup 的构造函数赋予 group 名字的同时又显式地指定了父 Group。

代码清单 6-1　ThreadGroupCreator.java

```java
package com.wangwenjun.concurrent.chapter06;

public class ThreadGroupCreator
{
    public static void main(String[] args)
    {
        //① 获取当前线程的 group
        ThreadGroup currentGroup = Thread.currentThread().getThreadGroup();
        //② 定义一个新的 group
        ThreadGroup group1 = new ThreadGroup("Group1");

        //③ 程序输出 true
        System.out.println(group1.getParent() == currentGroup);

        //④ 定义 group2，指定 group1 为其父 group
        ThreadGroup group2 = new ThreadGroup(group1, "Group2");

        //⑤ 程序输出 true
        System.out.println(group2.getParent() == group1);

    }
}
```

代码清单 6-1 定义了一个 group1，没有指定父 group，所以默认父 group 为当前线程所在的 group，在构造 group2 时，显式地指定了其父 group 为 group1。

6.3　复制 Thread 数组和 ThreadGroup 数组

在一个 ThreadGroup 中会加入若干个线程以及子 ThreadGroup，ThreadGroup 为我们提供了若干个方法，可以复制出线程和线程组。

6.3.1　复制 Thread 数组

先来看如下两个方法：

```
public int enumerate(Thread[] list)
```

```
public int enumerate(Thread[] list,boolean recurse)
```

上述两个方法，会将 ThreadGroup 中的 active 线程全部复制到 Thread 数组中，其中 recurse 参数如果为 true，则该方法会将所有子 group 中的 active 线程都递归到 Thread 数组中，enumerate（Thread[] list）实际上等价于 enumerate（Thread[] true），上面两个方法都调用了 ThreadGroup 的私有方法 enumerate：

```java
private int enumerate(Thread list[], int n, boolean recurse) {
    int ngroupsSnapshot = 0;
    ThreadGroup[] groupsSnapshot = null;
    synchronized (this) {
        if (destroyed) {
            return 0;
        }
        int nt = nthreads;
        if (nt > list.length - n) {
            nt = list.length - n;
        }
        for (int i = 0; i < nt; i++) {
            if (threads[i].isAlive()) {
                list[n++] = threads[i];
            }
        }
        if (recurse) {
            ngroupsSnapshot = ngroups;
            if (groups != null) {
                groupsSnapshot = Arrays.copyOf(groups, ngroupsSnapshot);
            } else {
                groupsSnapshot = null;
            }
        }
    }
    if (recurse) {
        for (int i = 0 ; i < ngroupsSnapshot ; i++) {
            n = groupsSnapshot[i].enumerate(list, n, true);
        }
    }
    return n;
}
```

下面，我们再写个简单的例子来演示一下如何使用这两个方法，首先定义一个 ThreadGroup，并且将该 group 加入到 main group 中，然后定义一个线程 thread 并将其加入到 myGroup 中，最后分别调用 enumerate 的递归和非递归方法，如代码清单 6-2 所示。

代码清单 6-2　ThreadGroupEnumerateThreads.java

```java
package com.wangwenjun.concurrent.chapter06;

import java.util.concurrent.TimeUnit;
```

```java
public class ThreadGroupEnumerateThreads
{
    public static void main(String[] args)
            throws InterruptedException
    {
        // 创建一个 ThreadGroup
        ThreadGroup myGroup = new ThreadGroup("MyGroup");
        // 创建线程传入 threadgroup
        Thread thread = new Thread(myGroup, () ->
        {
            while (true)
            {
                try
                {
                    TimeUnit.SECONDS.sleep(1);
                } catch (InterruptedException e)
                {
                }
            }
        }, "MyThread");
        thread.start();

        TimeUnit.MILLISECONDS.sleep(2);
        ThreadGroup mainGroup = Thread.currentThread().getThreadGroup();

        Thread[] list = new Thread[mainGroup.activeCount()];
        int recurseSize = mainGroup.enumerate(list);
        System.out.println(recurseSize);

        recurseSize = mainGroup.enumerate(list, false);
        System.out.println(recurseSize);
    }
}
```

上面的代码运行之后,最后一个输出会比第一个少 1,那是因为代码中将递归 recurse 设置为了 false,myGroup 中的线程将不会包含在内。

> **注意**
> - enumerate 方法获取的线程仅仅是个预估值,并不能百分之百地保证当前 group 的活跃线程,比如在调用复制之后,某个线程结束了生命周期或者新的线程加入了进来,都会导致数据的不准确。
> - enumerate 方法的返回值 int 相较 Thread[] 的长度更为真实,比如定义了数组长度的 Thread 数组,那么 enumerate 方法仅仅会将当前活跃的 thread 分别放进数组中,而返回值 int 则代表真实的数量,并非 Thread 数组的长度,可能是早期版本就有这个方法的缘故 (JDK1.0),其实用 List (JDK1.1 版本才引入) 会更好一些。

6.3.2 复制 ThreadGroup 数组

来看如下两种方法：

```java
public int enumerate(ThreadGroup[] list)
public int enumerate(ThreadGroup[] list,boolean recurse)
```

和复制 Thread 数组类似，上述两个方法，主要用于复制当前 ThreadGroup 的子 Group，同样 recurse 会决定是否以递归的方式复制。

代码清单 6-3　ThreadGroupEnumerateThreadGroups.java

```java
package com.wangwenjun.concurrent.chapter06;

import java.util.concurrent.TimeUnit;

public class ThreadGroupEnumerateThreadGroups
{
    public static void main(String[] args)
            throws InterruptedException
    {

        ThreadGroup myGroup1 = new ThreadGroup("MyGroup1");
        ThreadGroup myGroup2 = new ThreadGroup(myGroup1,"MyGroup2");


        TimeUnit.MILLISECONDS.sleep(2);
        ThreadGroup mainGroup = Thread.currentThread().getThreadGroup();


        ThreadGroup[] list = new ThreadGroup[mainGroup.activeGroupCount()];

        int recurseSize = mainGroup.enumerate(list);
        System.out.println(recurseSize);

        recurseSize = mainGroup.enumerate(list, false);
        System.out.println(recurseSize);
    }
}
```

在代码清单 6-3 中，myGroup1 的父 group 为 mainGroup，而 myGroup2 的父 group 为 myGroup1，因此上述的代码运行之后，递归复制的结果为 2，不递归的情况下为 1。

6.4　ThreadGroup 操作

ThreadGroup 并不能提供对线程的管理，ThreadGroup 的主要功能是对线程进行组织，在本节中，将详细介绍 ThreadGroup 的主要方法。

6.4.1 ThreadGroup 的基本操作

- activeCount() 用于获取 group 中活跃的线程，这只是个估计值，并不能百分之百地保证数字一定正确，原因前面已经分析过，该方法会递归获取其他子 group 中的活跃线程。
- activeGroupCount() 用于获取 group 中活跃的子 group，这也是一个近似估值，该方法也会递归获取所有的子 group。
- getMaxPriority() 用于获取 group 的优先级，默认情况下，Group 的优先级为 10，在该 group 中，所有线程的优先级都不能大于 group 的优先级。
- getName() 用于获取 group 的名字。
- getParent() 用于获取 group 的父 group，如果父 group 不存在，则会返回 null，比如 system group 的父 group 就为 null。
- list() 该方法没有返回值，执行该方法会将 group 中所有的活跃线程信息全部输出到控制台，也就是 System.out。
- parentOf（ThreadGroup g）会判断当前 group 是不是给定 group 的父 group，另外如果给定的 group 就是自己本身，那么该方法也会返回 true。
- setMaxPriority（int pri）会指定 group 的最大优先级，最大优先级不能超过父 group 的最大优先级，执行该方法不仅会改变当前 group 的最大优先级，还会改变所有子 group 的最大优先级。

下面我们给出一个简单的例子来测试一下上面的几个方法。

代码清单 6-4　ThreadGroupBasic.java

```java
package com.wangwenjun.concurrent.chapter06;

import java.util.concurrent.TimeUnit;

public class ThreadGroupBasic
{
    public static void main(String[] args) throws InterruptedException
    {
        /*
         * Create a thread group and thread.
         */
        ThreadGroup group = new ThreadGroup("group1");
        Thread thread = new Thread(group, () ->
        {
            while (true)
            {
                try
                {
                    TimeUnit.SECONDS.sleep(1);
```

```
                } catch (InterruptedException e)
                {
                    e.printStackTrace();
                }
            }
        }, "thread");
        thread.setDaemon(true);
        thread.start();

        //make sure the thread is started
        TimeUnit.MILLISECONDS.sleep(1);

        ThreadGroup mainGroup = Thread.currentThread().getThreadGroup();

        System.out.println("activeCount=" + mainGroup.activeCount());
        System.out.println("activeGroupCount=" + mainGroup.activeGroupCount());
        System.out.println("getMaxPriority=" + mainGroup.getMaxPriority());
        System.out.println("getName=" + mainGroup.getName());
        System.out.println("getParent=" + mainGroup.getParent());
        mainGroup.list();
        System.out.println("-------------------------");
        System.out.println("parentOf="+mainGroup.parentOf(group));
    System.out.println("parentOf="+mainGroup.parentOf(mainGroup));
    }
}
```

运行代码清单 6-4 所示的程序，执行结果如下，你的运行结果不一定与笔者的完全一样，大家可以自行分析一下。

```
activeCount=3
activeGroupCount=1
getMaxPriority=10
getName=main
getParent=java.lang.ThreadGroup[name=system,maxpri=10]
java.lang.ThreadGroup[name=main,maxpri=10]
    Thread[main,5,main]
    Thread[Monitor Ctrl-Break,5,main]
    java.lang.ThreadGroup[name=group1,maxpri=10]
        Thread[thread,5,group1]
-------------------------
parentOf=true
parentOf=true

Process finished with exit code 0
```

这里需要特别说明的是 **setMaxPriority**，在本书的 3.3 节中，我们通过分析源码得出结论，线程的最大优先级，不能高于所在线程组的最大优先级，但是如果我们把代码写成下面这样会怎么样呢？

```java
package com.wangwenjun.concurrent.chapter06;

import java.util.concurrent.TimeUnit;

public class ThreadGroupPriority
{
    public static void main(String[] args)
    {
        /*
         * Create a thread group and thread.
         */
        ThreadGroup group = new ThreadGroup("group1");
        Thread thread = new Thread(group, () ->
        {
            while (true)
            {
                try
                {
                    TimeUnit.SECONDS.sleep(1);
                } catch (InterruptedException e)
                {
                    e.printStackTrace();
                }
            }
        }, "thread");
        thread.setDaemon(true);
        thread.start();

        System.out.println("group.getMaxPriority()="+group.getMaxPriority());

        System.out.println("thread.getPriority()="+thread.getPriority());
        //① 改变 group 的最大优先级
        group.setMaxPriority(3);

        System.out.println("group.getMaxPriority()="+group.getMaxPriority());

        System.out.println("thread.getPriority()="+thread.getPriority());
    }
}
```

运行上面的程序,会出现 thread 的优先级大于所在 group 最大优先级的情况,如下所示:

```
group.getMaxPriority()=10
thread.getPriority()=5
group.getMaxPriority()=3
thread.getPriority()=5
```

虽然出现了已经加入该 group 的线程的优先级大于 group 最大优先级的情况,但是后面加入该 group 的线程再不会大于新设置的值:3,这一点需要大家注意。

6.4.2 ThreadGroup 的 interrupt

interrupt 一个 thread group 会导致该 group 中所有的 active 线程都被 interrupt，也就是说该 group 中每一个线程的 interrupt 标识都被设置了，下面是 ThreadGroup interrupt 方法的源码：

```java
public final void interrupt() {
    int ngroupsSnapshot;
    ThreadGroup[] groupsSnapshot;
    synchronized (this) {
        checkAccess();
        for (int i = 0 ; i < nthreads ; i++) {
            threads[i].interrupt();
        }
        ngroupsSnapshot = ngroups;
        if (groups != null) {
            groupsSnapshot = Arrays.copyOf(groups, ngroupsSnapshot);
        } else {
            groupsSnapshot = null;
        }
    }
    for (int i = 0 ; i < ngroupsSnapshot ; i++) {
        groupsSnapshot[i].interrupt();
    }
}
```

分析上述源码，我们可以看出在 interrupt 内部会执行所有 thread 的 interrupt 方法，并且会递归获取子 group，然后执行它们各自的 interrupt 方法，下面我们写个简单的程序测试一下：

```java
package com.wangwenjun.concurrent.chapter06;

import java.util.concurrent.TimeUnit;

public class ThreadGroupInterrupt
{
    public static void main(String[] args) throws InterruptedException
    {
        ThreadGroup group = new ThreadGroup("TestGroup");

        new Thread(group, () ->
        {
            while (true)
            {
                try
                {
                    TimeUnit.MILLISECONDS.sleep(2);
                } catch (InterruptedException e)
```

```
            {
                //received interrupt SIGNAL and clear quickly
                break;
            }
        }
        System.out.println("t1 will exit.");
    }, "t1").start();

    new Thread(group, () ->
    {
        while (true)
        {
            try
            {
                TimeUnit.MILLISECONDS.sleep(1);
            } catch (InterruptedException e)
            {
                //received interrupt SIGNAL and clear quickly
                break;
            }
        }
        System.out.println("t2 will exit.");
    }, "t2").start();

    //make sure all of above threads started.
    TimeUnit.MILLISECONDS.sleep(2);

    group.interrupt();
    }
}
```

上面的代码足够简单，不必多做解释，从运行结果中可以看出，group 中的 active thread 都将被 interrupt。

6.4.3 ThreadGroup 的 destroy

destroy 用于销毁 ThreadGroup，该方法只是针对一个没有任何 active 线程的 group 进行一次 destroy 标记，调用该方法的直接结果是在父 group 中将自己移除：

Destroys this thread group and all of its subgroups. This thread group must be empty, indicating that all threads that had been in this thread group have since stopped. （销毁 ThreadGroup 及其子 ThreadGroup，在该 ThreadGroup 中所有的线程必须是空的，也就是说 ThreadGroup 或者子 ThreadGroup 所有的线程都已经停止运行，如果有 Active 线程存在，调用 destroy 方法则会抛出异常。）

下面我们写一个简单的代码对其进行测试：

```
package com.wangwenjun.concurrent.chapter06;

public class ThreadGroupDestroy
```

```java
{
    public static void main(String[] args)
    {
        ThreadGroup group = new ThreadGroup("TestGroup");

        ThreadGroup mainGroup = Thread.currentThread().getThreadGroup();
        System.out.println("group.isDestroyed=" + group.isDestroyed());
        mainGroup.list();

        group.destroy();

        System.out.println("group.isDestroyed=" + group.isDestroyed());
        mainGroup.list();
    }
}
```

程序的运行结果如下所示，其中 isDestroyed 方法是判断 ThreadGroup 是否被 destroy 了：

```
ggroup.isDestroyed=false
java.lang.ThreadGroup[name=main,maxpri=10]
    Thread[main,5,main]
    Thread[Monitor Ctrl-Break,5,main]
    java.lang.ThreadGroup[name=TestGroup,maxpri=10]
group.isDestroyed=true
java.lang.ThreadGroup[name=main,maxpri=10]
    Thread[main,5,main]
    Thread[Monitor Ctrl-Break,5,main]
```

6.4.4 守护 ThreadGroup

线程可以设置为守护线程，ThreadGroup 也可以设置为守护 ThreadGroup，但是若将一个 ThreadGroup 设置为 daemon，也并不会影响线程的 daemon 属性，如果一个 ThreadGroup 的 daemon 被设置为 true，那么在 group 中没有任何 active 线程的时候该 group 将自动 destroy，下面我们给出一个简单的例子来对其进行说明：

```java
package com.wangwenjun.concurrent.chapter06;

import java.util.concurrent.TimeUnit;

public class ThreadGroupDaemon
{
    public static void main(String[] args)
            throws InterruptedException
    {

        ThreadGroup group1 = new ThreadGroup("Group1");
        new Thread(group1, () ->
```

```java
        {
            try
            {
                TimeUnit.SECONDS.sleep(1);
            } catch (InterruptedException e)
            {
                e.printStackTrace();
            }
        }, "group1-thread1").start();

        ThreadGroup group2 = new ThreadGroup("Group2");
        new Thread(group2, () ->
        {
            try
            {
                TimeUnit.SECONDS.sleep(1);
            } catch (InterruptedException e)
            {
                e.printStackTrace();
            }
        }, "group2-thread1").start();

        // 设置daemon为true
        group2.setDaemon(true);

        TimeUnit.SECONDS.sleep(3);
        System.out.println(group1.isDestroyed());
        System.out.println(group2.isDestroyed());
    }
}
```

在上面的代码中，第二个 group 的 daemon 被设置为 true，当其中没有 active 线程的时候，该 group 将会自动被 destroy，而第一个 group 则相反。

6.5 本章总结

在本章中，详细介绍了 ThreadGroup 与 Thread 的关系，ThreadGroup 并不是用来管理 Thread 的，而是针对 Thread 的一个组织；ThreadGroup 提供了比较多的 API，本章中几乎对所有的 API 都进行了介绍。

第 7 章
Hook 线程以及捕获线程执行异常

在本章中,将介绍如何获取线程在运行时期的异常信息,以及如何向 Java 程序注入 Hook 线程(Hook 通常也称钩子)。

7.1 获取线程运行时异常

在 Thread 类中,关于处理运行时异常的 API 总共有四个,如下所示:

- public void setUncaughtExceptionHandler(UncaughtExceptionHandler eh):为某个特定线程指定 UncaughtExceptionHandler。
- public static void setDefaultUncaughtExceptionHandler(UncaughtExceptionHandler eh):设置全局的 UncaughtExceptionHandler。
- public UncaughtExceptionHandler getUncaughtExceptionHandler():获取特定线程的 UncaughtExceptionHandler。
- public static UncaughtExceptionHandler getDefaultUncaughtExceptionHandler():获取全局的 UncaughtExceptionHandler。

7.1.1 UncaughtExceptionHandler 的介绍

线程在执行单元中是不允许抛出 checked 异常的,这一点前文中已经有过交代,而且线程运行在自己的上下文中,派生它的线程将无法直接获得它运行中出现的异常信息。对此,Java 为我们提供了一个 UncaughtExceptionHandler 接口,当线程在运行过程中出现异常时,会回调 UncaughtExceptionHandler 接口,从而得知是哪个线程在运行时出错,以及出现了什么样的错误,示例代码如下:

```java
@FunctionalInterface
public interface UncaughtExceptionHandler {
    /**
     * Method invoked when the given thread terminates due to the
     * given uncaught exception.
     * <p>Any exception thrown by this method will be ignored by the
     * Java Virtual Machine.
     * @param t the thread
     * @param e the exception
     */
    void uncaughtException(Thread t, Throwable e);
}
```

在上述代码中，UncaughtExceptionHandler 是一个 FunctionalInterface，只有一个抽象方法，该回调接口会被 Thread 中的 dispatchUncaughtException 方法调用，如下所示：

```java
/**
 * Dispatch an uncaught exception to the handler. This method is
 * intended to be called only by the JVM.
 */
private void dispatchUncaughtException(Throwable e) {
    getUncaughtExceptionHandler().uncaughtException(this, e);
}
```

当线程在运行过程中出现异常时，JVM 会调用 dispatchUncaughtException 方法，该方法会将对应的线程实例以及异常信息传递给回调接口。

7.1.2 UncaughtExceptionHandler 实例

本节中，我们将给出一个实例，来演示如何使用 UncaughtExceptionHandler。

代码清单 7-1 CaptureThreadException.java

```java
package com.wangwenjun.concurrent.chapter07;

import java.util.concurrent.TimeUnit;

public class CaptureThreadException
{
    public static void main(String[] args)
    {
        //① 设置回调接口
        Thread.setDefaultUncaughtExceptionHandler((t, e) ->
        {
            System.out.println(t.getName() + " occur exception");
            e.printStackTrace();
        });
        final Thread thread = new Thread(() ->
```

```
        {
            try
            {
                TimeUnit.SECONDS.sleep(2);
            } catch (InterruptedException e)
            {
            }

            //② 这里会出现unchecked异常
            //here will throw unchecked exception.
            System.out.println(1 / 0);
        }, "Test-Thread");

        thread.start();
    }
}
```

执行上面的程序，线程 Test-Thread 在运行两秒之后会抛出一个 unchecked 异常，我们设置的回调接口将获得该异常信息，程序的执行结果如下：

```
Test-Thread occur exception
java.lang.ArithmeticException: / by zero
    at com.wangwenjun.concurrent.chapter06.CaptureThreadException.lambda$main$1
(CaptureThreadException.java:26)
    at com.wangwenjun.concurrent.chapter06.CaptureThreadException$$Lambda$2/
531885035.run(Unknown Source)
    at java.lang.Thread.run(Thread.java:745)
```

在平时的工作中，这种设计方式是比较常见的，尤其是那种异步执行方法，比如Google 的 guava toolkit 就提供了 EventBus，在 EventBus 中事件源和实践的 subscriber 两者借助于 EventBus 实现了完全的解耦合，但是在 subscriber 执行任务时有可能会出现异常情况，EventBus 同样也是借助于一个 ExceptionHandler 进行回调处理的。

7.1.3 UncaughtExceptionHandler 源码分析

在没有向线程注入 UncaughtExceptionHandler 回调接口的情况下，线程若出现了异常又将如何处理呢？本节中我们将通过对 Thread 的源码进行剖析来追踪一下，示例代码如下：

```
public UncaughtExceptionHandler getUncaughtExceptionHandler() {
    return uncaughtExceptionHandler != null ?
        uncaughtExceptionHandler : group;
}
```

getUncaughtExceptionHandler 方法首先会判断当前线程是否设置了 handler，如果有则执行线程自己的 uncaughtException 方法，否则就到所在的 ThreadGroup 中获取，ThreadGroup 同样也实现了 UncaughtExceptionHandler 接口，下面再来看看 ThreadGroup 的 uncaughtException 方法。

```
public void uncaughtException(Thread t, Throwable e) {
    if (parent != null) {
        parent.uncaughtException(t, e);
    } else {
        Thread.UncaughtExceptionHandler ueh =
            Thread.getDefaultUncaughtExceptionHandler();
        if (ueh != null) {
            ueh.uncaughtException(t, e);
        } else if (!(e instanceof ThreadDeath)) {
            System.err.print("Exception in thread \""
                            + t.getName() + "\" ");
            e.printStackTrace(System.err);
        }
    }
}
```

- 该 ThreadGroup 如果有父 ThreadGroup，则直接调用父 Group 的 uncaughtException 方法。
- 如果设置了全局默认的 UncaughtExceptionHandler，则调用 uncaughtException 方法。
- 若既没有父 ThreadGroup，也没有设置全局默认的 UncaughtExceptionHandler，则会直接将异常的堆栈信息定向到 System.err 中。

代码清单 7-2　EmptyExceptionHandler.java

```java
package com.wangwenjun.concurrent.chapter07;

import java.util.concurrent.TimeUnit;

public class EmptyExceptionHandler
{
    public static void main(String[] args)
    {
        // get current thread's thread group
        ThreadGroup mainGroup = Thread.currentThread().getThreadGroup();
        System.out.println(mainGroup.getName());
        System.out.println(mainGroup.getParent());
        System.out.println(mainGroup.getParent().getParent());

        final Thread thread = new Thread(() ->
        {
            try
            {
                TimeUnit.SECONDS.sleep(2);
            } catch (InterruptedException e)
            {
            }

            //here will throw unchecked exception.
            System.out.println(1 / 0);
        }, "Test-Thread");
```

```
        thread.start();
    }
}
```

代码清单 7-2 中没有设置默认的 Handler，也没有对 thread 指定 Handler，因此当 thread 出现异常时，会向上寻找 Group 的 uncaughtException 方法，如图 7-1 所示。

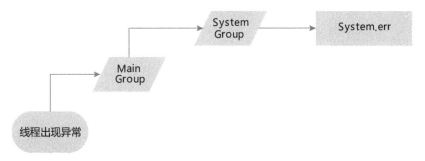

图 7-1 uncaughtException 执行调用关系图

7.2 注入钩子线程

7.2.1 Hook 线程介绍

JVM 进程的退出是由于 JVM 进程中没有活跃的非守护线程，或者收到了系统中断信号，向 JVM 程序注入一个 Hook 线程，在 JVM 进程退出的时候，Hook 线程会启动执行，通过 Runtime 可以为 JVM 注入多个 Hook 线程，下面就通过一个简单的例子来看一下如何向 Java 程序注入 Hook 线程。

代码清单 7-3 ThreadHook.java

```
package com.wangwenjun.concurrent.chapter07;

import java.util.concurrent.TimeUnit;

public class ThreadHook
{
    public static void main(String[] args)
    {
        //为应用程序注入钩子线程
        Runtime.getRuntime().addShutdownHook(new Thread()
        {
            @Override
            public void run()
            {
                try
                {
```

```
                System.out.println("The hook thread 1 is running.");
                TimeUnit.SECONDS.sleep(1);
            } catch (InterruptedException e)
            {
                e.printStackTrace();
            }
            System.out.println("The hook thread 1 will exit.");
        }
    });

    // 钩子线程可注册多个
    Runtime.getRuntime().addShutdownHook(new Thread()
    {
        @Override
        public void run()
        {
            try
            {
                System.out.println("The hook thread 2 is running.");
                TimeUnit.SECONDS.sleep(1);
            } catch (InterruptedException e)
            {
                e.printStackTrace();
            }
            System.out.println("The hook thread 2 will exit.");
        }
    });
    System.out.println("The program will is stopping.");
    }
}
```

在代码清单 7-3 中，给 Java 程序注入了两个 Hook 线程，在 main 线程中结束，也就是 JVM 中没有了活动的非守护线程，JVM 进程即将退出时，两个 Hook 线程会被启动并且运行，输出结果如下：

```
The program will is stopping.
The hook thread 1 is running.
The hook thread 2 is running.
The hook thread 1 will exit.
The hook thread 2 will exit.
```

7.2.2 Hook 线程实战

在我们的开发中经常会遇到 Hook 线程，比如为了防止某个程序被重复启动，在进程启动时会创建一个 lock 文件，进程收到中断信号的时候会删除这个 lock 文件，我们在 mysql 服务器、zookeeper、kafka 等系统中都能看到 lock 文件的存在，本节中，将利用 hook 线程的特点，模拟一个防止重复启动的程序，如代码清单 7-4 所示。

代码清单 7-4　PreventDuplicated.java

```java
package com.wangwenjun.concurrent.chapter07;

import java.io.IOException;
import java.nio.file.Files;
import java.nio.file.Path;
import java.nio.file.Paths;
import java.nio.file.attribute.PosixFilePermission;
import java.nio.file.attribute.PosixFilePermissions;
import java.util.Set;
import java.util.concurrent.TimeUnit;

public class PreventDuplicated
{
    private final static String LOCK_PATH = "/home/wangwenjun/locks/";

    private final static String LOCK_FILE = ".lock";

    private final static String PERMISSIONS = "rw-------";

    public static void main(String[] args) throws IOException
    {
        //② 检查是否存在 .lock 文件
        checkRunning();

        //① 注入 Hook 线程，在程序退出时删除 lock 文件
        Runtime.getRuntime().addShutdownHook(new Thread(() ->
        {
            System.out.println("The program received kill SIGNAL.");
            getLockFile().toFile().delete();
        }));

        //③ 简单模拟当前程序正在运行
        for (; ; )
        {
            try
            {
                TimeUnit.MILLISECONDS.sleep(1);
                System.out.println("program is running.");
            } catch (InterruptedException e)
            {
                e.printStackTrace();
            }
        }
    }

    private static void checkRunning() throws IOException
    {
        Path path = getLockFile();
```

```
        if (path.toFile().exists())
            throw new RuntimeException("The program already running.");

        Set<PosixFilePermission> perms = PosixFilePermissions.fromString(PERMISS-
IONS);
        Files.createFile(path, PosixFilePermissions.asFileAttribute(perms));
    }

    private static Path getLockFile()
    {
        return Paths.get(LOCK_PATH, LOCK_FILE);
    }
}
```

运行上面的程序，会发现在 /home/wangwenjun/locks 目录下多了一个 .lock 文件，如图 7-2 所示。

```
drwxrwxr-x  2 wangwenjun wangwenjun 4096 Oct 28 00:27 ./
drwxr-xr-x 27 wangwenjun wangwenjun 4096 Oct 28 00:23 ../
-rw-------  1 wangwenjun wangwenjun    0 Oct 28 00:27 .lock
```

图 7-2 程序运行则会创建 .lock 文件以防止重复启动

执行 kill pid 或者 kill -1 pid 命令之后，JVM 进程会收到中断信号，并且启动 hook 线程删除 .lock 文件，输出结果如下所示：

```
program is running.
program is running.
The program received kill SIGNAL.
program is running.
```

7.2.3 Hook 线程应用场景以及注意事项

- Hook 线程只有在收到退出信号的时候会被执行，如果在 kill 的时候使用了参数 -9，那么 Hook 线程不会得到执行，进程将会立即退出，因此 .lock 文件将得不到清理。
- Hook 线程中也可以执行一些资源释放的工作，比如关闭文件句柄、socket 链接、数据库 connection 等。
- 尽量不要在 Hook 线程中执行一些耗时非常长的操作，因为其会导致程序迟迟不能退出。

7.3 本章总结

本章介绍了如何通过 Handler 回调的方式获取线程运行期间的异常信息，并且分析了 Thread 的源码从追踪 uncaughtException 的执行顺序；Hook 线程是一个非常好的机制，可以帮助程序获得进程中断信号，有机会在进程退出之前做一些资源释放的动作，或者做一些告警通知。切记如果强制杀死进程，那么进程将不会收到任何中断信号。

CHAPTER8 · 第 8 章

线程池原理以及自定义线程池

自 JDK1.5 起，utils 包提供了 ExecutorService 线程池的实现，主要目的是为了重复利用线程，提高系统效率。通过前文的学习我们得知 Thread 是一个重量级的资源，创建、启动以及销毁都是比较耗费系统资源的，因此对线程的重复利用一种是非常好的程序设计习惯，加之系统中可创建的线程数量是有限的，线程数量和系统性能是一种抛物线的关系，也就是说当线程数量达到某个数值的时候，性能反倒会降低很多，因此对线程的管理，尤其是数量的控制更能直接决定程序的性能。

在本书中，不会讲解 JUC（JDK 的 Java utilities concurrent），而是从原理入手，设计一个线程池，其目的并不是重复地发明轮子，而是为了帮助读者弄清楚一个线程池应该具备哪些功能，线程池的实现需要注意哪些细节。

8.1 线程池原理

所谓线程池，通俗的理解就是有一个池子，里面存放着已经创建好的线程，当有任务提交给线程池执行时，池子中的某个线程会主动执行该任务。如果池子中的线程数量不够应付数量众多的任务时，则需要自动扩充新的线程到池子中，但是该数量是有限的，就好比池塘的水界线一样。当任务比较少的时候，池子中的线程能够自动回收，释放资源。为了能够异步地提交任务和缓存未被处理的任务，需要有一个任务队列，如图 8-1 所示。

通过上面的描述可知，一个完整的线程池应该具备如下要素。

❑ 任务队列：用于缓存提交的任务。

❑ 线程数量管理功能：一个线程池必须能够很好地管理和控制线程数量，可通过如下三个参数来实现，比如创建线程池时初始的线程数量 init；线程池自动扩充时最大的线程数量 max；在线程池空闲时需要释放线程但是也要维护一定数量的活跃数量或者核心数量 core。有了这三个参数，就能够很好地控制线程池中的线程数量，将其维护在一个合理的范围之内，三者之间的关系是 init<=core<=max。

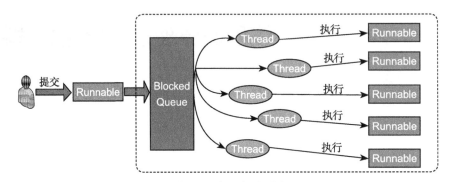

图 8-1　线程池原理图

- 任务拒绝策略：如果线程数量已达到上限且任务队列已满，则需要有相应的拒绝策略来通知任务提交者。
- 线程工厂：主要用于个性化定制线程，比如将线程设置为守护线程以及设置线程名称等。
- QueueSize：任务队列主要存放提交的 Runnable，但是为了防止内存溢出，需要有 limit 数量对其进行控制。
- Keepedalive 时间：该时间主要决定线程各个重要参数自动维护的时间间隔。

8.2　线程池实现

在本节中，实现一个比较简单的 ThreadPool，虽然比较简单，但是该有的功能基本上都具备，对读者学习和掌握 JUC 中的 ExecutorService 也有一定的帮助。图 8-2 所示的为线程池实现类图。

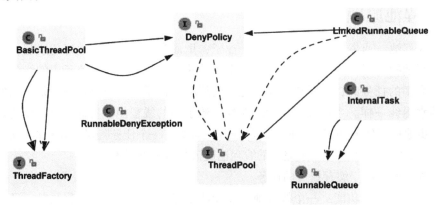

图 8-2　线程池实现类图

8.2.1 线程池接口定义

1. ThreadPool

ThreadPool 主要定义了一个线程池应该具备的基本操作和方法，下面是 ThreadPool 接口定义的方法：

代码清单 8-1　线程池接口 ThreadPool.java

```java
package com.wangwenjun.concurrent.chapter08;

public interface ThreadPool
{
    // 提交任务到线程池
    void execute(Runnable runnable);

    // 关闭线程池
    void shutdown();

    // 获取线程池的初始化大小
    int getInitSize();

    // 获取线程池最大的线程数
    int getMaxSize();

    // 获取线程池的核心线程数量
    int getCoreSize();

    // 获取线程池中用于缓存任务队列的大小
    int getQueueSize();

    // 获取线程池中活跃线程的数量
    int getActiveCount();

    // 查看线程池是否已经被 shutdown
    boolean isShutdown();
}
```

在上述代码中：

- execute（Runnable runnable）：该方法接受提交 Runnable 任务。
- shutdown()：关闭并且销毁线程池。
- getInitSize()：返回线程池的初始线程数量。
- getMaxSize()：返回线程池最大的线程数量。
- getCoreSize()：返回核心线程数量。
- getQueueSize()：返回当前线程池任务数量。
- getActiveCount()：返回线程池中当前活跃的线程数量。
- isShutdown()：判断线程池是否已被销毁。

2. RunnableQueue

RunanbleQueue 主要用于存放提交的 Runnable，该 Runnable 是一个 BlockedQueue，并且有 limit 的限制，示例代码如清单 8-2 所示。

代码清单 8-2　任务队列 RunnableQueue.java

```
package com.wangwenjun.concurrent.chapter08;

// 任务队列，主要用于缓存提交到线程池中的任务
public interface RunnableQueue
{
    // 当有新的任务进来时首先会 offer 到队列中
    void offer(Runnable runnable);

    // 工作线程通过 take 方法获取 Runnable
    Runnable take();

    // 获取任务队列中任务的数量
    int size();
}
```

在上述代码中：
- offer（Runnable runnable）方法主要用于将任务提交至队列中。
- Runnable take() 方法主要从队列中获取相应的任务。
- int size() 方法用于获取当前队列 Runnable 的数量。

3. ThreadFactory

ThreadFactory 提供了创建线程的接口，以便于个性化地定制 Thread，比如 Thread 应该被加到哪个 Group 中、优先级、线程名字以及是否为守护线程等，示例代码如清单 8-3 所示。

代码清单 8-3　创建 Thread 的工厂 ThreadFactory.java

```
package com.wangwenjun.concurrent.chapter08;

// 创建线程的工厂
@FunctionalInterface
public interface ThreadFactory
{
    Thread createThread(Runnable runnable);
}
```

其中，createThread（Runnable runnable）用于创建线程。

4. DenyPolicy

DenyPolicy 主要用于当 Queue 中的 runnable 达到了 limit 上限时，决定采用何种策略通知提交者。该接口中定义了三种默认的实现，具体如代码清单 8-4 所示。

代码清单 8-4　拒绝策略 DenyPolicy.java

```java
package com.wangwenjun.concurrent.chapter08;

@FunctionalInterface
public interface DenyPolicy
{

    void reject(Runnable runnable, ThreadPool threadPool);

    // 该拒绝策略会直接将任务丢弃
    class DiscardDenyPolicy implements DenyPolicy
    {

        @Override
        public void reject(Runnable runnable, ThreadPool threadPool)
        {
            //do nothing
        }
    }

    // 该拒绝策略会向任务提交者抛出异常
    class AbortDenyPolicy implements DenyPolicy
    {

        @Override
        public void reject(Runnable runnable, ThreadPool threadPool)
        {
            throw new RunnableDenyException("The runnable " + runnable + " will be abort.");
        }
    }
    // 该拒绝策略会使任务在提交者所在的线程中执行任务
    class RunnerDenyPolicy implements DenyPolicy
    {

        @Override
        public void reject(Runnable runnable, ThreadPool threadPool)
        {
            if (!threadPool.isShutdown())
            {
                runnable.run();
            }
        }
    }
}
```

在上述代码中：
- void reject（Runnable runnable，ThreadPool threadPool）为拒绝方法。
- DiscardDenyPolicy 策略会直接丢弃掉 Runnable 任务。

- AbortDenyPolicy 策略会抛出 RunnableDenyException 异常。
- RunnerDenyPolicy 策略，交给调用者的线程直接运行 runnable，而不会被加入到线程池中。

5. RunnableDenyException

RunnableDenyException 是 RuntimeException 的子类，主要用于通知任务提交者，任务队列已无法再接收新的任务。示例代码如清单 8-5 所示。

代码清单 8-5　异常 RunnableDenyException.java

```java
package com.wangwenjun.concurrent.chapter08;

public class RunnableDenyException extends RuntimeException
{
    public RunnableDenyException(String message)
    {
        super(message);
    }
}
```

6. InternalTask

InternalTask 是 Runnable 的一个实现，主要用于线程池内部，该类会使用到 RunnableQueue，然后不断地从 queue 中取出某个 runnable，并运行 runnable 的 run 方法，示例代码如代码清单 8-6 所示。

代码清单 8-6　InternalTask.java

```java
package com.wangwenjun.concurrent.chapter08;

public class InternalTask implements Runnable
{
    private final RunnableQueue runnableQueue;

    private volatile boolean running = true;

    public InternalTask(RunnableQueue runnableQueue)
    {
        this.runnableQueue = runnableQueue;
    }

    @Override
    public void run()
    {
        // 如果当前任务为 running 并且没有被中断，则其将不断地从 queue 中获取 runnable，然后执行 run 方法
        while (running && !Thread.currentThread().isInterrupted())
```

```java
        {
            try
            {
                Runnable task = runnableQueue.take();
                task.run();
            } catch (InterruptedException e)
            {
                running = false;
                break;
            }
        }
    }

    //停止当前任务,主要会在线程池的shutdown方法中使用
    public void stop()
    {
        this.running = false;
    }
}
```

除此之外,代码还对该类增加了一个开关方法 stop,主要用于停止当前线程,一般在线程池销毁和线程数量维护的时候会使用到。

8.2.2 线程池详细实现

本节将对线程池进行详细的实现,其中会涉及很多同步的技巧和资源竞争,我们将结合本书第一部分的大多数知识灵活使用,也算是一个综合练习。

1. LinkedRunnableQueue

LinkedRunnableQueue 的示例代码如清单 8-7 所示。

代码清单 8-7　LinkedRunnableQueue.java

```java
package com.wangwenjun.concurrent.chapter08;

import java.util.LinkedList;

public class LinkedRunnableQueue implements RunnableQueue
{
    //任务队列的最大容量,在构造时传入
    private final int limit;

    //若任务队列中的任务已经满了,则需要执行拒绝策略
    private final DenyPolicy denyPolicy;

    //存放任务的队列
    private final LinkedList<Runnable> runnableList = new LinkedList<>();

    private final ThreadPool threadPool;
```

```java
    public LinkedRunnableQueue(int limit, DenyPolicy denyPolicy, ThreadPool threadPool)
    {
        this.limit = limit;
        this.denyPolicy = denyPolicy;
        this.threadPool = threadPool;
    }
```

在 LinkedRunnableQueue 中有几个重要的属性,第一个是 limit,也就是 Runnable 队列的上限;当提交的 Runnable 数量达到 limit 上限时,则会调用 DenyPolicy 的 reject 方法;runnableList 是一个双向循环列表,用于存放 Runnable 任务,示例代码如下:

```java
@Override
public void offer(Runnable runnable)
{
    synchronized (runnableList)
    {
        if (runnableList.size() >= limit)
        {
            // 无法容纳新的任务时执行拒绝策略
            denyPolicy.reject(runnable, threadPool);
        } else
        {
            // 将任务加入到队尾,并且唤醒阻塞中的线程
            runnableList.addLast(runnable);
            runnableList.notifyAll();
        }
    }
}
```

offer 方法是一个同步方法,如果队列数量达到了上限,则会执行拒绝策略,否则会将 runnable 存放至队列中,同时唤醒 take 任务的线程:

```java
@Override
public Runnable take() throws InterruptedException
{
    synchronized (runnableList)
    {
        while (runnableList.isEmpty())
        {
            try
            {
                // 如果任务队列中没有可执行任务,则当前线程将会挂起,进入 runnableList 关联
                //   的 monitor waitset 中等待唤醒(新的任务加入)
                runnableList.wait();
            } catch (InterruptedException e)
            {
                // 被中断时需要将该异常抛出
```

```
            throw e;
        }
    }
    // 从任务队列头部移除一个任务
    return runnableList.removeFirst();
}
```

take 方法也是同步方法，线程不断从队列中获取 Runnable 任务，当队列为空的时候工作线程会陷入阻塞，有可能在阻塞的过程中被中断，为了传递中断信号需要在 catch 语句块中将异常抛出以通知上游（InternalTask），示例代码如下：

```
@Override
public int size()
{
    synchronized (runnableList)
    {
        // 返回当前任务队列中的任务数
        return runnableList.size();
    }
}
```

其中，size 方法用于返回 runnableList 的任务个数。

2. 初始化线程池

根据前面的讲解，线程池需要有数量控制属性、创建线程工厂、任务队列策略等功能，线程池初始化代码如清单 8-8 所示。

代码清单 8-8　BasicThreadPool.java

```java
package com.wangwenjun.concurrent.chapter08;

import java.util.ArrayDeque;
import java.util.Queue;
import java.util.concurrent.TimeUnit;
import java.util.concurrent.atomic.AtomicInteger;

public class BasicThreadPool extends Thread implements ThreadPool
{
    // 初始化线程数量
    private final int initSize;

    // 线程池最大线程数量
    private final int maxSize;

    // 线程池核心线程数量
    private final int coreSize;

    // 当前活跃的线程数量
    private int activeCount;
```

```java
// 创建线程所需的工厂
private final ThreadFactory threadFactory;

// 任务队列
private final RunnableQueue runnableQueue;

// 线程池是否已经被shutdown
private volatile boolean isShutdown = false;

// 工作线程队列
private final Queue<ThreadTask> threadQueue = new ArrayDeque<>();

    private final static DenyPolicy DEFAULT_DENY_POLICY = new DenyPolicy.DiscardDenyPolicy();

    private final static ThreadFactory DEFAULT_THREAD_FACTORY = new DefaultThreadFactory();

    private final long keepAliveTime;

    private final TimeUnit timeUnit;

// 构造时需要传递的参数: 初始的线程数量, 最大的线程数量, 核心线程数量, 任务队列的最大数量
public BasicThreadPool(int initSize, int maxSize, int coreSize,
                       int queueSize)
{
    this(initSize, maxSize, coreSize, DEFAULT_THREAD_FACTORY,
            queueSize, DEFAULT_DENY_POLICY, 10, TimeUnit.SECONDS);
}

// 构造线程池时需要传入的参数, 该构造函数需要的参数比较多
public BasicThreadPool(int initSize, int maxSize, int coreSize,
                       ThreadFactory threadFactory, int queueSize,
                       DenyPolicy denyPolicy, long keepAliveTime, TimeUnit
                       timeUnit)
{
    this.initSize = initSize;
    this.maxSize = maxSize;
    this.coreSize = coreSize;
    this.threadFactory = threadFactory;
        this.runnableQueue = new LinkedRunnableQueue(queueSize, denyPolicy,
                            this);
    this.keepAliveTime = keepAliveTime;
    this.timeUnit = timeUnit;
    this.init();
}
// 初始化时, 先创建initSize个线程
private void init()
{
    start();
```

```
        for (int i = 0; i < initSize; i++)
        {
            newThread();
        }
    }
```

一个线程池除了控制参数之外,最主要的是应该有活动线程,其中 Queue <Thread-Task> 主要用来存放活动线程,BasicThreadPool 同时也是 Thread 的子类,它在初始化的时候启动,在 keepalive 时间间隔到了之后再自动维护活动线程数量(采用继承 Thread 的方式其实不是一种好的方法,因为 BasicThreadPool 会暴露 Thread 的方法,建议将继承关系更改为组合关系,读者可以自行修改)。

3. 提交任务

提交任务非常简单,只是将 Runnable 插入 runnableQueue 中即可。示例代码如下:

```
@Override
public void execute(Runnable runnable)
{
    if (this.isShutdown)
        throw new IllegalStateException("The thread pool is destroy");
    // 提交任务只是简单地往任务队列中插入 Runnable
    this.runnableQueue.offer(runnable);
}
```

4. 线程池自动维护

线程池中线程数量的维护主要由 run 负责,这也是为什么 BasicThreadPool 继承自 Thread 了,不过笔者不推荐使用直接继承的方式,线程池自动维护代码如下:

```
private void newThread()
{
    // 创建任务线程,并且启动
    InternalTask internalTask = new InternalTask(runnableQueue);
    Thread thread = this.threadFactory.createThread(internalTask);
    ThreadTask threadTask = new ThreadTask(thread, internalTask);
    threadQueue.offer(threadTask);
    this.activeCount++;
    thread.start();
}
private void removeThread()
{
    // 从线程池中移除某个线程
    ThreadTask threadTask = threadQueue.remove();
    threadTask.internalTask.stop();
    this.activeCount--;
}
@Override
public void run()
```

```java
{
    //run方法继承自Thread，主要用于维护线程数量，比如扩容、回收等工作
    while (!isShutdown && !isInterrupted())
    {
        try
        {
            timeUnit.sleep(keepAliveTime);
        } catch (InterruptedException e)
        {
            isShutdown = true;
            break;
        }
        synchronized (this)
        {
            if (isShutdown)
                break;
            // 当前的队列中有任务尚未处理，并且activeCount< coreSize 则继续扩容
            if (runnableQueue.size() > 0&& activeCount < coreSize)
            {
                for (int i = initSize; i < coreSize; i++)
                {
                    newThread();
                }
                //continue 的目的在于不想让线程的扩容直接达到maxsize
                continue;
            }
            // 当前的队列中有任务尚未处理，并且activeCount< maxSize 则继续扩容
            if (runnableQueue.size() > 0&& activeCount < maxSize)
            {
                for (int i = coreSize; i < maxSize; i++)
                {
                    newThread();
                }
            }
            // 如果任务队列中没有任务，则需要回收，回收至 coreSize 即可
            if (runnableQueue.size()==0&& activeCount > coreSize)
            {
                for (int i = coreSize; i < activeCount; i++)
                {
                    removeThread();
                }
            }
        }
    }
}

//ThreadTask 只是 InternalTask 和 Thread 的一个组合
private static class ThreadTask
{
    public ThreadTask(Thread thread, InternalTask internalTask)
    {
```

```
            this.thread = thread;
            this.internalTask = internalTask;
    }
    Thread thread;
    InternalTask internalTask;
}
```

下面重点来解说线程自动维护的方法，自动维护线程的代码块是同步代码块，主要是为了阻止在线程维护过程中线程池销毁引起的数据不一致问题。

任务队列中若存在积压任务，并且当前活动线程少于核心线程数，则新建 coreSize-initSize 数量的线程，并且将其加入到活动线程队列中，为了防止马上进行 maxSize-coreSize 数量的扩充，建议使用 continue 终止本次循环。

任务队列中有积压任务，并且当前活动线程少于最大线程数，则新建 maxSize-coreSize 数量的线程，并且将其加入到活动队列中。

当前线程池不够繁忙时，则需要回收部分线程，回收到 coreSize 数量即可，回收时调用 removeThread() 方法，在该方法中需要考虑的一点是，如果被回收的线程恰巧从 Runnable 任务取出了某个任务，则会继续保持该线程的运行，直到完成了任务的运行为止，详见 InternalTask 的 run 方法。

5. 线程池销毁

线程池的销毁同样需要同步机制的保护，主要是为了防止与线程池本身的维护线程引起数据冲突，线程池销毁代码如下：

```
@Override
public void shutdown()
{
    synchronized (this)
    {
        if (isShutdown) return;
        isShutdown = true;
        threadQueue.forEach(threadTask ->
        {
            threadTask.internalTask.stop();
            threadTask.thread.interrupt();
        });
        this.interrupt();
    }
}
```

销毁线程池主要为了是停止 BasicThreadPool 线程，停止线程池中的活动线程并且将 isShutdown 开关变量更改为 true。

6. 线程池的其他方法

先来看一段代码：

```java
    @Override
    public int getInitSize()
    {
        if (isShutdown)
            throw new IllegalStateException("The thread pool is destroy");
        return this.initSize;
    }

    @Override
    public int getMaxSize()
    {
        if (isShutdown)
            throw new IllegalStateException("The thread pool is destroy");
        return this.maxSize;
    }

    @Override
    public int getCoreSize()
    {
        if (isShutdown)
            throw new IllegalStateException("The thread pool is destroy");
        return this.coreSize;
    }

    @Override
    public int getQueueSize()
    {
        if (isShutdown)
            throw new IllegalStateException("The thread pool is destroy");
        return runnableQueue.size();
    }

    @Override
    public int getActiveCount()
    {
        synchronized (this)
        {
            return this.activeCount;
        }
    }

    @Override
    public boolean isShutdown()
    {
        return this.isShutdown;
    }

    private static class DefaultThreadFactory implements ThreadFactory
    {
```

```java
        private static final AtomicInteger GROUP_COUNTER = new AtomicInteger(1);

        private static final ThreadGroup group = new ThreadGroup("MyThreadPool-"
+ GROUP_COUNTER.getAndDecrement());

        private static final AtomicInteger COUNTER = new AtomicInteger(0);

        @Override
        public Thread createThread(Runnable runnable)
        {
            return new Thread(group, runnable, "thread-pool-" + COUNTER.getAndDecrement());
        }
    }
```

上述代码为线程池的其他方法，主要是用来获取线程池各个参数的值（当线程池已经被关闭时，调用查询方法将会抛出异常信息），至此线程池的全部实现已经完成，在 8.3 节中我们将对线程池的应用进行简单的测试。

8.3 线程池的应用

本节将写一个简单的程序分别测试线程池的任务提交、线程池线程数量的动态扩展，以及线程池的销毁功能，代码如清单 8-9 所示。

代码清单 8-9　线程池测试代码

```java
package com.wangwenjun.concurrent.chapter08;

import java.util.concurrent.TimeUnit;

public class ThreadPoolTest
{
    public static void main(String[] args) throws InterruptedException
    {
// 定义线程池，初始化线程数为 2，核心线程数为 4，最大线程数为 6，任务队列最多允许 1000 个任务
        final ThreadPool threadPool = new BasicThreadPool(2, 6, 4, 1000);
// 定义 20 个任务并且提交给线程池
        for (int i = 0; i < 20; i++)
            threadPool.execute(() ->
            {
                try
                {
                    TimeUnit.SECONDS.sleep(10);
                    System.out.println(Thread.currentThread().getName() + " is running and done.");
                } catch (InterruptedException e)
                {
                    e.printStackTrace();
                }
```

```java
        });
        for (; ; )
        {
            // 不断输出线程池的信息
            System.out.println("getActiveCount:" + threadPool.getActiveCount());
            System.out.println("getQueueSize:" + threadPool.getQueueSize());
            System.out.println("getCoreSize:" + threadPool.getCoreSize());
            System.out.println("getMaxSize:" + threadPool.getMaxSize());
            System.out.println("=====================================");
            TimeUnit.SECONDS.sleep(5);
        }
    }
}
```

上述测试代码中，定义了一个 Basic 线程池，其中初始化线程数量为 2，核心线程数量为 4，最大线程数量为 6，最大任务队列数量为 1000，同时提交了 20 个任务到线程池中，然后在 main 线程中不断地输出线程池中的线程数量信息监控变化，运行上述代码，截取的部分输出信息如下：

```
getActiveCount:2
getQueueSize:18
getCoreSize:4
getMaxSize:6
...
thread-pool--2 is running and done.
thread-pool--3 is running and done.
thread-pool--1 is running and done.
thread-pool-0 is running and done.
getActiveCount:6
getQueueSize:2
getCoreSize:4
getMaxSize:6
...
getActiveCount:4
getQueueSize:0
getCoreSize:4
getMaxSize:6
```

通过上述输出信息可以看出，线程池中线程的动态扩展状况以及任务执行情况，在输出的最后会发现 active count 停留在了 core size 的位置，这也符合我们的设计，最后为了确定线程池中的活跃线程数量，可以使用 JVM 工具进行验证：

```
"thread-pool--5" #17 prio=5 os_prio=0 tid=0x000000000b3e8000 nid=0x1784 in Object.wait() [0x000000000c0bf000]
    java.lang.Thread.State: WAITING (on object monitor)
    at java.lang.Object.wait(Native Method)
    - waiting on <0x000000078b9599e0> (a java.util.LinkedList)
```

```
        at java.lang.Object.wait(Object.java:502)
        at com.wangwenjun.concurrent.chapter08.LinkedRunnableQueue.
take(LinkedRunnableQueue.java:49)
        - locked <0x000000078b9599e0> (a java.util.LinkedList)
        at com.wangwenjun.concurrent.chapter08.InternalTask.run(InternalTask.java:22)
        at java.lang.Thread.run(Thread.java:745)

"thread-pool--4" #16 prio=5 os_prio=0 tid=0x000000000b3e7000 nid=0x52c in Object.
wait() [0x000000000c81f000]
        java.lang.Thread.State: WAITING (on object monitor)
        at java.lang.Object.wait(Native Method)
        - waiting on <0x000000078b9599e0> (a java.util.LinkedList)
        at java.lang.Object.wait(Object.java:502)
        at com.wangwenjun.concurrent.chapter08.LinkedRunnableQueue.
take(LinkedRunnableQueue.java:49)
        - locked <0x000000078b9599e0> (a java.util.LinkedList)
        at com.wangwenjun.concurrent.chapter08.InternalTask.run(InternalTask.java:22)
        at java.lang.Thread.run(Thread.java:745)

"thread-pool--3" #15 prio=5 os_prio=0 tid=0x000000000b3e6800 nid=0x6b8 in Object.
wait() [0x000000000c62e000]
        java.lang.Thread.State: WAITING (on object monitor)
        at java.lang.Object.wait(Native Method)
        - waiting on <0x000000078b9599e0> (a java.util.LinkedList)
        at java.lang.Object.wait(Object.java:502)
        at com.wangwenjun.concurrent.chapter08.LinkedRunnableQueue.
take(LinkedRunnableQueue.java:49)
        - locked <0x000000078b9599e0> (a java.util.LinkedList)
        at com.wangwenjun.concurrent.chapter08.InternalTask.run(InternalTask.java:22)
        at java.lang.Thread.run(Thread.java:745)

"thread-pool--2" #14 prio=5 os_prio=0 tid=0x000000000b3e6000 nid=0x1d48 in
Object.wait() [0x000000000c41e000]
        java.lang.Thread.State: WAITING (on object monitor)
        at java.lang.Object.wait(Native Method)
        - waiting on <0x000000078b9599e0> (a java.util.LinkedList)
        at java.lang.Object.wait(Object.java:502)
        at com.wangwenjun.concurrent.chapter08.LinkedRunnableQueue.
take(LinkedRunnableQueue.java:49)
        - locked <0x000000078b9599e0> (a java.util.LinkedList)
        at com.wangwenjun.concurrent.chapter08.InternalTask.run(InternalTask.java:22)
        at java.lang.Thread.run(Thread.java:745)
```

再来应用一下线程池的销毁功能，将 for 循环中不断获取线程池线程数量的代码注释掉，用下面的代码取而代之：

```
TimeUnit.SECONDS.sleep(12);
// 线程池在 12 秒之后将被 shutdown
threadPool.shutdown();
```

```
// 使 main 线程 join，方便通过工具观察线程堆栈信息
Thread.currentThread().join();
```

程序运行 12 秒之后，线程池将被销毁，线程池中的线程都将被销毁，同样为了验证所有线程是否被成功销毁，也可以借助 JVM 工具查看堆栈信息，这也是在最后使用 currentThread join 进行阻塞以便于查看的原因。

8.4 本章总结

本章结合了线程的基础知识点编写了一个比较综合的实战 ThreadPool，在 JDK1.5 以前，笔者进行 J2SE 开发时就是采用这样的思路来开发线程池框架的，所以当 JDK1.5 提供了 ExecutorService 线程池 API 的时候就能马上理解和灵活使用。本章的目的并非是要写一个线程池在工作中取代 JUC 中的 ExecutorService，ExecutorService 的功能远比我们的线程池强大得多，但是万变不离其宗，二者的原理非常类似，读者可以自行实现一个线程池的功能，以便更加深入地理解 JDK 中的 ExecutorService。

笔者实现的线程池还存在诸多缺点，鉴于篇幅原因就不再优化，下面笔者将问题指出请读者参考自行优化。

1) BasicThreadPool 和 Thread 不应该是继承关系，采用组合关系更为妥当，这样就可以避免调用者直接使用 BasicThreadPool 中的 Thread 方法。

2) 线程池的销毁功能并未返回未被处理的任务，这样会导致未被处理的任务被丢弃。

3) BasicThreadPool 的构造函数太多，创建不太方便，建议采用 Builder 和设计模式对其进行封装或者提供工厂方法进行构造。

4) 线程池中的数量控制没有进行合法性校验，比如 initSize 数量不应该大于 maxSize 数量。

5) 其他缺点及相关优化请读者自行思考。

PART2 · 第二部分

Java ClassLoader

在 3.6 节中，我们没有对线程上下文类加载器相关的方法做太多解析，原因是要想说明白这个问题，必须掌握 JVM 的类加载器原理和整个类加载的过程，在详细地掌握了 ClassLoader 知识以后，再回过头来看线程上下文的类加载器肯定会更加透彻。

第 9 章"类的加载过程"：本章将着重阐述类的加载过程，循序渐进地理解一个 class 文件是如何被 JVM 加载并且最后得以运行的。

第 10 章"JVM 类加载器"：本章将介绍 JVM 自带的三大类加载器——根加载器、扩展类加载器、系统类加载器——不同的职责和各自的特点，然后重点讲解自定义类加载器、双亲委托机制，以及如何打破 JDK 的双亲委托模型。

第 11 章"线程上下文类加载器"：本章将重点介绍和线程关系非常密切的上下文类加载器，以及通过 JDBC 驱动的源码剖析，让读者理解数据库驱动的加载过程。

第 9 章 · CHAPTER9

类的加载过程

ClassLoader 的主要职责就是负责加载各种 class 文件到 JVM 中，ClassLoader 是一个抽象的 class，给定一个 class 的二进制文件名，ClassLoader 会尝试加载并且在 JVM 中生成构成这个类的各个数据结构，然后使其分布在 JVM 对应的内存区域中。

9.1 类的加载过程简介

类的加载过程一般分为三个比较大的阶段，分别是加载阶段、连接阶段和初始化阶段，如图 9-1 所示。

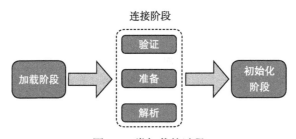

图 9-1　类加载的过程

- 加载阶段：主要负责查找并且加载类的二进制数据文件，其实就是 class 文件。
- 连接阶段：连接阶段所做的工作比较多，细分的话还可以分为如下三个阶段。
  - 验证：主要是确保类文件的正确性，比如 class 的版本，class 文件的魔术因子是否正确。
  - 准备：为类的静态变量分配内存，并且为其初始化默认值。
  - 解析：把类中的符号引用转换为直接引用。

❑ 初始化阶段：为类的静态变量赋予正确的初始值（代码编写阶段给定的值）。

当一个 JVM 在我们通过执行 Java 命令启动之后，其中可能包含的类非常多，是不是每一个类都会被初始化呢？答案是否定的，JVM 对类的初始化是一个延迟的机制，即使用的是 lazy 的方式，当一个类在首次使用的时候才会被初始化，在同一个运行时包下，一个 Class 只会被初始化一次（运行时包和类的包是有区别的，关于这点我们将在第 10 章中进行详细的讲解），那么什么是类的主动使用和被动使用呢？接下来我们通过一些实例来进行相应的总结。

9.2 类的主动使用和被动使用

JVM 虚拟机规范规定了，每个类或者接口被 Java 程序**首次主动使用**时才会对其进行初始化，当然随着 JIT 技术越来越成熟，JVM 运行期间的编译也越来越智能，不排除 JVM 在运行期间提前预判并且初始化某个类。

JVM 同时规范了以下 6 种主动使用类的场景，具体如下。

❑ 通过 new 关键字会导致类的初始化：这种是大家经常采用的初始化一个类的方式，它肯定会导致类的加载并且最终初始化。

❑ 访问类的静态变量，包括读取和更新会导致类的初始化，这种情况的示例代码如下：

```
public class Simple
{
    static
    {
        System.out.println("I will be initialized");
    }

    public static int x = 10;
}
```

这段代码中 x 是一个简单的静态变量，其他类即使不对 Simple 进行 new 的创建，直接访问变量 x 也会导致类的初始化。

❑ 访问类的静态方法，会导致类的初始化，这种情况的示例代码如下：

```
public class Simple
{
    static
    {
        System.out.println("I will be initialized");
    }

    //静态方法
    public static void test(){

    }
}
```

同样，在其他类中直接调用 test 静态方法也会导致类的初始化。
- 对某个类进行反射操作，会导致类的初始化，这种情况的示例代码如下：

```
public static void main(String[] args) throws ClassNotFoundException
{
Class.forName("com.wangwenjun.concurrent.chapter09.Simple");
}
```

运行上面的代码，同样会看到静态代码块中的输出语句执行。
- 初始化子类会导致父类的初始化，这种情况的示例代码如下：

```
public class Parent
{
    static
    {
        System.out.println("The parent is initialized");
    }

    public static int y=100;
}

public class Child extends Parent
{
    static
    {
        System.out.println("The child will be initialized");
    }

    public static int x = 10;
}

public class ActiveLoadTest
{
    public static void main(String[] args)
    {
        System.out.println(Child.x);
    }
}
```

在 ActiveLoadTest 中，我们调用了 Child 的静态变量，根据前面的知识可以得出 Child 类被初始化了，Child 类又是 Parent 类的子类，子类的初始化会进一步导致父类的初始化，当然这里需要注意的一点是，通过子类使用父类的静态变量只会导致父类的初始化，子类则不会被初始化，示例代码如下：

```
public class ActiveLoadTest
{
    public static void main(String[] args)
    {
        System.out.println(Child.y);
```

    }
}
```

改写后的 ActiveLoadTest，直接用 Child 访问父类的静态变量 y，并不会导致 Child 的初始化，仅仅会导致 Parent 的初始化。

- 启动类：也就是执行 main 函数所在的类会导致该类的初始化，比如使用 java 命令运行上文中的 ActiveLoadTest 类。

**除了上述 6 种情况，其余的都称为被动使用，不会导致类的加载和初始化。**

关于类的主动引用和被动引用，下面有几个容易引起大家混淆的例子，我们来看一看。

- 构造某个类的数组时并不会导致该类的初始化，比如下面的例子：

```java
public static void main(String[] args)
{
    Simple[] simples = new Simple[10];
    System.out.println(simples.length);
}
```

上面的代码中 new 方法新建了一个 Simple 类型的数组，但是它并不能导致 Simple 类的初始化，因此它是被动使用，不要被前面的 new 关键字所误导，事实上该操作只不过是在堆内存中开辟了一段连续的地址空间 4byte × 10。

- 引用类的静态常量不会导致类的初始化，请看下面的例子：

```java
package com.wangwenjun.concurrent.chapter09;

import java.util.Random;

public class GlobalConstants
{
    static
    {
        System.out.println("The GlobalConstants will be initialized.");
    }

    //在其他类中使用 MAX 不会导致 GlobalConstants 的初始化，静态代码块不会输出
    public final static int MAX = 100;

    //虽然 RANDOM 是静态常量，但是由于计算复杂，只有初始化之后才能得到结果，因此在其他类中使用
    //RANDOM 会导致 GlobalConstants 的初始化
    public final static int RANDOM = new Random().nextInt();
}
```

这段代码中，MAX 是一个被 final 修饰的静态变量，也就是一个静态常量，在其他类中直接访问 MAX 不会导致 GlobalConstants 的初始化，虽然它也是被 static 修饰的，但是如果在其他类中访问 RANDOM 则会导致类的初始化，因为 RANDOM 是需要进行随机函数计算的，在类的加载、连接阶段是无法对其进行计算的，需要进行初始化后才能对其赋予准确的值。
```

9.3 类的加载过程详解

在正式讲解类加载各个阶段的内容之前，请大家思考下面这段程序的输出结果，如果你不能准确计算出结果或者感觉有些模棱两可，那么在学习了本节的内容之后，相信你会很容易地得出结果，以及明白为什么是这样的结果：

```java
package com.wangwenjun.concurrent.chapter09;

public class Singleton
{
    //①
    private static int x = 0;

    private static int y;

    private static Singleton instance = new Singleton();//②

    private Singleton()
    {
        x++;
        y++;
    }

    public static Singleton getInstance()
    {
        return instance;
    }

    public static void main(String[] args)
    {
        Singleton singleton = Singleton.getInstance();
        System.out.println(singleton.x);
        System.out.println(singleton.y);
    }
}
```

运行上面的程序代码输出将是多少？如果将注释②的代码移到注释①的位置，输出结果又是什么呢？两种输出会产生不一样的结果，为何会发生这样的现象，下面我们就在本节中一起寻找答案。

9.3.1 类的加载阶段

简单来说，类的加载就是将 class 文件中的二进制数据读取到内存之中，然后将该字节流所代表的静态存储结构转换为方法区中运行时的数据结构，并且在堆内存中生成一个该类的 java.lang.Class 对象，作为访问方法区数据结构的入口，如图 9-2 所示。

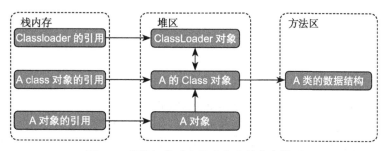

图9-2　类被加载后在栈内存中的分布情况

类加载的最终产物就是堆内存中的class对象，对同一个ClassLoader来讲，不管某个类被加载了多少次，对应到堆内存中的class对象始终是同一个。虚拟机规范中指出了类的加载是通过一个全限定名（包名＋类名）来获取二进制数据流，但是并没有限定必须通过某种方式去获得，比如我们常见的是class二进制文件的形式，但是除此之外还会有如下的几种形式。

- 运行时动态生成，比如通过开源的ASM包可以生成一些class，或者通过动态代理java.lang.Proxy也可以生成代理类的二进制字节流。
- 通过网络获取，比如很早之前的Applet小程序，以及RMI动态发布等。
- 通过读取zip文件获得类的二进制字节流，比如jar、war（其实，jar和war使用的是和zip同样的压缩算法）。
- 将类的二进制数据存储在数据库的BLOB字段类型中。
- 运行时生成class文件，并且动态加载，比如使用Thrift、AVRO等都是可以在运行时将某个Schema文件生成对应的若干个class文件，然后再进行加载。

注意我们在这里所说的加载是类加载过程中的第一个阶段，并不代表整个类已经加载完成了，在某个类完成加载阶段之后，虚拟机会将这些二进制字节流按照虚拟机所需的格式存储在方法区中，然后形成特定的数据结构，随之又在堆内存中实例化一个java.lang.Class类对象，在类加载的整个生命周期中，加载过程还没有结束，连接阶段是可以交叉工作的，比如连接阶段验证字节流信息的合法性，但是总体来讲加载阶段肯定是出现在连接阶段之前的。

9.3.2　类的连接阶段

类的连接阶段可以细分为三个小的过程，分别是验证、准备和解析，在本节中，我们将详细介绍每一个过程的具体细节。

1. 验证

验证在连接阶段中的主要目的是确保class文件的字节流所包含的内容符合当前JVM的规范要求，并且不会出现危害JVM自身安全的代码，当字节流的信息不符合要求时，则会抛出 VerifyError 这样的异常或者是其子异常。既然是验证,那么它到底验证了那些信息呢?

(1)验证文件格式
- 在很多二进制文件中,文件头部都存在着魔术因子,该因子决定了这个文件到底是什么类型,class 文件的魔术因子是 0xCAFEBABE。
- 主次版本号,Java 的版本是在不断升级的,JVM 规范同样也在不断升级,比如你用高版本的 JDK 编译的 class 就不能够被低版本的 JVM 所兼容,在验证的过程中,还需要查看当前 class 文件版本是否符合当前 JDK 所处理的范围。
- 构成 class 文件的字节流是否存在残缺或者其他附加信息,主要是看 class 的 MD5 指纹(每一个类在编译阶段经过 MD5 摘要算法计算之后,都会将结果一并附加给 class 字节流作为字节流的一部分)。
- 常量池中的常量是否存在不被支持的变量类型,比如 int64。
- 指向常量中的引用是否指到了不存在的常量或者该常量的类型不被支持。
- 其他信息。

当然了,JVM 对 class 字节流的验证远不止于此,由于获得 class 字节流的来源各种各样,甚至可以直接根据 JVM 规范编写出一个二进制字节流,对文件格式的验证可以在 class 被初始化之前将一些不符合规范的、恶意的字节流拒之门外,文件格式的验证充当了先锋关卡。

(2)元数据的验证

元数据的验证其实是对 class 的字节流进行语义分析的过程,整个语义分析就是为了确保 class 字节流符合 JVM 规范的要求。
- 检查这个类是否存在父类,是否继承了某个接口,这些父类和接口是否合法,或者是否真实存在。
- 检查该类是否继承了被 final 修饰的类,被 final 修饰的类是不允许被继承并且其中的方法是不允许被 override 的。
- 检查该类是否为抽象类,如果不是抽象类,那么它是否实现了父类的抽象方法或者接口中的所有方法。
- 检查方法重载的合法性,比如相同的方法名称、相同的参数但是返回类型不相同,这都是不被允许的。
- 其他语义验证。

(3)字节码验证

当经过了文件格式和元数据的语义分析过程之后,还要对字节码进行进一步的验证,该部分的验证是比较复杂的,主要是验证程序的控制流程,比如循环、分支等。
- 保证当前线程在程序计数器中的指令不会跳转到不合法的字节码指令中去。
- 保证类型的转换是合法的,比如用 A 声明的引用,不能用 B 进行强制类型转换。
- 保证任意时刻,虚拟机栈中的操作栈类型与指令代码都能正确地被执行,比如在压栈的时候传入的是一个 A 类型的引用，在使用的时候却将 B 类型载入了本地变量表。
- 其他验证。

（4）符号引用验证

我们说过，在类的加载过程中，有些阶段是交叉进行的，比如在加载阶段尚未结束之前，连接阶段可能已经开始工作了，这样做的好处是能够提高类加载的整体效率，同样符号引用的验证，其主要作用就是验证符号引用转换为直接引用时的合法性。
- 通过符号引用描述的字符串全限定名称是否能够顺利地找到相关的类。
- 符号引用中的类、字段、方法，是否对当前类可见，比如不能访问引用类的私有方法。
- 其他。

符号引用的验证目的是为了保证解析动作的顺利执行，比如，如果某个类的字段不存在，则会抛出 NoSuchFieldError，若方法不存在时则抛出 NoSuchMethodError 等，我们在使用反射的时候会遇到这样的异常信息。

2. 准备

当一个 class 的字节流通过了所有的验证过程之后，就开始为该对象的类变量，也就是静态变量，分配内存并且设置初始值了，类变量的内存会被分配到方法区中，不同于实例变量会被分配到堆内存之中。

所谓设置初始值，其实就是为相应的类变量给定一个相关类型在没有被设置值时的默认值，不同的数据类型及其初始值见表 9-1。

表 9-1 各种类型的默认初始值

数据类型	初始值
Byte	(byte) 0
Char	'\u0000'
Short	(short) 0
Int	0
Float	0.0F
Double	0.0D
Long	0L
Boolean	False
引用类型	null

为类变量设置初始值的代码如下：

```
public class LinkedPrepare
{
    private static int a = 10;           //①
    private final static int b = 10;     //②
}
```

其中 static int a = 10 在准备阶段不是 10，而是初始值 0，当然 final static int b 则还会是 10，为什么呢？因为 final 修饰的静态变量（可直接计算得出结果）不会导致类的初始化，是一种被动引用，因此就不存在连接阶段了。

当然了更加严谨的解释是 final static int b = 10 在类的编译阶段 javac 会将其 value 生成一个 ConstantValue 属性，直接赋予 10。

3. 解析

在连接阶段中经历了验证、准备之后，就可以顺利进入到解析过程了，当然在解析的过程中照样会交叉一些验证的过程，比如符号引用的验证，所谓解析就是在常量池中寻找类、接口、字段和方法的符号引用，并且将这些符号引用替换成直接引用的过程：

```java
public class ClassResolve
{
    static Simple simple = new Simple();

    public static void main(String[] args)
    {
        System.out.println(simple);
    }
}
```

在上面的代码中 ClassResolve 用到了 Simple 类，我们在编写程序的时候，可以直接使用 simple 这个引用去访问 Simple 类中可见的属性及方法，但是在 class 字节码中可不是这么简单，它会被编译成相应的助记符，这些助记符称为符号引用，在类的解析过程中，助记符还需要得到进一步的解析，才能正确地找到所对应的堆内存中的 Simple 数据结构，下面是一段 ClassResolve 字节码的信息片段：

```
public static void main(java.lang.String[]);
    flags: ACC_PUBLIC, ACC_STATIC
    Code:
        stack=2, locals=1, args_size=1
            0: getstatic    #2 // Field java/lang/System.out:Ljava/io/PrintStream;
            3: getstatic    #3 // Field simple:Lcom/wangwenjun/concurrent/chapter09/Simple;
            6: invokevirtual #4 // Method java/io/PrintStream.println:(Ljava/lang/Object;)V
            9: return
        LineNumberTable:
            line 15: 0
            line 16: 9
        LocalVariableTable:
            Start  Length  Slot  Name   Signature
                0      10     0   args   [Ljava/lang/String;

static {};
    flags: ACC_STATIC
    Code:
        stack=2, locals=0, args_size=0
            0: new          #5 // class com/wangwenjun/concurrent/chapter09/Simple
            3: dup
            4: invokespecial #6 // Method com/wangwenjun/concurrent/chapter09/Simple."<init>":()V
            7: putstatic    #3 // Field simple:Lcom/wangwenjun/concurrent/chapter09/Simple;
            10: return
        LineNumberTable:
            line 11: 0
```

在常量池中通过 getstatic 这个指令获取 PrintStream，同样 getstatic 也适用于获取 Simple，然后通过 invokevirtual 指令将 simple 传递给 PrintStream 的 println 方法，在字节码的执行过程中，getstatic 被执行之前，就需要进行解析。

虚拟机规范规定了，在 anewarray、checkcast、getfield、getstatic、instanceof、invokeinterface、invokespecial、invokestatic、invokevirtual、multianewarray、new、putfield、putstatic 这 13 个操作符号引用的字节码指令之前，必须对所有的符号提前进行解析。

解析过程主要是针对类接口、字段、类方法和接口方法这四类进行的，分别对应到常量池中的 CONSTANT_Class_info、CONSTANT_Fieldref_info、Constant_Methodref_info 和 Constant_InterfaceMethodred_info 这四种类型常量。

（1）类接口解析
- 假设前文代码中的 Simple，不是一个数组类型，则在加载的过程中，需要先完成对 Simple 类的加载，同样需要经历所有的类加载阶段。
- 如果 Simple 是一个数组类型，则虚拟机不需要完成对 Simple 的加载，只需要在虚拟机中生成一个能够代表该类型的数组对象，并且在堆内存中开辟一片连续的地址空间即可。
- 在类接口的解析完成之后，还需要进行符号引用的验证。

（2）字段的解析

所谓字段的解析，就是解析你所访问类或者接口中的字段，在解析类或者变量的时候，如果该字段不存在，或者出现错误，就会抛出异常，不再进行下面的解析。
- 如果 Simple 类本身就包含某个字段，则直接返回这个字段的引用，当然也要对该字段所属的类提前进行类加载。
- 如果 Simple 类中不存在该字段，则会根据继承关系自下而上，查找父类或者接口的字段，找到即可返回，同样需要提前对找到的字段进行类的加载过程。
- 如果 Simple 类中没有字段，一直找到了最上层的 java.lang.Object 还是没有，则表示查找失败，也就不再进行任何解析，直接抛出 NoSuchFieldError 异常。

这样也就解释了子类为什么重写了父类的字段之后能够生效的原因，因为找到子类的字段就直接初始化并返回了。

（3）类方法的解析

类方法和接口方法有所不同，类方法可以直接使用该类进行调用，而接口方法必须要有相应的实现类继承才能够进行调用。
- 若在类方法表中发现 class_index 中索引的 Simple 是一个接口而不是一个类，则直接返回错误。
- 在 Simple 类中查找是否有方法描述和目标方法完全一致的方法，如果有，则直接返回这个方法的引用，否则直接继续向上查找。
- 如果父类中仍然没有找到，则意味着查找失败，程序会抛出 NoSuchMethodError

异常。
- 如果在当前类或者父类中找到了和目标方法一致的方法，但是它是一个抽象类，则会抛出 AbstractMethodError 这个异常。

在查找的过程中也出现了大量的检查和验证。

（4）接口方法的解析

接口不仅可以定义方法，还可以继承其他接口。
- 在接口方法表中发现 class_index 中索引的 Simple 是一个类而不是一个接口，则会直接返回错误，因为方法接口表和类接口表所容纳的类型应该是不一样的，这也是为什么在常量池中必须要有 Constant_Methodref_info 和 Constant_InterfaceMethodred_info 两个不同的类型。
- 接下来的查找就和类方法的解析比较类似了，自下而上的查找，直到找到为止，或者没找到抛出 NoSuchMethodError 异常。

9.3.3 类的初始化阶段

经历了层层关卡，终于来到了类的初始化阶段，类的初始化阶段是整个类加载过程中的最后一个阶段，在初始化阶段做的最主要的一件事情就是执行 <clinit>() 方法的过程（clinit 是 class initialize 前面几个字母的简写）在 <clinit>() 方法中所有的类变量都会被赋予正确的值，也就是在程序编写的时候指定的值。

<clinit>() 方法是在编译阶段生成的，也就是说它已经包含在了 class 文件中了，<clinit>() 中包含了所有类变量的赋值动作和静态语句块的执行代码，编译器收集的顺序是由执行语句在源文件中的出现顺序所决定的（<clinit>() 是能够保证顺序性的，关于顺序性的话题会在本书的第三部分进行详细的介绍）。另外需要注意的一点是，静态语句块只能对后面的静态变量进行赋值，但是不能对其进行访问。如图 9-3 所示，静态代码块中对 x 的访问无法通过编译。

```
public class ClassInit
{
    static {
        System.out.println(x);
        x=100;
    }

    private static int x = 10;
}
```

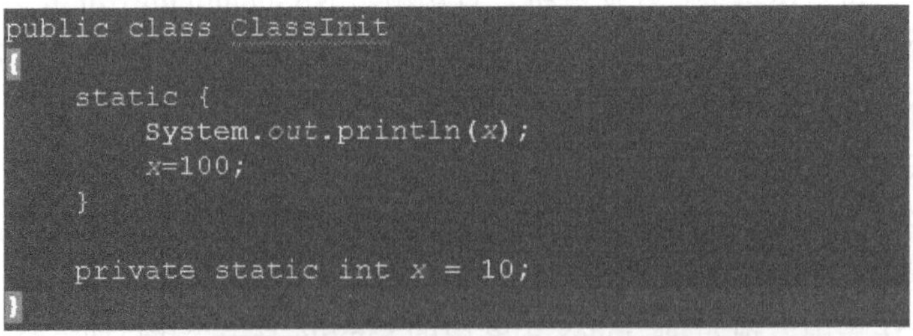

图 9-3 在静态代码块中对 x 的访问无法通过编译

另外 <clinit>() 方法与类的构造函数有所不同，它不需要显示的调用父类的构造器，虚拟机会保证父类的 <clinit>() 方法最先执行，因此父类的静态变量总是能够得到优先赋值，示例代码如下：

```java
package com.wangwenjun.concurrent.chapter09;

public class ClassInit
{
    // 父类中有静态变量value
    static class Parent
    {
        static int value = 10;

        static
        {
            value = 20;
        }

    }

    // 子类使用父类的静态变量为自己的静态变量赋值
    static class Child extends Parent
    {
        static int i = value;
    }

    public static void main(String[] args)
    {
        System.out.println(Child.i);
    }
}
```

上面程序的输出是20，而不是10，因为父类的 <clinit>() 方法优先得到了执行。

虽然说 Java 编译器会帮助 class 生成 <clinit>() 方法，但是该方法并不意味着总是会生成，比如某个类中既没有静态代码块，也没有静态变量，那么它就没有生成 <clinit>() 方法的必要了，接口中同样也是如此，由于接口天生不能定义静态代码块，因此只有当接口中有变量的初始化操作时才会生成 <clinit>() 方法。

<clinit>() 方法虽然是真实存在的，但是它只能被虚拟机执行，在主动使用触发了类的初始化之后就会调用这个方法，如果有多个线程同时访问这个方法，那么会不会引起线程安全问题呢？我们来看下面的这个例子：

```java
package com.wangwenjun.concurrent.chapter09;

import java.util.concurrent.TimeUnit;
import java.util.stream.IntStream;

public class ClassInit
{
    static
    {
        try
        {
```

```
            System.out.println("The ClassInit static code block will be invoke.");
            TimeUnit.MINUTES.sleep(10);
        } catch (InterruptedException e)
        {
            e.printStackTrace();
        }
    }

    public static void main(String[] args)
    {
        IntStream.range(0, 5)
                .forEach(i -> new Thread(ClassInit::new));
    }
}
```

运行上面的代码，你会发现在同一时间，只能有一个线程执行到静态代码块中的内容，并且静态代码块仅仅只会被执行一次，JVM 保证了 <clinit>() 方法在多线程的执行环境下的同步语义，因此在单例设计模式下，采用 Holder 的方式是一种最佳的设计方案（本书的第 14 章会进行详细的介绍）。

9.4 本章总结

好了，至此我们非常详细地介绍了类加载的整个过程和其中的细节，随着 JVM 的不断升级，其中的一些细节可能会不断变化，但是总体来讲，类的加载过程还是会围绕着二进制文件的加载、二进制数据的连接以及类的初始化这样的过程去进行，从第 10 章开始我们将进入编程实战阶段，深入 ClassLoader 这个类进行各个知识点的讲解。

不过别忘了我们在 9.3 节开始时留下的那个看似诡异的程序，我们通过前面学习和掌握的内容一起来解释一下：

```
private static int x = 0;
private static int y;
private static Singleton instance = new Singleton();
```

在连接阶段的准备过程中，每一个类变量都被赋予了相应的初始值：

```
x = 0, y = 0, instance=null
```

下面跳过解析过程，来看类的初始化阶段，初始化阶段会为每一个类变量赋予正确的值，也就是执行 <clinit>() 方法的过程：

```
x = 0, y=0, instance = new Singleton()
```

在 new Singleton 的时候会执行类的构造函数，而在构造函数中分别对 x 和 y 进行了自增，因此结果为：

```
x = 1, y = 1
```

我们再来看调换顺序之后的程序输出：

```
private static Singleton instance = new Singleton();
private static int x = 0;
private static int y;
```

在连接阶段的准备过程中，每一个类变量都被赋予了相应的初始值：

```
instance=null, x = 0, y = 0
```

在类的初始化阶段，需要为每一个类赋予程序编写时期所期待的正确的初始值，首先会进入 instance 的构造函数中（构造函数其实在 class 中也有相应的函数名，init() 少了一个 class 的前缀），执行完 instance 的构造函数之后，各个静态变量的值如下：

```
instance=Singleton@3f99bd52, x = 1, y = 1
```

然后，为 x 初始化，由于 x 没有显式地进行赋值，因此 0 才是所期望的正确赋值，而 y 由于没有给定初始值，在构造函数中计算所得的值就是所谓的正确赋值，因此结果又会变成：

```
instance=Singleton@3f99bd52, x = 0, y = 1
```

第 10 章 · CHAPTER10

JVM 类加载器

顾名思义，类的加载器就是负责类的加载职责，对于任意一个 class，都需要由加载它的类加载器和这个类本身确立其在 JVM 中的唯一性，这也就是运行时包，关于运行时包我们在后文中会讲到。在前文中我们提到过，任何一个对象的 class 在 JVM 中只存在唯一的一份，比如 String.class、Object.class 在堆内存以及方法区中肯定是唯一的，但是不能绝对地理解为我们自定义的类在 JVM 中同样也是这样。本章我们将站在开发者的角度，通过实战编程的方式得出关于类加载器的知识总结。

10.1 JVM 内置三大类加载器

JVM 为我们提供了三大内置的类加载器，不同的类加载器负责将不同的类加载到 JVM 内存之中，并且它们之间严格遵守着父委托的机制，如图 10-1 所示。

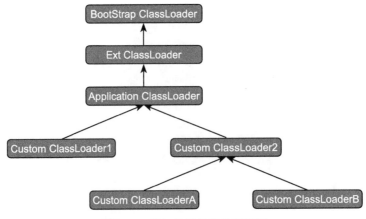

图 10-1　类加载器的父委托机制

10.1.1 根类加载器介绍

根加载器又称为 Bootstrap 类加载器，该类加载器是最为顶层的加载器，其没有任何父加载器，它是由 C++ 编写的，主要负责虚拟机核心类库的加载，比如整个 java.lang 包都是由根加载器所加载的，可以通过 -Xbootclasspath 来指定根加载器的路径，也可以通过系统属性来得知当前 JVM 的根加载器都加载了哪些资源，示例代码如清单 10-1 所示。

代码清单 10-1　根加载器测试 BootStrapClassLoader.java

```
package com.wangwenjun.concurrent.chapter10;

public class BootStrapClassLoader
{
    public static void main(String[] args)
    {
        System.out.println("Bootstrap:" +
String.class.getClassLoader());
System.out.println(System.getProperty("sun.boot.class.path"));

    }
}
```

上述程序运行输出的结果如下所示，其中 String.class 的类加载器是根加载器，根加载器是获取不到引用的，因此输出为 null，而根加载器所在的加载路径可以通过 sun.boot.class.path 这个系统属性来获得。

```
Bootstrap:null
D:\Program Files\Java\jdk1.8.0_40\jre\lib\resources.jar;
D:\Program Files\Java\jdk1.8.0_40\jre\lib\rt.jar;
D:\Program Files\Java\jdk1.8.0_40\jre\lib\sunrsasign.jar;
D:\Program Files\Java\jdk1.8.0_40\jre\lib\jsse.jar;
D:\Program Files\Java\jdk1.8.0_40\jre\lib\jce.jar;
D:\Program Files\Java\jdk1.8.0_40\jre\lib\charsets.jar;
D:\Program Files\Java\jdk1.8.0_40\jre\lib\jfr.jar;
D:\Program Files\Java\jdk1.8.0_40\jre\classes
```

10.1.2 扩展类加载器介绍

扩展类加载器的父加载器是根加载器，它主要用于加载 JAVA_HOME 下的 jre\lb\ext 子目录里面的类库。扩展类加载器是由纯 Java 语言实现的，它是 java.lang.URLClassLoader 的子类，它的完整类名是 sun.misc.Launcher$ExtClassLoader。扩展类加载器所加载的类库可以通过系统属性 java.ext.dirs 获得，示例代码如清单 10-2 所示。

代码清单 10-2　扩展类载器测试 ExtClassLoader.java

```
package com.wangwenjun.concurrent.chapter10;

public class ExtClassLoader
```

```
{
    public static void main(String[] args)
    {
        System.out.println(System.getProperty("java.ext.dirs"));
    }
}
```

运行上面的程序会得到扩展类加载器加载资源的路径，输出如下：

```
D:\Program Files\Java\jdk1.8.0_40\jre\lib\ext;
C:\Windows\Sun\Java\lib\ext
```

当然你也可以将自己的类打包成 jar 包，放到扩展类加载器所在的路径中，扩展类加载器会负责加载你所需要的类，比如笔者简单做了一个 jar 包，其中就一个 Hello.class，然后放置在扩展类加载器的加载路径之中，代码如下：

```
public static void main(String[] args)
        throws ClassNotFoundException
{
    Class<?> helloClass = Class.forName("Hello");
    System.out.println(helloClass.getClassLoader());
}
```

运行程序会发现加载 Hello 这个 Class 的类加载器是扩展类加载器，输出如下：

```
sun.misc.Launcher$ExtClassLoader@3a71f4dd
```

10.1.3　系统类加载器介绍

系统类加载器是一种常见的类加载器，其负责加载 classpath 下的类库资源。我们在进行项目开发的时候引入的第三方 jar 包，系统类加载器的父加载器是扩展类加载器，同时它也是自定义类加载器的默认父加载器，系统类加载器的加载路径一般通过 -classpath 或者 -cp 指定，同样也可以通过系统属性 java.class.path 进行获取，示例代码如清单 10-3 所示。

代码清单 10-3　系统类加载器测试 ApplicationClassLoader.java

```
package com.wangwenjun.concurrent.chapter10;

public class ApplicationClassLoader
{
    public static void main(String[] args)
    {
        System.out.println(System.getProperty("java.class.path"));
System.out.println(ApplicationClassLoader.class.getClassLoader());
    }
}
```

JDK 除了提供上述三大类内置类加载器之外，还允许开发人员进行类加载器的扩展，也就是自定义类加载器，很多开源项目借助于自定义类加载器开发出了很多伟大的系统，比如 OSGI、Tomcat 的容器隔离等。在 10.2 节中，我们将详细介绍如何自定义类的加载器，以及详细剖析类的父委托机制等话题。

10.2 自定义类加载器

在本节中，我们开始用程序实现自定义的类加载器，所有的自定义类加载器都是 ClassLoader 的直接子类或者间接子类，java.lang.ClassLoader 是一个抽象类，它里面并没有抽象方法，但是有 findClass 方法，务必实现该方法，否则将会抛出 Class 找不到的异常，示例代码如下：

```
protected Class<?> findClass(String name) throws ClassNotFoundException {
    throw new ClassNotFoundException(name);
}
```

10.2.1 自定义类加载器，问候世界

在本节中，我们自定义一个简单的 ClassLoader，然后使用该类加载器加载一个简单的类，示例代码如清单 10-4 所示。

代码清单 10-4　自定义类加载器 MyClassLoader.java

```java
package com.wangwenjun.concurrent.chapter10;

import java.io.ByteArrayOutputStream;
import java.io.IOException;
import java.nio.file.Files;
import java.nio.file.Path;
import java.nio.file.Paths;

// 自定义类加载器必须是 ClassLoader 的直接或者间接子类
public class MyClassLoader extends ClassLoader
{
    // 定义默认的 class 存放路径
    private final static Path DEFAULT_CLASS_DIR = Paths.get("G:", "classloader1");

    private final Path classDir;

    // 使用默认的 class 路径
    public MyClassLoader()
    {
        super();
```

```java
            this.classDir = DEFAULT_CLASS_DIR;
        }

        // 允许传入指定路径的class路径
        public MyClassLoader(String classDir)
        {
            super();
            this.classDir = Paths.get(classDir);
        }

    // 指定class路径的同时,指定父类加载器
        public MyClassLoader(String classDir, ClassLoader parent)
        {
            super(parent);
            this.classDir = Paths.get(classDir);
        }

        // 重写父类的findClass方法,这是至关重要的步骤
        @Override
        protected Class<?> findClass(String name)
                throws ClassNotFoundException
    {
        // 读取class的二进制数据
            byte[] classBytes = this.readClassBytes(name);
            // 如果数据为null,或者没有读到任何信息,则抛出ClassNotFoundException异常
            if (null == classBytes || classBytes.length == 0)
            {
                throw new ClassNotFoundException("Can not load the class " + name);
            }

            // 调用defineClass方法定义class
            return this.defineClass(name, classBytes, 0, classBytes.length);
        }

        // 将class文件读入内存
        private byte[] readClassBytes(String name)
                throws ClassNotFoundException
        {
            // 将包名分隔符转换为文件路径分隔符
            String classPath = name.replace(".", "/");
            Path classFullPath = classDir.resolve(Paths.get(classPath + ".class"));
            if (!classFullPath.toFile().exists())
                    throw new ClassNotFoundException("The class " + name + " not found.");
            try (ByteArrayOutputStream baos = new ByteArrayOutputStream())
            {
                Files.copy(classFullPath, baos);
                return baos.toByteArray();
            } catch (IOException e)
```

```
            {
                throw new ClassNotFoundException("load the class " + name + " occur error.", e);
            }
        }

        @Override
        public String toString()
        {
            return "My ClassLoader";
        }
    }
```

至此我们完成了一个非常简单的基于磁盘的 ClassLoader, 几个关键的地方都已经做了标注, 第一个构造函数使用默认的文件路径, 第二个构造函数允许外部指定一个特定的磁盘目录, 第三个构造函数除了可以指定磁盘目录以外还可以指定该类加载器的父加载器。

在我们定义的类加载器中, 通过将类的全名称转换成文件的全路径重写 findClass 方法, 然后读取 class 文件的字节流数据, 最后使用 ClassLoader 的 defineClass 方法对 class 完成了定义。

在开始使用我们定义的 ClassLoader 之前, 有几个地方需要特别强调一下。第一, 关于类的全路径格式, 一般情况下我们的类都是类似于 java.lang.String 这样的格式, 但是有时候不排除内部类, 匿名内部类等; 全路径格式有如下几种情况。

- java.lang.String: 包名 . 类名
- javax.swing.JSpinner$DefaultEditor: 包名 . 类名 $ 内部类
- java.security.KeyStore$Builder$FileBuilder$1: 包名 . 类名 $ 内部类 $ 内部类 $ 匿名内部类
- java.net.URLClassLoader$3$1: 包名 . 类名 $ 匿名内部类 $ 匿名内部类

第二个需要强调的是 defineClass 方法, 该方法的完整方法描述是 defineClass (String name, byte[] b, int off, int len), 其中, 第一个是要定义类的名字, 一般与 findClass 方法中的类名保持一致即可; 第二个是 class 文件的二进制字节数组, 这个也不难理解; 第三个是字节数组的偏移量; 第四个是从偏移量开始读取多长的 byte 数据。大家思考一下, 在类的加载过程中, 第一个阶段的加载主要是获取 class 的字节流信息, 那么我们将整个字节流信息交给 defineClass 方法不就行了吗, 为什么还要画蛇添足地指定偏移量和读取长度呢? 原因是因为 class 字节数组不一定是从一个 class 文件中获得的, 有可能是来自网络的, 也有可能是用编程的方式写入的, 由此可见, 一个字节数组中很有可能存储多个 class 的字节信息。

好了, 下面我们写一个简单的程序, 使用定义的 ClassLoader 对其进行加载, 代码如清单 10-5 所示。

代码清单 10-5　HelloWorld.java

```java
package com.wangwenjun.concurrent.chapter10;

public class HelloWorld
{
    static
    {
        System.out.println("Hello World Class is Initialized.");
    }

    public String welcome()
    {
        return "Hello World";
    }
}
```

Java 文件编译之后，将 class 文件复制到你想要的目录，比如笔者将编译后的完整路径复制到了 G:\classloader1 目录，同时在程序运行的 class path 中删除掉 HelloWorld 这个 class。如果你用的是集成开发环境，那么应该连 HelloWorld.java 也一并删除，否则 HelloWorld 将会被系统类加载器加载，这是由于类加载器的委托机制所导致的，关于委托机制在接下来章节中会进行详细的介绍。

ClassLoader 定义好了，需要加载的类也准备好了，我们可以简单地对其进行一下测试，测试代码如清单 10-6 所示。

代码清单 10-6　测试 MyClassLoader.java

```java
package com.wangwenjun.concurrent.chapter10;

import java.lang.reflect.InvocationTargetException;
import java.lang.reflect.Method;
public class MyClassLoaderTest
{
    public static void main(String[] args)
            throws ClassNotFoundException,
            IllegalAccessException,
            InstantiationException, NoSuchMethodException,
            InvocationTargetException
    {
        // 声明一个 MyClassLoader
        MyClassLoader classLoader = new MyClassLoader();
        // 使用 MyClassLoader 加载 HelloWorld
        Class<?> aClass = classLoader
                .loadClass("com.wangwenjun.concurrent.chapter10.HelloWorld");


        System.out.println(aClass.getClassLoader());

        //① 注释
```

```
        Object helloWorld = aClass.newInstance();
        System.out.println(helloWorld);
        Method welcomeMethod = aClass.getMethod("welcome");
        String result = (String) welcomeMethod.invoke(helloWorld);
        System.out.println("Result:" + result);
    }
}
```

如果没有错误发生,程序会正常输出类加载器以及对世界的一句问候,在测试代码中注释掉①以下所有的代码你会发现,虽然 aClass 被成功加载并且输出了类加载器的信息,但是 HelloWorld 的静态代码块并没有得到输出,那是因为使用类加载器 loadClass 并不会导致类的主动初始化,它只是执行了加载过程中的加载阶段而已,这一点需要读者注意。

10.2.2 双亲委托机制详细介绍

了解了如何自定义一个类加载器之后,我们来研究一下类加载器最重要的机制——双亲委托机制,有时候也称为父委托机制。当一个类加载器被调用了 loadClass 之后,它并不会直接将其加载,而是先交给当前类加载器的父加载器尝试加载直到最顶层的父加载器,然后再依次向下进行加载,这也是为什么在 10.2.1 节中要求将 HelloWorld.java 和 HelloWorld.class 删除的原因,图 10-2 显示了类加载器的委托流程。

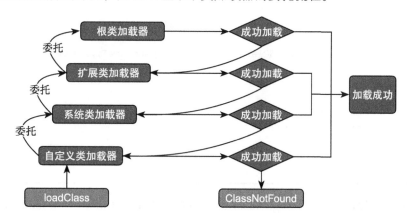

图 10-2 类加载器的父委托机制

在开始分析 loadClass 源码之前,请大家思考一个问题,由于我们担心 HelloWorld.class 被系统类加载器加载,所以删除了 HelloWorld 的相关文件,那么有什么办法可以不用删除又可以使用 MyClassLoader 对 HelloWorld 进行加载的吗?来看个代码片段,如清单 10-7 所示。

代码清单 10-7　JDK ClassLoader 源码片段

```
public Class<?> loadClass(String name) throws ClassNotFoundException
{
        return loadClass(name, false);
}
```

```java
protected Class<?> loadClass(String name, boolean resolve)
    throws ClassNotFoundException
{
    synchronized (getClassLoadingLock(name)) {
        // First, check if the class has already been loaded
        Class<?> c = findLoadedClass(name);
        if (c == null) {
            long t0 = System.nanoTime();
            try {
                if (parent != null) {
                    c = parent.loadClass(name, false);
                } else {
                    c = findBootstrapClassOrNull(name);
                }
            } catch (ClassNotFoundException e) {
                // ClassNotFoundException thrown if class not found
                // from the non-null parent class loader
            }
            if (c == null) {
                // If still not found, then invoke findClass in order
                // to find the class.
                long t1 = System.nanoTime();
                c = findClass(name);
                // this is the defining class loader; record the stats
sun.misc.PerfCounter.getParentDelegationTime().addTime(t1 - t0);
sun.misc.PerfCounter.getFindClassTime().addElapsedTimeFrom(t1);
                sun.misc.PerfCounter.getFindClasses().increment();
            }
        }
        if (resolve) {
            resolveClass(c);
        }
        return c;
    }
}
```

上面的代码片段是 java.lang.ClassLoader 的 loadClass（name）和 loadClass（name，resolve）方法，由于 loadClass（name）调用的是 loadClass（name，false），因此我们重点解释后者即可。

- 从当前类加载器的已加载类缓存中根据类的全路径名查询是否存在该类，如果存在则直接返回。
- 如果当前类存在父类加载器，则调用父类加载器的 loadClass（name，false）方法对其进行加载。
- 如果当前类加载器不存在父类加载器，则直接调用根类加载器对该类进行加载。
- 如果当前类的所有父类加载器都没有成功加载 class，则尝试调用当前类加载器的

findClass 方法对其进行加载，该方法就是我们自定义加载器需要重写的方法。
- 最后如果类被成功加载，则做一些性能数据的统计。
- 由于 loadClass 指定了 resolve 为 false，所以不会进行连接阶段的继续执行，这也就解释了为什么通过类加载器加载类并不会导致类的初始化。

分析完了 loadClass 方法的执行过程，再结合父委托机制的类加载流程图，相信读者应该对父委托机制有一个比较清晰的认识，回到本节开始时的问题，如何在不删除 HelloWorld.class 文件的情况下使用 MyClassLoader 而不是系统类加载器进行 HelloWorld 的加载，有如下两种方法可以做到。

- 第一种方式是绕过系统类加载器，直接将扩展类加载器作为 MyClassLoader 的父加载器，示例代码如下：

```
ClassLoader extClassLoader = MyClassLoaderTest.class.getClassLoader().getParent();
MyClassLoader classLoader = new MyClassLoader("G:\\classloader1",extClassLoader);
Class<?> aClass = classLoader.loadClass("com.wangwenjun.concurrent.chapter10.HelloWorld");
System.out.println(aClass);
System.out.println(aClass.getClassLoader());
```

首先我们通过 MyClassLoaderTest.class 获取系统类加载器，然后在获取系统类加载器的父类加载器中扩展类加载器，使其成为 MyClassLoader 的父类加载器，这样一来，根加载器和扩展类加载器都无法对 G:\\classloader1 中的类文件进行加载，自然而然就交给了 MyClassLoader 对 HelloWorld 进行加载了，这种方式也是充分利用了类加载器父委托机制的特性。

- 第二种方式是在构造 MyClassLoader 的时候指定其父类加载器为 null，示例代码如下：

```
MyClassLoader classLoader = new MyClassLoader("G:\\classloader1",null);
        Class<?> aClass = classLoader.loadClass("com.wangwenjun.concurrent.chapter10.HelloWorld");
System.out.println(aClass);
System.out.println(aClass.getClassLoader());
```

根据对 loadClass 方法的源码分析，当前类在没有父类加载器的情况下，会直接使用根加载器对该类进行加载，很显然，HelloWorld 在根加载器的加载路径下是无法找到的，那么它自然而然地就交给当前类加载器进行加载了。

10.2.3 破坏双亲委托机制

通过 10.2.2 节对 loadClass 源码的分析，我们发现类加载器的父委托机制的逻辑主要是由 loadClass 来控制的，有些时候我们需要打破这种双亲委托的机制，比如 HelloWorld 这个类就是不希望通过系统类加载器对其进行加载。虽然在 10.2.2 节我们给出了两种解决方案，但是采取的都是绕过 ApplicationClassLoader 的方式去实现的，并没有避免一层一层的委托，那么有没有办法可以绕过这种双亲委托的模型呢？

值得庆幸的一点是，JDK 提供的双亲委托机制并非一个强制性的模型，程序开发人员是可以对其进行灵活发挥破坏这种委托机制的，比如我们想要在程序运行时进行某个模块功能的升级，甚至是在不停止服务的前提下增加新的功能，这就是我们常说的热部署。热部署首先要卸载掉加载该模块所有 Class 的类加载器，卸载类加载器会导致所有类的卸载，很显然我们无法对 JVM 三大内置加载器进行卸载，我们只有通过控制自定义类加载器才能做到这一点。

我们可以通过破坏父委托机制的方式来实现对 HelloWorld 类的加载，而不需要在工程中删除该文件，清单 10-8 中的代码和 MyClassLoader 非常类似，只不过增加了一个 loadClass 方法的重写过程，读者可以直接继承 MyClassLoader 然后重写 loadClass 方法。

代码清单 10-8　BrokerDelegateClassLoader.java

```java
package com.wangwenjun.concurrent.chapter10;

import java.io.ByteArrayOutputStream;
import java.io.IOException;
import java.nio.file.Files;
import java.nio.file.Path;
import java.nio.file.Paths;

public class BrokerDelegateClassLoader extends ClassLoader
{
...省略代码
@Override
protected Class<?> loadClass(String name, boolean resolve)
        throws ClassNotFoundException
{
    //①
    synchronized (getClassLoadingLock(name))
    {
        //②
        Class<?> klass = findLoadedClass(name);
        //③
        if (klass == null)
        {
            //④
            if (name.startsWith("java.") || name.startsWith("javax"))
            {
                try
                {
                    klass = getSystemClassLoader().loadClass(name);
                } catch (Exception e)
                {
                    //ignore
                }
            } else
```

```
            {
                //⑤
                try
                {
                    klass = this.findClass(name);
                } catch (ClassNotFoundException e)
                {
                    //ignore
                }
                //⑥
                if (klass == null)
                {
                    if (getParent() != null)
                    {
                        klass = getParent().loadClass(name);
                    } else
                    {
                        klass = getSystemClassLoader().loadClass(name);
                    }
                }
            }
            //⑦
            if (null == klass)
            {
                    throw new ClassNotFoundException("The class " + name + " not found.");
            }
            if (resolve)
            {
                resolveClass(klass);
            }
            return klass;
        }
    }
    ...
    }
```

在上述代码中：

①根据类的全路径名称进行加锁，确保每一个类在多线程的情况下只被加载一次。

②到已加载类的缓存中查看该类是否已经被加载，如果已加载则直接返回。

③④若缓存中没有被加载的类，则需要对其进行首次加载，如果类的全路径以 java 和 javax 开头，则直接委托给系统类加载器对其进行加载。

⑤如果类不是以 java 和 javax 开头，则尝试用我们自定义的类加载进行加载。

⑥若自定义类加载仍旧没有完成对类的加载，则委托给其父类加载器进行加载或者系统类加载器进行加载。

⑦经过若干次的尝试之后，如果还是无法对类进行加载，则抛出无法找到类的异常。

好了，至此我们的 BrokerDelegateClassLoader 已经完成，使用它对 HelloWorld 进行加载可以不用在工程文件中删除 HelloWorld。既然我们可以通过这种打破双亲委托的方式进行类加载，那么我们是否可以使用我们自定义的类加载器加载一个和 String 类全路径名称完全一致的 class 呢？读者可以提前思考一下，我们在后文中将会对其进行验证。

10.2.4 类加载器命名空间、运行时包、类的卸载等

1. 类加载器命名空间

每一个类加载器实例都有各自的命名空间，命名空间是由该加载器及其所有父加载器所构成的，因此在每个类加载器中同一个 class 都是独一无二的，类加载器命令空间代码如清单 10-9 所示。

代码清单 10-9　NameSpace.java

```
package com.wangwenjun.concurrent.chapter10;
public class NameSpace
{
    public static void main(String[] args)
            throws ClassNotFoundException
    {
        //获取系统类加载器
        ClassLoader classLoader = NameSpace.class.getClassLoader();
        Class<?> aClass = classLoader.loadClass("com.wangwenjun.concurrent.chapter10.Test");
        Class<?> bClass = classLoader.loadClass("com.wangwenjun.concurrent.chapter10.Test");
        System.out.println(aClass.hashCode());
        System.out.println(bClass.hashCode());
        System.out.println(aClass == bClass);
    }
}
```

运行上面的代码，不论 load 多少次 Test，你都将会发现他们始终是同一份 class 对象，这也完全符合我们在本书 9.1 节中的描述。类被加载后的内存情况如图 10-3 所示。

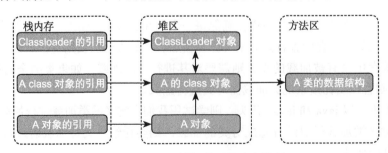

图 10-3　class 被加载后的内存情况

但是，使用不同的类加载器，或者同一个类加载器的不同实例，去加载同一个 class，

则会在堆内存和方法区产生多个 class 的对象。

（1）不同类加载器加载同一个 class

```
public static void main(String[] args)
        throws ClassNotFoundException
{
    MyClassLoader classLoader1 = new MyClassLoader("G:\\classloader1", null);
    BrokenDelegateClassLoader classLoader2 = new BrokenDelegateClassLoader("G:\\classloader1", null);
    Class<?> aClass = classLoader1.loadClass("com.wangwenjun.concurrent.chapter10.Test");
    Class<?> bClass = classLoader2.loadClass("com.wangwenjun.concurrent.chapter10.Test");
    System.out.println(aClass.getClassLoader());
    System.out.println(bClass.getClassLoader());
    System.out.println(aClass.hashCode());
    System.out.println(bClass.hashCode());
    System.out.println(aClass == bClass);
}
```

程序的输出结果显示，aClass 和 bClass 不是同一个 class 实例。

（2）相同类加载器加载同一个 class

```
public static void main(String[] args)
        throws ClassNotFoundException
{
    MyClassLoader classLoader1 = new MyClassLoader("G:\\classloader1", null);
    MyClassLoader classLoader2 = new MyClassLoader("G:\\classloader1", null);
    Class<?> aClass = classLoader1.loadClass("com.wangwenjun.concurrent.chapter10.Test");
    Class<?> bClass = classLoader2.loadClass("com.wangwenjun.concurrent.chapter10.Test");
    System.out.println(aClass.getClassLoader());
    System.out.println(bClass.getClassLoader());
    System.out.println(aClass.hashCode());
    System.out.println(bClass.hashCode());
    System.out.println(aClass == bClass);
}
```

程序的输出结果显示，aClass 和 bClass 不是同一个 class 实例。

我们再来分析 JDK 中关于 ClassLoader 的相关源代码，具体如下：

```
protected Class<?> loadClass(String name, boolean resolve)
    throws ClassNotFoundException
{
    synchronized (getClassLoadingLock(name)) {
        // First, check if the class has already been loaded
        Class<?> c = findLoadedClass(name);
...
```

```
protected final Class<?> findLoadedClass(String name) {
    if (!checkName(name))
        return null;
    return findLoadedClass0(name);
}
private native final Class<?> findLoadedClass0(String name);
```

在类加载器进行类加载的时候，首先会到加载记录表也就是缓存中，查看该类是否已经被加载过了，如果已经被加载过了，就不会重复加载，否则将会认为其是首次加载，图10-4 是同一个 class 被不同类加载器加载之后的内存情况。

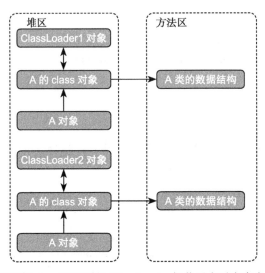

图 10-4　相同的 class 被不同的 ClassLoader 加载后会对应多个 class 实例

经过对类加载器命名空间的了解之后，同一个 class 实例只能在 JVM 中存在一份这样的说法是不够严谨的，更准确的说法应该是同一个 class 实例在同一个类加载器命名空间之下是唯一的。

2. 运行时包

我们在编写代码的时候通常会给一个类指定一个包名，包的作用是为了组织类，防止不同包下同样名称的 class 引起冲突，还能起到封装的作用，包名和类名构成了类的全限定名称。在 JVM 运行时 class 会有一个运行时包，运行时的包是由类加载器的命名空间和类的全限定名称共同组成的，比如 Test 的运行时包如下所示：

BootstrapClassLoader.ExtClassLoader.AppClassloader.MyClassLoader.com.wangwenjun.concurrent.chapter10.Test

这样做的好处同样是出于安全和封装的考虑，在 java.lang.String 中存在仅包可见的方法 void getChars（char[] var1，int var2），java.lang 包以外的 class 是无法直接对其访问的。假设用户想自己定义一个类 java.lang.HackString，并且由自定义的类加载器进行加载，尝试访问 getChars 方法，由于 java.lang.HackString 和 java.lang.String 是由不同的类加载器进行加载的，它们拥有各自不同的运行时包，因此 HackString 是无法访问 java.lang.String 的包可见方法以及成员变量的。

3. 初始类加载器

由于运行时包的存在，JVM 规定了不同的运行时包下的类彼此之间是不可以进行访问的，那么问题来了，为什么我们在开发的程序中可以访问 java.lang 包下的类呢？根据前面所学的知识，我们知道 java.lang 包是由根加载器进行加载的，而我们开发的程序或者第三方类库一般是由系统类加载器进行加载的，为什么我们在程序中能够 new Object() 或者 new String() 等任意的 java.lang 包下的类呢？

为了能让读者更好地理解这一点，我们写了一个简单的例子对此加以说明，新建一个磁盘目录 G:\\classloader2，用于存放我们编译后的 class 文件，代码如清单 10-10 所示。

代码清单 10-10　SimpleClass.java

```java
package com.wangwenjun.concurrent.chapter10;

import java.util.ArrayList;
import java.util.List;

public class SimpleClass
{
    // 在 SimpleClass 中使用 byte[]
    private static byte[] buffer = new byte[8];

    // 在 SimpleClass 中使用 String
    private static String str = "";

    // 在 SimpleClass 中使用 List
    private static List<String> list = new ArrayList<>();

    static
    {
        buffer[0] = (byte) 1;
        str = "Simple";
        list.add("element");
        System.out.println(buffer[0]);
        System.out.println(str);
        System.out.println(list.get(0));
    }
}
```

编译 SimpleClass.java 文件，将 SimpleClass.class 复制到 G:\\classloader2 目录之下，在 SimpleClass 中，我们访问了 java.lang.String、java.utils.List 以及 java.lang.Object 等类，这些类都存在于 rt.jar 包下，在 JVM 启动的时候这些类是由根加载器进行加载的。下面使用前文中创建的 BrokerDelegateClassLoader 对 SimpleClass 进行加载，代码如清单 10-11 所示。

代码清单 10-11　LoadSimpleClass.java

```
package com.wangwenjun.concurrent.chapter10;

public class LoadSimpleClass
{
    public static void main(String[] args) throws ClassNotFoundException,
            IllegalAccessException, InstantiationException
    {
        BrokerDelegateClassLoader classLoader = new BrokerDelegateClassLoader
("G\\classloader2");
        Class<?> aClass = classLoader.loadClass("com.wangwenjun.concurrent.
chapter10.SimpleClass");
        System.out.println(classLoader.getParent());
        aClass.newInstance();
    }
}
```

在上面的程序中，SimpleClass 是由我们自定义的 ClassLoader 加载的，但是其能够访问不同的运行时包下的类，比如 String，下面我们来分析一下原因。

每一个类在经过 ClassLoader 的加载之后，在虚拟机中都会有对应的 Class 实例，如果某个类 C 被类加载器 CL 加载，那么 CL 就被称为 C 的初始类加载器。JVM 为每一个类加载器维护了一个列表，该列表中记录了将该类加载器作为初始类加载器的所有 class，在加载一个类时，JVM 使用这些列表来判断该类是否已经被加载过了，是否需要首次加载。

根据 JVM 规范的规定，在类的加载过程中，所有参与的类加载器，即使没有亲自加载过该类，也都会被标识为该类的初始类加载器，比如 java.lang.String 首先经过了 BrokerDelegateClassLoader 类加载器，依次又经过了系统类加载器、扩展类加载器、根类加载器，这些类加载器都是 java.lang.String 的初始类加载器，JVM 会在每一个类加载器维护的列表中添加该 class 类型，如图 10-5 所示。

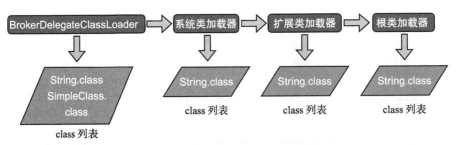

图 10-5　初始类加载器及 class 列表

虽然 SimpleClass 和 java.lang.String 由不同的类加载器加载，但是在 BrokerDelegate-ClassLoader 的 class 列表中维护了 SimpleClass.class 和 String.class，因此在 SimpleClass 中是可以正常访问 rt.jar 中的 class 的。

4. 类的卸载

在 JVM 的启动过程中，JVM 会加载很多的类，在运行期间同样也会加载很多的类，比如用自定义的类加载器进行类的加载，或者像 Apache Drools 框架一样会在每一个 DSL 文件解析成功之后生成相应的类文件。关于 JVM 在运行期间到底加载了多少 class，可以在启动 JVM 时指定 -verbose:class 参数观察得到，我们知道某个对象在堆内存中如果没有其他地方引用则会在垃圾回收器线程进行 GC 的时候被回收掉，那么该对象在堆内存中的 Class 对象以及 Class 在方法区中的数据结构何时被回收呢？

JVM 规定了一个 Class 只有在满足下面三个条件的时候才会被 GC 回收，也就是类被卸载。

- 该类所有的实例都已经被 GC，比如 Simple.class 的所有 Simple 实例都被回收掉。
- 加载该类的 ClassLoader 实例被回收。
- 该类的 class 实例没有在其他地方被引用。

10.3 本章总结

在本章中我们不仅介绍了 JVM 内置的三大类加载器，并且通过继承 ClassLoader 重写 findClass 方法自定义了 MyClassLoader。通过对 loadClass 方法的源码剖析我们详细分析了双亲委托机制的原理，双亲委托机制是一种包含关系，而并非继承关系，通过对自定义类加载器的分析，读者对此应该深有体会。

好了，下面我们来看 10.2.3 节中留下的关于自定义 java.lang.String 类，并且使用自定义类加载进行加载的问题，代码如下：

```java
package java.lang;

// 自定义与 java.lang.String 同名的类
public class String
{
    static
    {
        System.out.println("自定义java.lang.String");
    }
}
```

将编译好的 java.lang.String.class 放置到 G:\\classloader3 中，然后使用自定义类加载器 BrokerDelegateClassLoader 对其进行加载，代码如下：

```java
package com.wangwenjun.concurrent.chapter10;

public class LoadString
{
    public static void main(String[] args)
```

```
            throws ClassNotFoundException
    {
            BrokerDelegateClassLoader classLoader = new BrokerDelegateClassLoader
("G:\\classloader3");
            Class<?> aClass = classLoader.loadClass("java.lang.String");
    }
}
```

想要 BrokerDelegateClassLoader 加载 G:\\classloader 下的 String.class，需要对 Broker-DelegateClassLoader 源码进行修改，去掉对 java 和 javax 前缀的判断，虽然能够准确找到 G:\\classloader3 目录下的 java.lang.String class 文件，但 JVM 是不允许你这样做的，运行上面的程序会出现 java.lang.SecurityException: Prohibited package name: java.lang 错误，打开 ClassLoader 源码会发现 JVM 在 defineClass 的时候做了安全性检查，代码如下：

```
        private ProtectionDomain preDefineClass(String name,
                                                ProtectionDomain pd)
        {
            if (!checkName(name))
                throw new NoClassDefFoundError("IllegalName: " + name);

            if ((name != null) && name.startsWith("java.")) {
                throw new SecurityException
                    ("Prohibited package name: " +
                     name.substring(0, name.lastIndexOf('.')));
            }
            if (pd == null) {
                pd = defaultDomain;
            }

            if (name != null) checkCerts(name, pd.getCodeSource());

            return pd;
        }
```

几乎没有人会定义与 JDK 核心类库完全同限定名称的类，我们这么做的主要目的是为了让读者能够更进一步地了解 JDK 类加载的内部细节而已。

第 11 章

线程上下文类加载器

本书的内容主要是为了介绍多线程以及高并发程序的设计,但本书的第二部分却花费了大量的篇幅介绍 Java 类加载器的知识,其主要是为了解释线程的上下文类加载器原理和使用场景。

11.1 为什么需要线程上下文类加载器

根据 Thread 类的文档你会发现线程上下文方法是从 JDK1.2 开始引入的,getContextClassLoader() 和 setContextClassLoader（ClassLoader cl）分别用于获取和设置当前线程的上下文类加载器,如果当前线程没有设置上下文类加载器,那么它将和父线程保持同样的类加载器。站在开发者的角度,其他线程都是由 Main 线程,也就是 main 函数所在的线程派生的,它是其他线程的父线程或者祖先线程。下面进行线程上下文类的简单测试,代码如清单 11-1 所示。

代码清单 11-1　线程上下文类加载器简单测试

```
package com.wangwenjun.concurrent.chapter11;
import static java.lang.Thread.currentThread;
public class MainThreadClassLoader
{
    public static void main(String[] args)
    {
        System.out.println(currentThread().getContextClassLoader());
    }
}
```

为什么要有线程上下文类加载器呢？这就与 JVM 类加载器双亲委托机制自身的缺陷

是分不开的，JDK 的核心库中提供了很多 SPI（Service Provider Interface），常见的 SPI 包括 JDBC、JCE、JNDI、JAXP 和 JBI 等，JDK 只规定了这些接口之间的逻辑关系，但不提供具体的实现，具体的实现需要由第三方厂商来提供，作为 Java 程序员或多或少地都写过 JDBC 的程序，在编写 JDBC 程序时几乎百分之百的都在与 java.sql 包下的类打交道。

如图 11-1 所示，Java 使用 JDBC 这个 SPI 完全透明了应用程序和第三方厂商数据库驱动的具体实现，不管数据库类型如何切换，应用程序只需要替换 JDBC 的驱动 jar 包以及数据库的驱动名称即可，而不用进行任何更新。

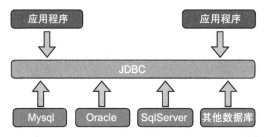

图 11-1　JDBC 作为 SPI 透明了各大数据库厂商的驱动细节

这样做的好处是 JDBC 提供了高度抽象，应用程序则只需要面向接口编程即可，不用关心各大数据库厂商提供的具体实现，但问题在于 java.lang.sql 中的所有接口都由 JDK 提供，加载这些接口的类加载器是根加载器，第三方厂商提供的类库驱动则是由系统类加载器加载的，由于 JVM 类加载器的双亲委托机制，比如 Connections、Statement、RowSet 等皆由根加载器加载，第三方的 JDBC 驱动包中的实现不会被加载，那么又是如何解决这个问题的呢？在 11.2 节中我们将通过对 Mysql 数据库的源码分析来一探究竟。

11.2　数据库驱动的初始化源码分析

在编写所有的 JDBC 程序时，首先都需要调用 Class.forName（"xxx.xxx.xxx.Driver"）对数据库驱动进行加载，打开 Mysql 驱动 Driver 源码，代码如清单 11-2 所示。

代码清单 11-2　Mysql 数据库驱动 Driver 源码

```
public class Driver extends NonRegisteringDriver
            implements java.sql.Driver {
    public Driver() throws SQLException {
    }

    static {
        try {
            // 在 Mysql 的静态方法中将 Driver 实例注册到 DriverManager 中
            DriverManager.registerDriver(new Driver());
        } catch (SQLException var1) {
            throw new RuntimeException("Can't register driver!");
```

            }
        }
    }
```

Driver 类的静态代码块主要是将 Mysql 的 Driver 实例注册给 DriverManager，因此直接使用 DriverManager.registerDriver（new com.mysql.jdbc.Driver()）其作用与 Class.forName（"xxx.xxx.xxx.Driver"）是完全等价的。

下面我们继续看 DriverManager 的源码，毕竟数据库的连接就是从它而来的，代码如下：

```
    private static Connection getConnection(
            String url, java.util.Properties info, Class<?> caller) throws SQLException {
    //注释①
        ClassLoader callerCL = caller != null ? caller.getClassLoader() : null;
        synchronized(DriverManager.class) {
            if (callerCL == null) {
                callerCL = Thread.currentThread().getContextClassLoader();
            }
        }
    //...
    //注释②
        for(DriverInfo aDriver : registeredDrivers) {
            if(isDriverAllowed(aDriver.driver, callerCL)) {
                try {
                    println("   trying " + aDriver.driver.getClass().getName());
                    Connection con = aDriver.driver.connect(url, info);
                    if (con != null) {
                        println("getConnection returning " + aDriver.driver.getClass().getName());
                        return (con);
    //...
    }

    //注释③
    private static boolean isDriverAllowed(Driver driver, ClassLoader classLoader) {
        boolean result = false;
        if(driver != null) {
            Class<?> aClass = null;
            try {
                aClass = Class.forName(driver.getClass().getName(), true, classLoader);
            } catch (Exception ex) {
                result = false;
            }

            result = (aClass == driver.getClass()) ? true : false;
```
```

```
        }

        return result;
    }
```

上面的代码片段我只截取了部分，主要是用于说明与线程上下文类加载器有关的内容。在注释①处获取当前线程的上下文类加载器，该类就是调用 Class.forName（"X"）所在线程的线程上下文类加载器，通常是系统类加载器。

注释②中通过递归 DriverManager 中已经注册的驱动类，然后验证该数据库驱动是否可以被指定的类加载器加载（线程上下文类加载器），如果验证通过则返回 Connection，此刻返回的 Connection 则是数据库厂商提供的实例。

注释③中的关键地方就在于 Class.forName（driver.getClass().getName()，true，classLoader）；其使用线程上下文类加载器进行数据库驱动的加载以及初始化。

下面就来回顾数据库驱动加载的整个过程，由于 JDK 定义了 SPI 的标准接口，加之这些接口被作为 JDK 核心标准类库的一部分，既想完全透明标准接口的实现，又想与 JDK 核心库进行捆绑，由于 JVM 类加载器双亲委托机制的限制，启动类加载器不可能加载得到第三方厂商提供的具体实现。为了解决这个困境，JDK 只好提供一种不太优雅的设计——线程上下文类加载器，有了线程上文类加载器，启动类加载器（根加载器）反倒需要委托子类加载器去加载厂商提供的 SPI 具体实现。

父委托变成了子委托的方式，这也打破了双亲委托机制的模型，而且是由 JDK 官方亲自打破的，自此几乎所有涉及 SPI 加载的动作采用的都是这种方式，比如 JNDI、JDBC、JCE、JAXB 和 JBI，等等，当然这可能是由于早期 Java 开发者没有考虑那么周全的原因所导致的，在现在的开源社区中也经常会遇到标准的接口和第三方实现独立设计的情况，比如 slf4j 只是 log 的标准接口库，而 slf4j-log4j 则是其中的一个实现，在真实项目中，两者皆由同一个类加载器进行加载，就不至于像数据库驱动的加载一样，绕了一大圈。

11.3 本章总结

本章通过分析 Mysql 驱动加载过程的源码，帮助读者清晰地理解线程上下文类加载器所发挥的作用了。

在 Thread 类中增加 getContextClassLoader() 和 setContextClassLoader（ClassLoader cl）方法实属无奈之举，它不仅破坏了类加载器的父委托机制，而且还反其道而行之，允许"子委托机制"，关于线程上下文类加载器方法的设计在各大论坛的争论还是比较大的，有人甚至认为它是 Java 设计中存在的一个缺陷。

PART3 · 第三部分

深入理解 volatile 关键字

自 Java 1.5 版本起，volatile 关键字所扮演的角色越来越重要，该关键字也成为并发包的基础，所有的原子数据类型都以此作为修饰，相比 synchronized 关键字，volatile 被称为"轻量级锁"，能实现部分 synchronized 关键字的语义。

第 12 章 "volatile 关键字介绍" 本章先通过一个简单的例子引入 volatile 关键字的使用场景，然后详细介绍 Java 的内存模型以及 CPU Cache 模型等，很好地掌握 JMM 以及 CPU Cache 等知识，对理解 volatile 关键字会有很大的帮助。

第 13 章 "深入 volatile 关键字" 本章先从并发编程所要注意的三个主要特性：原子性、可见性、有序性入手，然后分析了在 Java 内存模型中应通过何种方式来保证这三个特性，最后深入分析了 volatile 关键字的语义、原理和使用场景。

第 14 章 "7 种单例设计模式的设计" 将本章内容安排到本书并且放到第三部分确实有些牵强，但是其中有一种单例的设计方法能够很好地说明 volatile 的使用场景，因此将其安排到本部分中，另外单例模式也许并不像你想象得那样简单，其中蕴含了很多对于类加载、高并发等细节的思考。

第 12 章 · CHAPTER12

volatile 关键字的介绍

正如我们前面所说的，volatile 是一个非常重要的关键字，虽然看起来很简单，但是想要彻底弄清楚 volatile 的来龙去脉还是需要具备 Java 内存模型、CPU 缓存模型等知识的，在本章中，我们将首先介绍一个最能说明 volatile 特征的例子，然后对 Java 内存模型、CPU 缓存模型等知识进行展开讲解，这样对理解 volatile 关键字是非常有帮助的。

12.1 初识 volatile 关键字

清单 12-1 所示的这段程序分别启动了两个线程，一个线程负责对变量进行修改，一个线程负责对变量进行输出，根据本书第一部分的知识讲解，该变量就是共享资源（数据），那么在多线程操作的情况下，很有可能会引起数据不一致等线程安全的问题。

代码清单 12-1　VolatileFoo.java

```java
package com.wangwenjun.concurrent.chapter12;

import java.util.concurrent.TimeUnit;

public class VolatileFoo
{
    //init_value 的最大值
    final static int MAX = 5;
    //init_value 的初始值
    static int init_value = 0;

    public static void main(String[] args)
    {
        // 启动一个 Reader 线程，当发现 local_value 和 init_value 不同时，则输出 init_value 被修改的信息
```

```java
        new Thread(() ->
        {
            int localValue = init_value;
            while (localValue < MAX)
            {
                if (init_value != localValue)
                {
                    System.out.printf("The init_value is updated to [%d]\n", init_value);
                    // 对 localValue 进行重新赋值
                    localValue = init_value;
                }
            }
        }, "Reader").start();

        // 启动 Updater 线程，主要用于对 init_value 的修改，当 local_value>=5 的时候则退出生命周期
        new Thread(() ->
        {
            int localValue = init_value;
            while (localValue < MAX)
            {
                // 修改 init_value
                System.out.printf("The init_value will be changed to [%d]\n", ++localValue);
                init_value = localValue;
                try
                {
                    // 短暂休眠，目的是为了使 Reader 线程能够来得及输出变化内容
                    TimeUnit.SECONDS.sleep(2);
                } catch (InterruptedException e)
                {
                    e.printStackTrace();
                }
            }
        }, "Updater").start();
    }
}
```

在运行上面的程序之前，想象一下程序的输出将会是怎样的呢？Updater 线程的每一次修改都会使得 Reader 线程进行一次输出？如果你是这样认为的，那么事实会让你大跌眼镜，输出如下：

```
The init_value will be changed to [1]
The init_value will be changed to [2]
The init_value will be changed to [3]
The init_value will be changed to [4]
The init_value will be changed to [5]
```

通过控制台的输出信息我们不难发现，Reader 线程压根就没有感知到 init_value 的变化

而进入了死循环，这是为什么呢？我们将 init_value 的定义做一次小小的调整，代码如下：

```
static volatile int init_value = 0;
```

这里为 init_value 变量增加 volatile 关键字的修饰。再次运行修改后的程序，你会发现 Reader 线程会感知到 init_value 变量的变化，并且在条件不满足时退出运行，输出如下：

```
The init_value will be changed to [1]
The init_value is updated to [1]
The init_value will be changed to [2]
The init_value is updated to [2]
The init_value will be changed to [3]
The init_value is updated to [3]
The init_value will be changed to [4]
The init_value is updated to [4]
The init_value will be changed to [5]
The init_value is updated to [5]
```

为什么会出现这样的情况呢，其实这一切都是 volatile 关键字所起的作用，在本章接下来的内容以及第 13 章 "深入 volatile 关键字" 中都会介绍和 volatile 有关的内容，下面就请带着疑问阅读接下来的内容吧。

> **注意** volatile 关键字只能修饰类变量和实例变量，对于方法参数、局部变量以及实例常量、类常量都不能进行修饰，比如上面代码中的 MAX 就不能使用 volatile 关键字进行修饰。

12.2 机器硬件 CPU

在计算机中，所有的运算操作都是由 CPU 的寄存器来完成的，CPU 指令的执行过程需要涉及数据的读取和写入操作，CPU 所能访问的所有数据只能是计算机的主存（通常是指RAM），虽然 CPU 的发展频率不断地得到提升，但受制于制造工艺以及成本等的限制，计算机的内存反倒在访问速度上并没有多大的突破，因此 CPU 的处理速度和内存的访问速度之间的差距越拉越大，通常这种差距可以达到上千倍，极端情况下甚至会在上万倍以上。

12.2.1 CPU Cache 模型

由于两边速度严重的不对等，通过传统 FSB 直连内存的访问方式很明显会导致 CPU 资源受到大量的限制，降低 CPU 整体的吞吐量，于是就有了在 CPU 和主内存之间增加缓存的设计，现在缓存的数量都可以增加到 3 级了，最靠近 CPU 的缓存称为 L1，然后依次是 L2、L3 和主内存，CPU 缓存模型如图 12-1 所示。

由于程序指令和程序数据的行为和热点分布差异很大，因此 L1 Cache 又被划分成了 L1i（i 是 instruction 的首字母）和 L1d（d 是 data 的首字母）这两种有各自专门用途的缓存，

CPU Cache 又是由很多个 Cache Line 构成的，Cache Line 可以认为是 CPU Cache 中的最小缓存单位，目前主流 CPU Cache 的 Cache Line 大小都是 64 字节，图 12-2 是一张主存以及各级缓存之间的响应时间对比图。

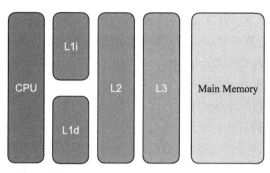

图 12-1　CPU Cache 模型

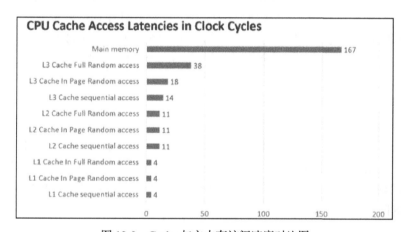

图 12-2　Cache 与主内存访问速度对比图

通过图 12-2，我们可以发现主内存的读写速度远远落后于 CPU Cache 的速度，更别说 CPU 本身的计算速度了。

Cache 的出现是为了解决 CPU 直接访问内存效率低下问题的，程序在运行的过程中，会将运算所需要的数据从主存复制一份到 CPU Cache 中，这样 CPU 进行计算时就可以直接对 CPU Cache 中的数据进行读取和写入，当运算结束之后，再将 CPU Cache 中的最新数据刷新到主内存当中，CPU 通过直接访问 Cache 的方式替代直接访问主存的方式极大地提高了 CPU 的吞吐能力，有了 CPU Cache 之后，整体的 CPU 和主内存之间交互的架构大致如图 12-3 所示。

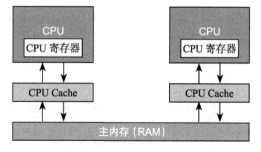

图 12-3　CPU 通过 Cache 与主内存进行交互

12.2.2 CPU 缓存一致性问题

由于缓存的出现，极大地提高了 CPU 的吞吐能力，但是同时也引入了缓存不一致的问题，比如 i++ 这个操作，在程序的运行过程中，首先需要将主内存中的数据复制一份存放到 CPU Cache 中，那么 CPU 寄存器在进行数值计算的时候就直接到 Cache 中读取和写入，当整个过程运算结束之后再将 Cache 中的数据刷新到主存当中，具体过程如下。

1）读取主内存的 i 到 CPU Cache 中。
2）对 i 进行加一操作。
3）将结果写回到 CPU Cache 中。
4）将数据刷新到主内存中。

i++ 在单线程的情况下不会出现任何问题，但是在多线程的情况下就会有问题，每个线程都有自己的工作内存（本地内存，对应于 CPU 中的 Cache），变量 i 会在多个线程的本地内存中都存在一个副本。如果同时有两个线程执行 i++ 操作，假设 i 的初始值为 0，每一个线程都从主内存中获取 i 的值存入 CPU Cache 中，然后经过计算再写入主内存中，很有可能 i 在经过了两次自增之后结果还是 1，这就是典型的缓存不一致性问题。

为了解决缓存不一致性问题，通常主流的解决方法有如下两种。

❑ 通过总线加锁的方式。
❑ 通过缓存一致性协议。

第一种方式常见于早期的 CPU 当中，而且是一种悲观的实现方式，CPU 和其他组件的通信都是通过总线（数据总线、控制总线、地址总线）来进行的，如果采用总线加锁的方式，则会阻塞其他 CPU 对其他组件的访问，从而使得只有一个 CPU（抢到总线锁）能够访问这个变量的内存。这种方式效率低下，所以就有了第二种通过缓存一致性协议的方式来解决不一致的问题（见图 12-4）。

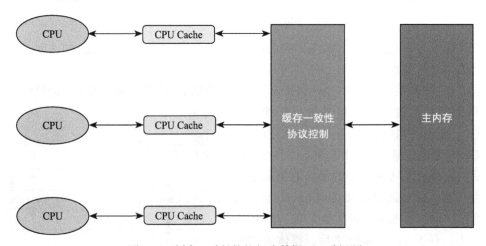

图 12-4 缓存一致性协议解决数据不一致问题

在缓存一致性协议中最为出名的是 Intel 的 MESI 协议，MESI 协议保证了每一个缓存中使用的共享变量副本都是一致的，它的大致思想是，当 CPU 在操作 Cache 中的数据时，如果发现该变量是一个共享变量，也就是说在其他的 CPU Cache 中也存在一个副本，那么进行如下操作：

1）读取操作，不做任何处理，只是将 Cache 中的数据读取到寄存器。

2）写入操作，发出信号通知其他 CPU 将该变量的 Cache line 置为无效状态，其他 CPU 在进行该变量读取的时候不得不到主内存中再次获取。

12.3　Java 内存模型

Java 的内存模型（Java Memory Mode，JMM）指定了 Java 虚拟机如何与计算机的主存（RAM）进行工作，如图 12-5 所示，理解 Java 内存模型对于编写行为正确的并发程序是非常重要的。在 JDK1.5 以前的版本中，Java 内存模型存在着一定的缺陷，在 JDK1.5 的时候，JDK 官方对 Java 内存模型重新进行了修订，JDK1.8 及最新的 JDK 版本都沿用了 JDK1.5 修订的内存模型。

Java 的内存模型决定了一个线程对共享变量的写入何时对其他线程可见，Java 内存模型定义了线程和主内存之间的抽象关系，具体如下。

❑ 共享变量存储于主内存之中，每个线程都可以访问。
❑ 每个线程都有私有的工作内存或者称为本地内存。
❑ 工作内存只存储该线程对共享变量的副本。
❑ 线程不能直接操作主内存，只有先操作了工作内存之后才能写入主内存。
❑ 工作内存和 Java 内存模型一样也是一个抽象的概念，它其实并不存在，它涵盖了缓存、寄存器、编译器优化以及硬件等。

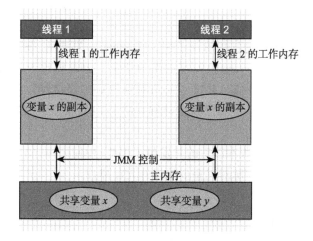

图 12-5　Java 内存模型（JMM）

假设主内存的共享变量为 0，线程 1 和线程 2 分别拥有共享变量 X 的副本，假设线程 1 此时将工作内存中的 x 修改为 1，同时刷新到主内存中，当线程 2 想要去使用副本 x 的时候，就会发现该变量已经失效了，必须到主内存中再次获取然后存入自己的工作内容中，这一点和 CPU 与 CPU Cache 之间的关系非常类似。

Java 的内存模型是一个抽象的概念，其与计算机硬件的结构并不完全一样，比如计算机物理内存不会存在栈内存和堆内存的划分，无论是堆内存还是虚拟机栈内存都会对应到物理的主内存，当然也有一部分堆栈内存的数据有可能会存入 CPU Cache 寄存器中。图 12-6 所示的是 Jave 内存模型与 CPU 硬件架构的交互图。

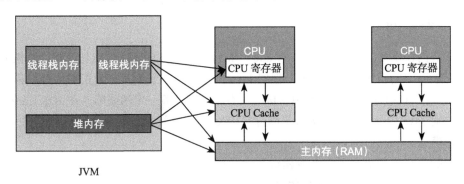

图 12-6　Java 内存模型与 CPU 硬件架构交互图

当同一个数据被分别存储到了计算机的各个内存区域时，势必会导致多个线程在各自的工作区域中看到的数据有可能是不一样的，在 Java 语言中如何保证不同线程对某个共享变量的可见性？以及又该如何解释 12.1 节中增加了 volatile 修饰之后不一样的运行效果？第 13 章将会为读者详细讲解。

12.4　本章总结

本章通过一个非常具有代表性的例子引入了 volatile 关键字，通过该示例不难发现，volatile 关键字具备了 synchronized 关键字的部分语义，但是相比于 synchronized 关键字，要理解 volatile 关键字会困难很多，需要了解机器硬件 CPU 的架构以及 Java 的内存模型等其他知识，因此我们在 12.2 节和 12.3 节中分别介绍了 CPU 模型、CPU 缓存一致性的解决方案以及 Java 的内存模型。

volatile 关键字在 JDK1.5 版本以后的并发包（JUC）中使用得非常广泛，因此彻底理解和掌握 volatile 关键字的来龙去脉对进一步的提升大有裨益。

CHAPTER13 · 第 13 章

深入 volatile 关键字

在第 12 章 "volatile 关键字的介绍"中，我们写了一个关于 volatile 的简单实例，然后介绍了 Java 内存模型以及 CPU 缓存相关的内容，这些内容对深入理解 volatile 关键字乃至并发编程都有很大的帮助，希望读者能够掌握。

13.1 并发编程的三个重要特性

并发编程有三个至关重要的特性，分别是原子性、有序性和可见性，理解这三个特性对于开发正确的高并发程序会有很大的帮助，在本节中，我们将逐一对其进行介绍。

13.1.1 原子性

所谓原子性是指在一次的操作或者多次操作中，要么所有的操作全部都得到了执行并且不会受到任何因素的干扰而中断，要么所有的操作都不执行。说起原子性一般都会用银行转账来进行举例说明，比如从 Alex 的账号往 Tina 的账号转入 1000 元，这个动作将包含两个最基本的操作：从 Alex 的账号上扣除 1000 元；给 Tina 的账号增加 1000 元。这两个操作必须符合原子性的要求，要么都成功要么都失败，总之不能出现 Alex 的账号扣除了 1000元，但是 Tina 的账号并未增加 1000 元或者 Alex 账号未扣除 1000 元，Tina 的账号反倒增加了 1000 元的情况。

同样在我们编写代码的过程中，比如一个简单的赋值语句：

```
Object o = new Object();
```

引用类型 o 占用四个字节（32 位），假设这样的赋值语句不能够保证原子性的话，那么会导致赋值出现错误的数据。

> **注意**：两个原子性的操作结合在一起未必还是原子性的，比如 i++（其中 get i，i+1 和 set i = x 三者皆是原子性操作，但是不代表 i++ 就是原子性操作）。

volatile 关键字不保证数据的原子性，synchronized 关键字保证，自 JDK1.5 版本起，其提供的原子类型变量也可以保证原子性。

13.1.2 可见性

可见性是指，当一个线程对共享变量进行了修改，那么另外的线程可以立即看到修改后的最新值，来回顾一下我们在 12.1 节中写的 Reader 线程始终看不到 init_value 的例子：

```
new Thread(() ->
    {
        int localValue = init_value;
        while (localValue < MAX)
        {
            if (init_value != localValue)
            {
                System.out.printf("The init_value is updated to [%d]\n", init_value);
                // 重新对 localValue 进行赋值
                localValue = init_value;
            }
        }
    }, "Reader").start();
}
```

根据 12.2 节的分析，Reader 线程会将 init_value 从主内存缓存到 CPU Cache 中，也就是从主内存缓存到线程的本地内存中，Updater 线程对 init_value 的修改对 Reader 线程是不可见的。

13.1.3 有序性

所谓有序性是指程序代码在执行过程中的先后顺序，由于 Java 在编译器以及运行期的优化，导致了代码的执行顺序未必就是开发者编写代码时的顺序，比如：

```
int x = 10;
int y =0;
x++;
y= 20;
```

上面这段代码定义了两个 int 类型的变量 x 和 y，对 x 进行自增操作，对 y 进行赋值操作，从编写程序的角度来看上面的代码肯定是顺序执行下来的，但是在 JVM 真正地运行这段代码的时候未必会是这样的顺序，比如 y=20 语句有可能会在 x++ 语句的前面得到执行，这种情况就是我们通常所说的指令重排序（Instruction Recorder）。

一般来说，处理器为了提高程序的运行效率，可能会对输入的代码指令做一定的优化，它不会百分之百的保证代码的执行顺序严格按照编写代码中的顺序来进行，但是它会保证程序的最终运算结果是编码时所期望的那样，比如上文中的 x++ 与 y=20 不管它们的执行顺序如何，执行完上面的四行代码之后得到的结果肯定都是 x=11，y=20。

当然对指令的重排序要严格遵守指令之间的数据依赖关系，并不是可以任意进行重排序的，比如下面的代码片段：

```
int x = 10;
int y = 0;
x++;
y=x+1;
```

对于这段代码有可能它的执行顺序就是代码本身的顺序，有可能发生了重排序导致 int y=0 优先于 int x=10 执行，但是绝对不可能出现 y=x+1 优先于 x++ 执行的执行情况，如果一个指令 x 在执行的过程中需要用到指令 y 的执行结果，那么处理器会保证指令 y 在指令 x 之前执行，这就好比 y=x+1 执行之前肯定要先执行 x++ 一样。

在单线程情况下，无论怎样的重排序最终都会保证程序的执行结果和代码顺序执行的结果是完全一致的，但是在多线程的情况下，如果有序性得不到保证，那么很有可能就会出现非常大的问题，比如下面的代码片段：

```
private boolean initialized = false;
private Context context;
public Context load(){
    if(!initialized){
        context=loadContext();
        initialized = true;
    }
    return context;
}
```

上述这段代码使用 boolean 变量 initialized 来控制 context 是否已经被加载过了，在单线程下无论怎样的重排序，最终返回给使用者的 context 都是可用的。如果在多线程的情况下发生了重排序，比如 context=loadContext() 的执行被重排序到了 initialized = true 的后面，那么这将是灾难性的了。比如第一个线程首先判断到 initialized=false，因此准备执行 context 的加载，但是它在执行 loadContext() 方法之前二话不说先将 initialized 置为 true 然后再执行 loadContext() 方法，那么如果另外一个线程也执行 load 方法，发现此时 initialized 已经为 true 了，则直接返回一个还未被加载成功的 context，那么在程序的运行过程中势必会出现错误。

13.2　JMM 如何保证三大特性

我们在 13.1 节中比较详细地介绍了高并发编程需要保证的三个主要特性，这对并发程序正确的运行有着至关重要的作用。在本节中，我们将结合 Java 的内存模型来看看在 Java 的世界中，是通过何种方式来保证原子性、可见性和顺序性这三大特性的。

JVM 采用内存模型的机制来屏蔽各个平台和操作系统之间内存访问的差异，以实现让 Java 程序在各种平台下达到一致的内存访问效果，比如 C 语言中的整型变量，在某些平台下占用了两个字节的内存，在某些平台下则占用了四个字节的内存，Java 则在任何平台下，Int 类型就是四个字节，这就是所谓的一致内存访问效果。

Java 的内存模型规定了所有的变量都是存在于主内存（RAM）当中的，而每个线程都有自己的工作内存或者本地内存（这一点很像 CPU 的 Cache），线程对变量的所有操作都必须在自己的工作内存中进行，而不能直接对主内存进行操作，并且每一个线程都不能访问其他线程的工作内存或者本地内存。

比如在某个线程中对变量 i 的赋值操作 i=1，该线程必须在本地内存中对 i 进行修改之后才能将其写入主内存之中。

13.2.1 JMM 与原子性

在 Java 语言中，对基本数据类型的变量读取赋值操作都是原子性的，对引用类型的变量读取和赋值的操作也是原子性的，因此诸如此类的操作是不可被中断的，要么执行，要么不执行，正所谓一荣俱荣一损俱损。

不过话虽如此简单，但是理解起来未必不会出错，下面我们就来看几个例子。

（1）x=10；赋值操作

x=10 的操作是原子性的，执行线程首先会将 x=10 写入工作内存中，然后再将其写入主内存（有可能在往主内存进行数值刷新的过程中其他线程也在对其进行刷新操作，比如另外一个线程将其写为 11，但是最终的结果肯定要么是 10，要么是 11，不可能出现其他情况，单就赋值语句这一点而言其是原子性的）。

（2）y=x；赋值操作

这条操作语句是非原子性的，因为它包含如下两个重要的步骤。

1）执行线程从主内存中读取 x 的值（如果 x 已经存在于执行线程的工作内存中，则直接获取）然后将其存入当前线程的工作内存之中。

2）在执行线程的工作内存中修改 y 的值为 x，然后将 y 的值写入主内存之中。

虽然第一步和第二步都是原子类型的操作，但是合在一起就不是原子操作了。

（3）y++；自增操作

这条操作语句是非原子性的，因为它包含三个重要的步骤，具体如下。

1）执行线程从主内存中读取 y 的值（如果 y 已经存在于执行线程的工作内存中，则直接获取），然后将其存入当前线程的工作内存之中。

2）在执行线程工作内存中为 y 执行加 1 操作。

3）将 y 的值写入主内存。

（4）z = z + 1；加一操作（与自增操作等价）

这条操作语句是非原子性的，因为它包含三个重要的步骤，具体如下。

1）执行线程从主内存中读取 z 的值（如果 z 已经存在于执行线程的工作内存中，则直接获取），然后将其存入当前线程的工作内存之中。

2）在执行线程工作内存中为 z 执行加 1 操作。

3）将 z 的值写入主内存。

综合上面的四个例子，我们可以发现只有第一种操作即赋值操作具备原子性，其余的均不具备原子性，由此我们可以得出以下几个结论。

- 多个原子性的操作在一起就不再是原子性操作了。
- 简单的读取与赋值操作是原子性的，将一个变量赋给另外一个变量的操作不是原子性的。
- Java 内存模型（JMM）只保证了基本读取和赋值的原子性操作，其他的均不保证，如果想要使得某些代码片段具备原子性，需要使用关键字 synchronized，或者 JUC 中的 lock。如果想要使得 int 等类型自增操作具备原子性，可以使用 JUC 包下的原子封装类型 java.util.concurrent.atomic.*

总结：**volatile 关键字不具备保证原子性的语义。**

13.2.2　JMM 与可见性

在多线程的环境下，如果某个线程首次读取共享变量，则首先到主内存中获取该变量，然后存入工作内存中，以后只需要在工作内存中读取该变量即可。同样如果对该变量执行了修改的操作，则先将新值写入工作内存中，然后再刷新至主内存中。但是什么时候最新的值会被刷新至主内存中是不太确定的，这也就解释了为什么 VolatileFoo 中的 Reader 线程始终无法获取到 init_value 最新的变化。

Java 提供了以下三种方式来保证可见性。

- 使用关键字 volatile，当一个变量被 volatile 关键字修饰时，对于共享资源的读操作会直接在主内存中进行（当然也会缓存到工作内存中，当其他线程对该共享资源进行了修改，则会导致当前线程在工作内存中的共享资源失效，所以必须从主内存中再次获取），对于共享资源的写操作当然是先要修改工作内存，但是修改结束后会立刻将其刷新到主内存中。
- 通过 synchronized 关键字能够保证可见性，synchronized 关键字能够保证同一时刻只有一个线程获得锁，然后执行同步方法，并且还会确保在锁释放之前，会将对变量的修改刷新到主内存当中。
- 通过 JUC 提供的显式锁 Lock 也能够保证可见性，Lock 的 lock 方法能够保证在同一时刻只有一个线程获得锁然后执行同步方法，并且会确保在锁释放（Lock 的 unlock 方法）之前会将对变量的修改刷新到主内存当中。

总结：volatile 关键字具有保证可见性的语义。

13.2.3 JMM 与有序性

在 Java 的内存模型中，允许编译器和处理器对指令进行重排序，在单线程的情况下，重排序并不会引起什么问题，但是在多线程的情况下，重排序会影响到程序的正确运行，Java 提供了三种保证有序性的方式，具体如下。

- 使用 volatile 关键字来保证有序性。
- 使用 synchronized 关键字来保证有序性。
- 使用显式锁 Lock 来保证有序性。

后两者采用了同步的机制，同步代码在执行的时候与在单线程情况下一样自然能够保证顺序性（最终结果的顺序性），对于 volatile 关键字我们在 13.3 节中还会进行更加详细的介绍。

此外，Java 的内存模型具备一些天生的有序性规则，不需要任何同步手段就能够保证有序性，这个规则被称为 Happens-before 原则。如果两个操作的执行次序无法从 happens-before 原则推导出来，那么它们就无法保证有序性，也就是说虚拟机或者处理器可以随意对它们进行重排序处理。

下面我们来具体看看都有哪些 happens-before 原则。

- 程序次序规则：在一个线程内，代码按照编写时的次序执行，编写在后面的操作发生于编写在前面的操作之后。

这句话的意思看起来是程序按照编写的顺序来执行，但是虚拟机还是可能会对程序代码的指令进行重排序，只要确保在一个线程内最终的结果和代码顺序执行的结果一致即可。

- 锁定规则：一个 unlock 操作要先行发生于对同一个锁的 lock 操作。

这句话的意思是，无论是在单线程还是在多线程的环境下，如果同一个锁是锁定状态，那么必须先对其执行释放操作之后才能继续进行 lock 操作。

- volatile 变量规则：对一个变量的写操作要早于对这个变量之后的读操作。

根据字面的意思来理解是，如果一个变量使用 volatile 关键字修饰，一个线程对它进行读操作，一个线程对它进行写操作，那么写入操作肯定要先行发生于读操作，关于这个规则我们在 13.3 节中还会继续介绍。

- 传递规则：如果操作 A 先于操作 B，而操作 B 又先于操作 C，则可以得出操作 A 肯定要先于操作 C，这一点说明了 happens-before 原则具备传递性。
- 线程启动规则：Thread 对象的 start() 方法先行发生于对该线程的任何动作，这也是我们在第一部分中讲过的，只有 start 之后线程才能真正运行，否则 Thread 也只是一个对象而已。
- 线程中断规则：对线程执行 interrupt() 方法肯定要优先于捕获到中断信号，这句话的意思是指如果线程收到了中断信号，那么在此之前势必要有 interrupt()。

- 线程的终结规则：线程中所有的操作都要先行发生于线程的终止检测，通俗地讲，线程的任务执行、逻辑单元执行肯定要发生于线程死亡之前。
- 对象的终结规则：一个对象初始化的完成先行发生于 finalize() 方法之前，这个更没什么好说的了，先有生后有死。

总结：volatile 关键字具有保证顺序性的语义。

13.3　volatile 关键字深入解析

将近两个章节的内容介绍，其实都是为 volatile 关键字做铺垫，没有完全孤立存在的知识，只有通过某个点将一些相关性的知识串联起来才能够形成一个知识体系，以助于理解和记忆，这也是为什么我们长篇大论了一番才切入正题讲解 volatile 关键字的原因。

13.3.1　volatile 关键字的语义

被 volatile 修饰的实例变量或者类变量具备如下两层语义。
- 保证了不同线程之间对共享变量操作时的可见性，也就是说当一个线程修改 volatile 修饰的变量，另外一个线程会立即看到最新的值。
- 禁止对指令进行重排序操作。

（1）理解 volatile 保证可见性

关于共享变量在多线程间的可见性，在 VolatileFoo 例子中已经体现得非常透彻了，Updater 线程对 init_value 变量的每一次更改都会使得 Reader 线程能够看到（在 happens-before 规则中，第三条 volatile 变量规则：对一个变量的写操作要早于对这个变量之后的读操作），其步骤具体如下。

1）Reader 线程从主内存中获取 init_value 的值为 0，并且将其缓存到本地工作内存中。

2）Updater 线程将 init_value 的值在本地工作内存中修改为 1，然后立即刷新至主内存中。

3）Reader 线程在本地工作内存中的 init_value 失效（反映到硬件上就是 CPU 的 L1 或者 L2 的 Cache Line 失效）。

4）由于 Reader 线程工作内存中的 init_value 失效，因此需要到主内存中重新读取 init_value 的值。

经过上面几个步骤的分析，相信读者对 volatile 关键字保证可见性有了一个更加清晰的认识了。

（2）理解 volatile 保证顺序性

volatile 关键字对顺序性的保证就比较霸道一点，直接禁止 JVM 和处理器对 volatile 关键字修饰的指令重排序，但是对于 volatile 前后无依赖关系的指令则可以随便怎么排序，比如

```
int x = 0;
int y = 1;
volatile int z = 20;
x++;
y--;
```

在语句 volatile int z = 20 之前,先执行 x 的定义还是先执行 y 的定义,我们并不关心,只要能够百分之百地保证在执行到 z=20 的时候 x=0、y=1,同理关于 x 的自增以及 y 的自减操作都必须在 z=20 以后才能发生。

再回到关于 load context 的那个例子,如果对 initialize 布尔变量增加了 volatile 的修饰,那就意味着 initialize =true 的时候一定是执行且完成了对 loadContext() 的方法调用,代码如下:

```
private volatile boolean initialized = false;
private Context context;
public Context load(){
    if(!initialized){
        context=loadContext();
        initialized = true;// 阻止重排序
    }
    return context;
}
```

(3) 理解 volatile 不保证原子性

我们已经不止一次地说过了,volatile 关键字不保证操作的原子性,在这里我们结合 JMM 的知识分析一下其中的原因,比如下面的这段代码:

```
package com.wangwenjun.concurrent.chapter13;

import com.wangwenjun.concurrent.chapter23.CountDownLatch;
import com.wangwenjun.concurrent.chapter23.Latch;

public class VolatileTest
{
    // 使用 volatile 修饰共享资源 i
    private static volatile int i = 0;
    // 使用在第 23 章中开发的 CountDownLatch
    private static final Latch latch = new CountDownLatch(10);

    private static void inc()
    {
        i++;
    }

    public static void main(String[] args) throws InterruptedException
    {
        for (int i = 0; i < 10; i++)
        {
```

```
            new Thread(() ->
            {
                for (int x = 0; x < 1000; x++)
                {
                    inc();
                }
                // 使计数器减1
                latch.countDown();
            }).start();
        }
        // 等待所有的线程完成工作
        latch.await();
        System.out.println(i);
    }
}
```

上面这段代码创建了 10 个线程，每个线程执行 1000 次对共享变量 i 的自增操作，但是最终的结果 i 肯定不是 10000，而且每次运行的结果也是各不相同，下面来分析一下其中的原因。

i++ 的操作其实是由三步组成的，具体如下。

1）从主内存中获取 i 的值，然后缓存至线程工作内存中。

2）在线程工作内存中为 i 进行加 1 的操作。

3）将 i 的最新值写入主内存中。

上面三个操作单独的每一个操作都是原子性操作，但是合起来就不是，因为在执行的中途很有可能会被其他线程打断，例如如下操作情况。

1）假设此时 i 的值为 100，线程 A 要对变量 i 执行自增操作，首先它需要到主内存中读取 i 的值，可是此时由于 CPU 时间片调度的关系，执行权切换到了线程 B，A 线程进入了 RUNNABLE 状态而不是 RUNNING 状态。

2）线程 B 同样需要从主内存中读取 i 的值，由于线程 A 没有对 i 做过任何修改操作，因此此时 B 获取到的 i 仍然是 100。

3）线程 B 工作内存中为 i 执行了加 1 操作，但是未刷新至主内存中。

4）CPU 时间片的调度又将执行权给了线程 A，A 线程直接对工作线程中的 100 进行加 1 运算（因为 A 线程已经从主内存中读取了 i 的值），由于 B 线程并未写入 i 的最新值，因此 A 线程工作空间中的 100 不会被失效。

5）线程 A 将 i=101 写入主内存之中。

6）线程 B 将 i=101 写入到主内存中。

这样，两次运算实际上只对 i 进行了一次数值的修改变化。

13.3.2 volatile 的原理和实现机制

经过前面内容的学习，我们知道了 volatile 关键字可以保证可见性以及顺序性，那么它到底是如何做到的呢？通过对 OpenJDK 下 unsafe.cpp 源码的阅读，会发现被 volatile 修饰的变量存在于一个"lock;"的前缀，源码如下：

```
// Adding a lock prefix to an instruction on MP machine
#define LOCK_IF_MP(mp) "cmp $0, " #mp "; je 1f; lock; 1: "
...
inline jint Atomic::cmpxchg (jint exchange_value, volatile jint* dest, jint compare_value) {
    int mp = os::is_MP();
    __asm__ volatile (LOCK_IF_MP(%4) "cmpxchgl %1,(%3)"
                  : "=a" (exchange_value)
                  : "r" (exchange_value), "a" (compare_value), "r" (dest), "r" (mp)
                  : "cc", "memory");
    return exchange_value;
}
```

"lock;"前缀实际上相当于是一个内存屏障，该内存屏障会为指令的执行提供如下几个保障。

- 确保指令重排序时不会将其后面的代码排到内存屏障之前。
- 确保指令重排序时不会将其前面的代码排到内存屏障之后。
- 确保在执行到内存屏障修饰的指令时前面的代码全部执行完成。
- 强制将线程工作内存中值的修改刷新至主内存中。
- 如果是写操作，则会导致其他线程工作内存（CPU Cache）中的缓存数据失效。

13.3.3 volatile 的使用场景

虽然 volatile 有部分 synchronized 关键字的语义，但是 volatile 不可能完全替代 synchronized 关键字，因为 volatile 关键字不具备原子性操作语义，我们在使用 volatile 关键字的时候也是充分利用它的可见性以及有序性（防止重排序）特点。

（1）开关控制利用可见性的特点

开关控制中最常见的就是进行线程的关闭操作，在 3.9.1 节中已经对其有过介绍，这里再简单回顾一下：

```
package com.wangwenjun.concurrent.chapter13;

public class ThreadCloseale extends Thread
{
    //volatile关键字保证了started线程的可见性
    private volatile boolean started = true;

    @Override
    public void run()
    {
        while (started)
```

```
        {
            //do work
        }
    }
    public void shutdown()
    {
        this.started = false;
    }
}
```

当外部线程执行 ThreadCloseable 的 shutdown 方法时，ThreadCloseable 会立刻看到 started 发生了变化（原因是因为 ThreadCloseable 工作内存中的 started 失效了，不得不到主内存中重新获取）。

如果 started 没有被 volatile 关键字修饰，那么很有可能外部线程在其工作内存中修改了 started 之后不及时刷新到主内存中，或者 ThreadCloseable 一直到自己的工作内存中读取 started 变量，都有可能导致 started=true 不生效，线程就会无法关闭。

（2）状态标记利用顺序性特点

关于使用状态标记说明顺序性的例子，我们之前用 context 进行过了举例，这里就不再赘述了。

（3）Singleton 设计模式的 double-check 也是利用了顺序性特点

原本不打算在本书中介绍 singleton 设计模式的，但是笔者发现它与类的主动加载，以及在高并发情况下如何获取唯一实例且保持高效等内容息息相关，尤其是使用 volatile 关键字对 double-check 的改进，因此将 singleton 设计模式加入到本书并且将其作为第三部分的一个章节来讲解。

13.3.4 volatile 和 synchronized

通过对 volatile 关键字的学习和之前对 synchronized 关键字的学习，我们在这里总结一下两者之间的区别。

（1）使用上的区别
- volatile 关键字只能用于修饰实例变量或者类变量，不能用于修饰方法以及方法参数和局部变量、常量等。
- synchronized 关键字不能用于对变量的修饰，只能用于修饰方法或者语句块。
- volatile 修饰的变量可以为 null，synchronized 关键字同步语句块的 monitor 对象不能为 null。

（2）对原子性的保证
- volatile 无法保证原子性。
- 由于 synchronized 是一种排他的机制，因此被 synchronized 关键字修饰的同步代码是无法被中途打断的，因此其能够保证代码的原子性。

（3）对可见性的保证
- 两者均可以保证共享资源在多线程间的可见性，但是实现机制完全不同。
- synchronized 借助于 JVM 指令 monitor enter 和 monitor exit 对通过排他的方式使得同步代码串行化，在 monitor exit 时所有共享资源都将会被刷新到主内存中。
- 相比较于 synchronized 关键字 volatile 使用机器指令（偏硬件）"lock；"的方式迫使其他线程工作内存中的数据失效，不得到主内存中进行再次加载。

（4）对有序性的保证
- volatile 关键字禁止 JVM 编译器以及处理器对其进行重排序，所以它能够保证有序性。
- 虽然 synchronized 关键字所修饰的同步方法也可以保证顺序性，但是这种顺序性是以程序的串行化执行换来的，在 synchronized 关键字所修饰的代码块中代码指令也会发生指令重排序的情况，比如：

```
synchronized(this){
    int x = 10;
    int y = 20;
    x++;
    y = y+1;
}
```

x 和 y 谁最先定义以及谁最先进行运算，对程序来说没有任何的影响，另外 x 和 y 之间也没有依赖关系，但是由于 synchronized 关键字同步的作用，在 synchronized 的作用域结束时 x 必定是 11，y 必定是 21，也就是说达到了最终的输出结果和代码编写顺序的一致性。

（5）其他
- volatile 不会使线程陷入阻塞。
- synchronized 关键字会使线程进入阻塞状态。

13.4 本章总结

本章首先非常详细地介绍了并发编程的三个重要特性：原子性、可见性和有序性，并且结合 Java 的内存模型 JMM 分别解释了 Java 是如何保证这三个重要特性的，在本章的最后深入解析了本部分的主人公"volatile 关键字"的原理和实现，以及 volatile 关键字的使用场景。

CHAPTER14 · 第 14 章

7 种单例设计模式的设计

单例设计模式是 GoF23 种设计模式中最常用的设计模式之一，无论是第三方类库，还是我们在日常的开发中，几乎都可以看到单例的影子，单例设计模式提供了一种在多线程情况下保证实例唯一性的解决方案，单例设计模式的实现虽然非常简单，但是实现方式却多种多样，在本章中笔者列举了 7 种单例设计模式的实现方法，为了比较这 7 种实现方式的优劣，我们从三个维度对其进行评估：线程安全、高性能、懒加载。

14.1 饿汉式

饿汉式单例设计模式示例代码如清单 14-1 所示。

代码清单 14-1　饿汉式的单例模式的设计

```java
package com.wangwenjun.concurrent.chapter14;

//final 不允许被继承
public final class Singleton
{
    // 实例变量
    private byte[] data = new byte[1024];

    // 在定义实例对象的时候直接初始化
    private static Singleton instance = new Singleton();

    // 私有构造函数，不允许外部 new
    private Singleton()
    {
    }

    public static Singleton getInstance()
```

```
        {
            return instance;
        }
    }
```

饿汉式的关键在于 instance 作为类变量并且直接得到了初始化,根据本书第二部分的知识我们可以得知,如果主动使用 Singleton 类,那么 instance 实例将会直接完成创建,包括其中的实例变量都会得到初始化,比如 1K 空间的 data 将会同时被创建。

instance 作为类变量在类初始化的过程中会被收集进 <clinit>() 方法中,该方法能够百分之百地保证同步,也就是说 instance 在多线程的情况下不可能被实例化两次,但是 instance 被 ClassLoader 加载后可能很长一段时间才被使用,那就意味着 instance 实例所开辟的堆内存会驻留更久的时间。

如果一个类中的成员属性比较少,且占用的内存资源不多,饿汉的方式也未尝不可,相反,如果一个类中的成员都是比较重的资源,那么这种方式就会有些不妥。

总结起来,饿汉式的单例设计模式可以保证多个线程下的唯一实例,getInstance 方法性能也比较高,但是无法进行懒加载。

14.2 懒汉式

所谓懒汉式就是在使用类实例的时候再去创建(用时创建),这样就可以避免类在初始化时提前创建,懒汉式示例代码如清单 14-2 所示。

代码清单 14-2　懒汉式单例模式的设计与实现

```
package com.wangwenjun.concurrent.chapter14;

//final 不允许被继承
public final class Singleton
{
    // 实例变量
    private byte[] data = new byte[1024];

    // 定义实例,但是不直接初始化
    private static Singleton instance = null;

    private Singleton()
    {
    }

    public static Singleton getInstance()
    {
        if (null == instance)
            instance = new Singleton();
        return instance;
    }
}
```

Singleton 的类变量 instance=null，因此当 Singleton.class 被初始化的时候 instance 并不会被实例化，在 getInstance 方法中会判断 instance 实例是否被实例化，看起来没有什么问题，但是将 getInstance 方法放在多线程环境下进行分析，则会导致 instance 被实例化一次以上，并不能保证单例的唯一性，如图 14-1 所示。

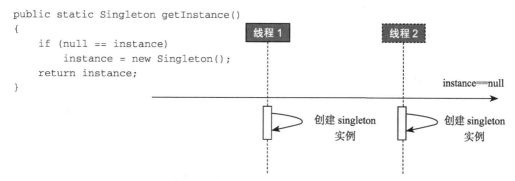

图 14-1　两个线程同时看到 instance 为 null

如图 14-1 所示，两个线程同时看到 instance==null，那么 instance 将无法保证单例的唯一性。

14.3　懒汉式 + 同步方法

懒汉式的方式可以保证实例的懒加载，但无法保证实例的唯一性，根据本书第 4 章的内容，在多线程的情况下，instance 又称为共享资源（数据），当多个线程对其访问使用时，需要保证数据的同步性，对 14.2 节中的程序稍加修改，增加同步的约束即可，修改后的代码如清单 14-3 所示。

代码清单 14-3　懒汉式 + 同步实现单例模式

```
package com.wangwenjun.concurrent.chapter14;
//final 不允许被继承
public final class Singleton
{
    // 实例变量
    private byte[] data = new byte[1024];

    private static Singleton instance = null;

    private Singleton()
    {
    }
    // 向 getInstance 方法加入同步控制，每次只能有一个线程能够进入
    public static synchronized Singleton getInstance()
    {
        if (null == instance)
            instance = new Singleton();
```

```
        return instance;
    }
}
```

采用懒汉式+数据同步的方式既满足了懒加载又能够百分之百地保证 instance 实例的唯一性,但是 synchronized 关键字天生的排他性导致了 getInstance 方法只能在同一时刻被一个线程所访问,性能低下。

14.4 Double-Check

Double-Check 是一种比较聪明的设计方式,他提供了一种高效的数据同步策略,那就是首次初始化时加锁,之后则允许多个线程同时进行 getInstance 方法的调用来获得类的实例,Double-Check 的示例代码如清单 14-4 所示。

代码清单 14-4 Double Check 实现单例模式

```
package com.wangwenjun.concurrent.chapter14;

import java.net.Socket;
import java.sql.Connection;

//final 不允许被继承

public final class Singleton
{
    // 实例变量
    private byte[] data = new byte[1024];

    private static Singleton instance = null;

    Connection conn;

    Socket socket;

    private Singleton()
    {
        this.conn   // 初始化 conn
        this.socket // 初始化 socket
    }

    public static Singleton getInstance()
    {
        // 当 instance 为 null 时,进入同步代码块,同时该判断避免了每次都需要进入同步代码块,可以提高效率
        if (null == instance)
        {
            // 只有一个线程能够获得 Singleton.class 关联的 monitor
            synchronized (Singleton.class)
            {
                // 判断如果 instance 为 null 则创建
```

```
            if (null == instance)
            {
                instance = new Singleton();
            }
        }
    }
    return instance;
}
```

当两个线程发现 null==instance 成立时，只有一个线程有资格进入同步代码块，完成对 instance 的实例化，随后的线程发现 null==instance 不成立则无须进行任何动作，以后对 getInstance 的访问就不需要数据同步的保护了。

这种方式看起来是那么的完美和巧妙，既满足了懒加载，又保证了 instance 实例的唯一性，Double-Check 的方式提供了高效的数据同步策略，可以允许多个线程同时对 getInstance 进行访问，但是这种方式在多线程的情况下有可能会引起空指针异常，下面我们来分析一下引发异常的原因。

在 Singleton 的构造函数中，需要分别实例化 conn 和 socket 两个资源，还有 Singleton 自身，根据 JVM 运行时指令重排序和 Happens-Before 规则，这三者之间的实例化顺序并无前后关系的约束，那么极有可能是 instance 最先被实例化，而 conn 和 socket 并未完成实例化，未完成初始化的实例调用其方法将会抛出空指针异常，下面我们结合图 14-2 来进行分析。

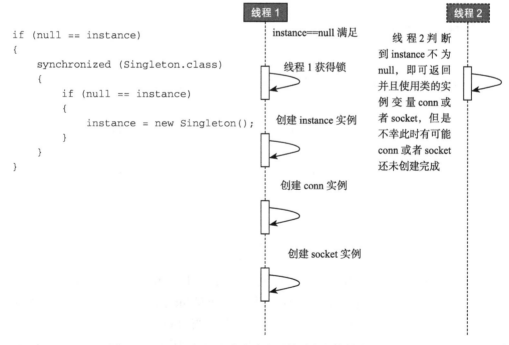

图 14-2　Double-Check 的方式有可能引起空指针异常

14.5 Volatile+Double-Check

Double-Check 虽然是一种巧妙的程序设计，但是有可能会引起类成员变量的实例化 conn 和 socket 发生在 instance 实例化之后，这一切均是由于 JVM 在运行时指令重排序所导致的，而 volatile 关键字则可以防止这种重排序的发生，因此代码稍作修改即可满足多线程下的单例、懒加载以及获取实例的高效性，代码修改如下：

```
private volatile static Singleton instance = null;
```

14.6 Holder 方式

Holder 的方式完全是借助了类加载的特点，下面我们对整个单例模式进行重构，然后结合类加载器的知识点分析这样做的好处在哪里，如代码清单 14-5 所示。

代码清单 14-5　采用 holder 的方式实现单例设计模式

```
package com.wangwenjun.concurrent.chapter14;

// 不允许被继承
public final class Singleton
{
    // 实例变量
    private byte[] data = new byte[1024];

    private Singleton()
    {
    }

    // 在静态内部类中持有 Singleton 的实例，并且可被直接初始化
    private static class Holder
    {
        private static Singleton instance = new Singleton();
    }

    // 调用 getInstance 方法，事实上是获得 Holder 的 instance 静态属性
    public static Singleton getInstance()
    {
        return Holder.instance;
    }
}
```

在 Singleton 类中并没有 instance 的静态成员，而是将其放到了静态内部类 Holder 之中，因此在 Singleton 类的初始化过程中并不会创建 Singleton 的实例，Holder 类中定义了 Singleton 的静态变量，并且直接进行了实例化，当 Holder 被主动引用的时候则会创建 Singleton 的实例，Singleton 实例的创建过程在 Java 程序编译时期收集至 <clinit>() 方法

中，该方法又是同步方法，同步方法可以保证内存的可见性、JVM 指令的顺序性和原子性、Holder 方式的单例设计是最好的设计之一，也是目前使用比较广的设计之一。

14.7　枚举方式

使用枚举的方式实现单例模式是《Effective Java》作者力推的方式，在很多优秀的开源代码中经常可以看到使用枚举方式实现单例模式的（身影），枚举类型不允许被继承，同样是线程安全的且只能被实例化一次，但是枚举类型不能够懒加载，对 Singleton 主动使用，比如调用其中的静态方法则 INSTANCE 会立即得到实例化，读者可以自行测试，示例代码如清单 14-6 所示。

代码清单 14-6　使用枚举实现单例设计模式

```
package com.wangwenjun.concurrent.chapter14;

// 枚举类型本身是 final 的，不允许被继承
public enum Singleton
{
    INSTANCE;
    // 实例变量
    private byte[] data = new byte[1024];

    Singleton()
    {
        System.out.println("INSTANCE will be initialized immediately");
    }

    public static void method()
    {
        // 调用该方法则会主动使用 Singleton，INSTANCE 将会被实例化
    }

    public static Singleton getInstance()
    {
        return INSTANCE;
    }
}
```

但是也可以对其进行改造，增加懒加载的特性，类似于 Holder 的方式，改进后的代码如清单 14-7 所示。

代码清单 14-7　使用枚举实现单例设计模式

```
package com.wangwenjun.concurrent.chapter14;

public class Singleton
{
```

```java
    // 实例变量
    private byte[] data = new byte[1024];

    private Singleton()
    {
    }
    // 使用枚举充当holder
    private enum EnumHolder
    {
        INSTANCE;
        private Singleton instance;

        EnumHolder()
        {
            this.instance = new Singleton();
        }

        private Singleton getSingleton()
        {
            return instance;
        }
    }

    public static Singleton getInstance()
    {
        return EnumHolder.INSTANCE.getSingleton();
    }
}
```

14.8 本章总结

虽然单例设计模式非常简单，但是在多线程的情况下，你设计的单例程序未必就能够满足单实例、懒加载以及高性能的特点。在本章中，我们分别介绍了7种单例模式的设计并且结合前文中所学的知识进行了深入的分析，笔者本人在日常的开发中也经常使用Holder和枚举方式来设计单例。

PART4 · 第四部分

多线程设计架构模式

第四部分主要介绍多线程的系统架构与设计模式,虽然代表线程的类只有一个 java.lang.Thread,但是针对它的应用场景却是形形色色、五花八门的,在不同的业务场景下,使用合适的设计方法可以提高系统的吞吐量,减少业务受理的延迟时间。

由于我们在第 5 章"线程间通信"中已经非常详细地介绍过了生产者与消费者模式,因此在本部分中将不会重复介绍,希望读者能够掌握和理解生产者消费者模式原理,它是多线程对资源读写比较典型的应用场景,也是使用最多的一种设计模式之一。另外学习本部分的内容需要读者深入地掌握线程相关的 API,尤其是本书第 4 章"线程安全与数据同步"和第 5 章"线程间通信"的内容,如果在阅读过程中有什么不解之处,希望回到前面的章节中进行回顾。

❏ 通过第一部分内容的学习,我们知道了线程的生命周期,以及各个状态之间的切换和导致切换可能产生的方法,那么在线程的业务逻辑执行单元中,我们又该如何监控任务的生命周期呢?在第 15 章"监控任务的生命周期"中我们通过将监听者模式与线程相结合很好地观察到了任务执行的周期,而且还能灵活方便地解决 run 方法无法获得线程运行后无返回值的问题。

❏ 第 16 章"Single Thread Execution 设计模式":本章重点介绍了在多线程环境下对资源竞争的保护措施,通过机场安检的例子让读者掌握引起线程不安全问题的原因及其解决方法;当采用同步的方式对资源进行保护时,还需要注意同步的方式是否引起了锁的交叉使用,锁在交叉使用的情况下会导致出现死锁的情况,如何通过设计的方式进行避免,在本章中都进行了比较详细的讲解。

❑ 对一个共享资源的操作大致上可以分为两类,读(读取)操作和写(增加、修改、删除)操作,读-写之间会引起冲突,写-写之间也会引起冲突,不假思索地对所有的操作都进行同步控制可以解决冲突问题,但是同时也带来了效率的降低,当多个线程同时对资源进行读-读操作时则无须锁的保护,在本部分的第 17 章"读写锁分离设计模式"中,我们进行了读写锁的分离设计,旨在提高读场景明显多过写场景的程序效率。

❑ 除了对共享资源进行锁的同步可以解决多线程间数据冲突的问题之外,使用不可变对象的设计方式同样可以既高效又安全地解决多线程间数据冲突的问题,将一个对象设计成不可变,任何线程都没有机会去修改它,这样就可以避免读-写、写-写之间数据不一致的情况出现,java.lang.String(线程安全的类)为我们提供了很好的设计思路,第 18 章"不可变对象设计模式"通过不可变 Int 类型的累加器的设计,可以让读者掌握如何才能设计一个线程安全的不可变对象。

❑ 第 19 章"Future 设计模式":本章中主要讲解 Future 设计模式及其使用场景,同步的 API 调用比较耗时的时候可以先选择获取一个 Future(结果凭据,立即返回的),调用者所在的线程不用陷入未知时长的阻塞中,使用 Future 可以使调用者立即返回进行其他任务的处理,在"未来"某个时间再凭借凭据获取结果,在本章的最后部分还增加了回调的机制,调用者无须通过阻塞方法 get() 获取结果,取而代之的是注入一个回调接口实现被动地等待结果,这样极大地提高了系统的响应时间,从而充分利用 CPU 资源。

❑ 第 20 章"Guarded Suspension 设计模式":本章中我们介绍了 Guarded(担保)Suspension(挂起)设计模式,该设计模式是很多设计模式的基础,比如生产者消费者设计模式,当某个对象的状态此刻不满足处理调用者请求的时候,选择将调用者正确的挂起,这样可以保证数据的正确性和完整性。

❑ 在大多数的系统设计中我们都会遇到上下文(content)的概念,比如 Servlet 中的 Application 是系统级别的上下文,而 Session 则是用户级别的上下文,在第 21 章"线程上下文设计模式"中,我们通过线程保险箱(ThreadLocal)设计实现了每一个线程的上下文,线程之间的上下文彼此独立,互相之间不受影响,确保了上下文只能被一个线程访问,从而避免了资源的竞争引起的冲突。

❑ 与 Guarded Suspension 模式不同的是,当某个对象此刻的状态不满足的时候,Bulking 设计模式选择的是一种放弃的方案而不是挂起,在第 22 章"Balking 设计模式"中,我们通过 word 文档自动保存与主动保存的例子,为读者详细介绍了 Bulking 设计模式的核心思想以及使用场景。

❑ 第 23 章"Latch 设计模式":本章中我们将会学习到 Latch(门阀、门闩)设计模式,该设计模式旨在解决当所有涉及的线程完成各自的任务之后才能开始下一步动作的场景。在本章中,我们通过 Alex、Jack、Gavin、Dillon 四个码农周末出游的例子，为读者详细剖析了无限等待 Latch 的设计以及具备超时能力的 Latch 设计。
- 第 24 章 "Thread-Per-Message 设计模式"：本章中我们通过一个多客户端的 TCP 聊天程序，为大家介绍了 Thread Per Message 设计模式，该设计模式主要是为了提高系统的并发能力，在同一时刻可以更多地受理更多的业务。
- 当线程正常结束或者错误结束的时候，如何才能最大可能地保证线程能够回收所使用的资源，在第 25 章 "Two Phase Termination 设计模式" 中 Two Phase Termination 为我们提供了很好的解决方案。除此之外，在本章的知识扩展部分中，我们为读者详细介绍了 Strong Reference、Soft Reference、Weak Reference 以及 Phantom Reference 的特点和使用场景。
- 第 26 章 "Worker-Thread 设计模式"：本章中我们模拟车间流水线产品加工的场景，为读者详细介绍了 Work Thread 设计模式的核心思想，并且通过图解的方式为读者区分了生产者消费者模式与 Worker Thread 模式之间的不同。
- 第 27 章 "Active Objects 设计模式"：本章中我们通过分析垃圾回收方法 System.gc() 为读者介绍了什么是 Active Objects，以及如何设计一个可接受异步消息的主动对象，Active Objects 设计模式是比较复杂的一种多线程设计方法，也是对诸如 Future 设计模式、Proxy 设计模式、Worker-Thread 设计模式等其他设计模式的综合应用。
- 在分布式系统中，我们往往会选择消息中间件（Apache ActiveMQ、Apache Kafka）降低系统之间的耦合，提高系统的吞吐量，在第 28 章 "Event Bus（消息总线）设计模式"中我们通过类似于 MQ 的设计——Event Bus 来解决如何在一个进程内部降低多个线程之间耦合的问题，被动接受来自广播的推送消息。在本章中，我们还结合 NIO2.0 中的 WatchService 实现了对文件目录变化事件的异步广播推送。
- 除了 Event Bus 的架构设计之外，Event Driven（事件驱动）的架构设计也可以降低模块之间耦合，提高系统的响应速度，是容易扩展的是设计方法，在本书的第 29 章 "Event Driven（事件驱动）设计模式"中，我们将实现一个基础的 Event Driven 框架，并且使用该框架来处理数据状态不断变化产生的事件。

第 15 章 · CHAPTER15

监控任务的生命周期

15.1 场景描述

虽然 Thread 为我们提供了可获取状态，以及判断是否 alive 的方法，但是这些方法均是针对线程本身的，而我们提交的任务 Runnable 在运行过程中所处的状态如何是无法直接获得的，比如它什么时候开始，什么时候结束，最不好的一种体验是无法获得 Runnable 任务执行后的结果。一般情况下想要获得最终结果，我们不得不为 Thread 或者 Runnable 传入共享变量，但是在多线程的情况下，共享变量将导致资源的竞争从而增加了数据不一致性的安全隐患。

15.2 当观察者模式遇到 Thread

当某个对象发生状态改变需要通知第三方的时候，观察者模式就特别适合胜任这样的工作。观察者模式需要有事件源，也就是引发状态改变的源头，很明显 Thread 负责执行任务的逻辑单元，它最清楚整个过程的始末周期，而事件的接收者则是通知接受者一方，严格意义上的观察者模式是需要 Observer 的集合的，我们在这里不需要完全遵守这样的规则，只需将执行任务的每一个阶段都通知给观察者即可。

15.2.1 接口定义

1. Observable 接口定义

Observable 接口定义的代码如清单 15-1 所示。

代码清单 15-1　Observable.java

```java
package com.wangwenjun.concurrent.chapter15;

public interface Observable
{
    // 任务生命周期的枚举类型
    enum Cycle
    {
        STARTED, RUNNING, DONE, ERROR
    }

    // 获取当前任务的生命周期状态
    Cycle getCycle();

    // 定义启动线程的方法，主要作用是为了屏蔽 Thread 的其他方法
    void start();

    // 定义线程的打断方法，作用与 start 方法一样，也是为了屏蔽 Thread 的其他方法
    void interrupt();
}
```

该接口主要是暴露给调用者使用的，其中四个枚举类型分别代表了当前任务执行生命周期的各个阶段，具体如下。

- getCycle() 方法用于获取当前任务处于哪个执行阶段。
- start() 方法的目的主要是为了屏蔽 Thread 类其他的 API，可通过 Observable 的 start 对线程进行启动。
- interrupt() 方法的作用与 start 一样，可通过 Observable 的 interrupt 对当前线程进行中断。

2. TaskLifecycle 接口定义

TaskLifecycle 接口定义的代码如清单 15-2 所示。

代码清单 15-2　TaskLifecycle.java

```java
package com.wangwenjun.concurrent.chapter15;

public interface TaskLifecycle<T>
{
    // 任务启动时会触发 onStart 方法
    void onStart(Thread thread);

    // 任务正在运行时会触发 onRunning 方法
    void onRunning(Thread thread);

    // 任务运行结束时会触发 onFinish 方法，其中 result 是任务执行结束后的结果
    void onFinish(Thread thread, T result);

    // 任务执行报错时会触发 onError 方法
```

```java
        void onError(Thread thread, Exception e);

        // 生命周期接口的空实现（Adapter）
        class EmptyLifecycle<T> implements TaskLifecycle<T>
        {
            @Override
            public void onStart(Thread thread)
            {
                //do nothing
            }

            @Override
            public void onRunning(Thread thread)
            {
                //do nothing
            }

            @Override
            public void onFinish(Thread thread, T result)
            {
                //do nothing
            }

            @Override
            public void onError(Thread thread, Exception e)
            {
                //do nothing
            }
        }
    }
```

TaskLifecycle 接口定义了在任务执行的生命周期中会被触发的接口，其中 EmptyLifycycle 是一个空的实现，主要是为了让使用者保持对 Thread 类的使用习惯。

- onStart（Thread thread）当任务开始执行时会被回调的方法。
- onRunning（Thread thread）任务运行时被回调的方法，由于我们针对的是任务的生命周期，不同于线程生命周期中的 RUNNING 状态，如果当前线程进入了休眠或者阻塞，那么任务都是 running 状态。
- onFinish（Thread thread，T result）任务正确执行结束后会被回调，其中 result 是任务执行后的结果，可允许为 null。
- onError（Thread thread，Exception e）任务在运行过程中出现任何异常抛出时，onError 方法都将被回调，并将异常信息一并传入。

3. Task 函数接口定义

Task 函数接口定义的代码如清单 15-3 所示。

代码清单 15-3　函数式接口 Task.java

```
package com.wangwenjun.concurrent.chapter15;

@FunctionalInterface
public interface Task<T>
{
    // 任务执行接口，该接口允许有返回值
    T call();
}
```

由于我们需要对线程中的任务执行增加可观察的能力，并且需要获得最后的计算结果，因此 Runnable 接口在可观察的线程中将不再使用，取而代之的是 Task 接口，其作用与 Runnable 类似，主要用于承载任务的逻辑执行单元。

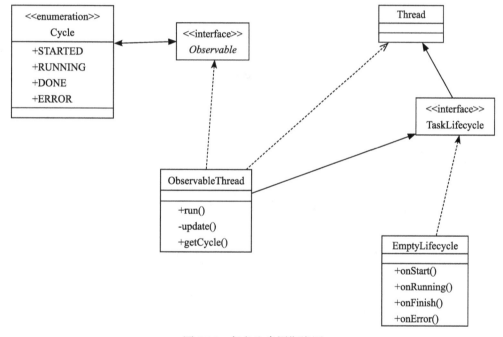

图 15-1　任务生命周期类图

15.2.2　ObservableThread 实现

ObservableThread 是任务监控的关键，它继承自 Thread 类和 Observable 接口，并且在构造期间需要传入 Task 的具体实现，代码如清单 15-4 所示。

代码清单 15-4　ObservableThread.java

```
package com.wangwenjun.concurrent.chapter15;
public class ObservableThread<T> extends Thread
```

```java
                implements Observable
{
    private final TaskLifecycle<T> lifecycle;

    private final Task<T> task;

    private Cycle cycle;

    // 指定 Task 的实现，默认情况下使用 EmptyLifecycle
    public ObservableThread(Task<T> task)
    {
        this(new TaskLifecycle.EmptyLifecycle<>(), task);
    }

    // 指定 TaskLifecycle 的同时指定 Task
    public ObservableThread(TaskLifecycle<T> lifecycle, Task<T> task)
    {
        super();
        //Task 不允许为 null
        if (task == null)
            throw new IllegalArgumentException("The task is required.");
        this.lifecycle = lifecycle;
        this.task = task;
    }

    @Override
    public final void run()
    {
        // 在执行线程逻辑单元的时候，分别触发相应的事件
        this.update(Cycle.STARTED, null, null);
        try
        {
            this.update(Cycle.RUNNING, null, null);

            T result = this.task.call();
            this.update(Cycle.DONE, result, null);
        } catch (Exception e)
        {
            this.update(Cycle.ERROR, null, e);
        }
    }

    private void update(Cycle cycle, T result, Exception e)
    {
        this.cycle = cycle;
        if (lifecycle == null)
            return;
        try
        {
            switch (cycle)
```

```java
        {
            case STARTED:
                this.lifecycle.onStart(currentThread());
                break;
            case RUNNING:
                this.lifecycle.onRunning(currentThread());
                break;
            case DONE:
                this.lifecycle.onFinish(currentThread(), result);
                break;
            case ERROR:
                this.lifecycle.onError(currentThread(), e);
                break;
        }
    } catch (Exception ex)
    {
        if (cycle == Cycle.ERROR)
        {
            throw ex;
        }
    }
}
@Override
public Cycle getCycle()
{
    return this.cycle;
}
```

重写父类的 run 方法,并且将其修饰为 final 类型,不允许子类再次对其进行重写,run 方法在线程的运行期间,可监控任务在执行过程中的各个生命周期阶段,任务每经过一个阶段相当于发生了一次事件。

update 方法用于通知时间的监听者,此时任务在执行过程中发生了什么,最主要的通知是异常的处理。如果监听者也就是 TaskLifecycle,在响应某个事件的过程中出现了意外,则会导致任务的正常执行受到影响,因此需要进行异常捕获,并忽略这些异常信息以保证 TaskLifecycle 的实现不影响任务的正确执行,但是如果任务执行过程中出现错误并且抛出了异常,那么 update 方法就不能忽略该异常,需要继续抛出异常,保持与 call 方法同样的意图。

15.3 本章总结

15.3.1 测试运行

关于 ObservableThread 我们已经完成实现,在这里进行简单测试,其中需要读者重点关注的是，ObservableThread 是否保持了与 Thread 相同的使用习惯，其次读者可以通过实现 TaskLifecycle 监听感兴趣的事件，比如获取最终的计算结果等，代码如下：

```java
public static void main(String[] args)
{
    Observable observableThread = new ObservableThread<>(() ->
    {
        try
        {
            TimeUnit.SECONDS.sleep(10);
        } catch (InterruptedException e)
        {
            e.printStackTrace();
        }
        System.out.println(" finished done.");
        return null;
    });
    observableThread.start();
}
```

这段程序与你平时使用 Thread 并没有太大的区别，只不过 ObservableThread 是一个泛型类，我们将其定义为 Void 类型，表示不关心返回值，默认的 EmptyLifecycle 同样表示不关心生命周期的每一个阶段，代码如下：

```java
public static void main(String[] args)
{
    final TaskLifecycle<String> lifecycle = new TaskLifecycle.EmptyLifecycle<String>()
    {
        public void onFinish(Thread thread, String result)
        {
            System.out.println("The result is " + result);
        }
    };
    Observable observableThread = new ObservableThread<>(lifecycle, () ->
    {
        try
        {
            TimeUnit.SECONDS.sleep(10);
        } catch (InterruptedException e)
        {
            e.printStackTrace();
        }
        System.out.println(" finished done.");
        return "Hello Observer";
    });
    observableThread.start();
}
```

上面这段程序代码定义了一个需要返回值的 ObservableThread，并且通过重写 EmptyLifecycle 的 onFinsh 方法输出最终的返回结果。

15.3.2 关键点总结

- 在接口 Observable 中定义与 Thread 同样的方法用于屏蔽 Thread 的其他 API，在使用的过程中使用 Observable 声明 ObservableThread 的类型，如果使用者还想知道更多的关于 Thread 的 API，只需要在 Observable 接口中增加即可。
- 将 ObservableThread 中的 run 方法修饰为 final，或者将 ObservableThread 类修饰为 final，防止子类继承重写，导致整个生命周期的监控失效，我们都知道，任务的逻辑执行单元是存在于 run 方法之中的，而在 ObservableThread 中我们摒弃了这一点，让它专门监控业务执行单元的生命周期，而将真正的业务逻辑执行单元交给了一个可返回计算结果的接口 Task。
- ObservableThread 本身的 run 方法充当了事件源的发起者，而 TaskLifecycle 则扮演了事件回调的响应者。

第 16 章 · CHAPTER16

Single Thread Execution 设计模式

Single Thread Execution 模式是指在同一时刻只能有一个线程去访问共享资源,就像独木桥一样每次只允许一人通行,简单来说,Single Thread Execution 就是采用排他式的操作保证在同一时刻只能有一个线程访问共享资源。

16.1 机场过安检

相信大家都有乘坐飞机的经历,在进入登机口之前必须经过安全检查,安检口类似于独木桥,每次只能通过一个人,工作人员除了检查你的登机牌以外,还要联网检查身份证信息以及是否携带危险物品,如图 16-1 所示。

图 16-1　Single Thread Execution 设计模式

16.1.1 非线程安全

先模拟一个非线程安全的安检口类,旅客(线程)分别手持登机牌和身份证接受工作人员的检查,示例代码如清单 16-1 所示。

代码清单 16-1　FlightSecurity.java

```
package com.wangwenjun.concurrent.chapter16;

public class FlightSecurity
{
    private int count = 0;
    //登机牌
    private String boardingPass = "null";
    //身份证
    private String idCard = "null";

    public void pass(String boardingPass, String idCard)
    {
        this.boardingPass = boardingPass;
        this.idCard = idCard;
        this.count++;
        check();
    }

    private void check()
    {
        //简单的测试,当登机牌和身份证首字母不相同时则表示检查不通过
        if (boardingPass.charAt(0) != idCard.charAt(0))
            throw new RuntimeException("====Exception====" + toString());
    }

    public String toString()
    {
        return "The " + count + " passengers,boardingPass [" + boardingPass +
"],idCard [" + idCard + "]";
    }
}
```

FlightSecurity 比较简单,提供了一个 pass 方法,将旅客的登机牌和身份证传递给 pass 方法,在 pass 方法中调用 check 方法对旅客进行检查,检查的逻辑也足够的简单,只需要检测登机牌和身份证首字母是否相等(当然这样在现实中非常不合理,但是为了使测试简单我们约定这么做),我们看代码清单 16-2 所示的测试。

代码清单 16-2　测试代码 FlightSecurityTest.java

```
package com.wangwenjun.concurrent.chapter16;

public class FlightSecurityTest
```

```java
{
    // 旅客线程
    static class Passengers extends Thread
    {
        // 机场安检类
        private final FlightSecurity flightSecurity;

        // 旅客的身份证
        private final String idCard;

        // 旅客的登机牌
        private final String boardingPass;

        // 构造旅客时传入身份证、登机牌以及机场安检类
        public Passengers(FlightSecurity flightSecurity,
                    String idCard, String boardingPass)
        {
            this.flightSecurity = flightSecurity;
            this.idCard = idCard;
            this.boardingPass = boardingPass;
        }

        @Override
        public void run()
        {
            while (true)
            {
                // 旅客不断地过安检
                flightSecurity.pass(boardingPass, idCard);
            }
        }
    }

    public static void main(String[] args)
    {
        // 定义三个旅客，身份证和登机牌首字母均相同
        final FlightSecurity flightSecurity = new FlightSecurity();
        new Passengers(flightSecurity, "A123456", "AF123456").start();
        new Passengers(flightSecurity, "B123456", "BF123456").start();
        new Passengers(flightSecurity, "C123456", "CF123456").start();
    }
}
```

看起来每一个客户都是合法的，因为每一个客户的身份证和登机牌首字母都一样，运行上面的程序却出现了错误，而且错误的情况还不太一样，笔者运行了 5 次左右，发现了两种类型的错误信息，程序输出如下：

====Exception====The 930 passengers,boardingPass [BF123456],idCard [B123456]
====Exception====The 771672 passengers,boardingPass [CF123456],idCard [B123456]

首字母相同检查不能通过和首字母不相同检查不能通过，为什么会出现这样的情况呢？首字母相同却不能通过？更加奇怪的是传入的参数明明全都是首字母相同的，为什么会出现首字母不相同的错误呢？

16.1.2　问题分析

在本节中我们将尝试分析 16.1.1 节出现的两种错误情况，在多线程的情况下调用 pass 方法，如果传入"A123456""AF123456"，虽然参数的传递百分之百能保证就是这两个值，但是在 pass 方法中对 boardingPass 和 idCard 的赋值很有可能交叉的，不能保证原子性操作。

（1）首字母相同却未通过检查

图 16-2 所示的为首字母相同却无法通过安检的分析过程。

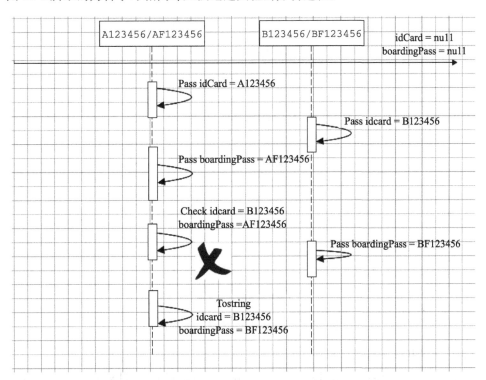

图 16-2　首字母相同却无法通过安检的分析过程

这种情况的执行步骤顺序如下。

1）线程 A 调用 pass 方法，传入"A123456""AF123456"并且对 idCard 赋值成功，由于 CPU 调度器时间片的轮转，CPU 的执行权归 B 线程所有。

2）线程 B 调用 pass 方法，传入"B123456""BF123456"并且对 idCard 赋值成功，覆盖 A 线程赋值的 idCard。

3）线程 A 重新获得 CPU 的执行权，将 boardingPass 赋予 AF123456，因此 check 无法通过。

4）在输出 toString 之前，B 线程成功将 boardingPass 覆盖为 BF123456。

（2）为何出现首字母不相同的情况

明明传入的身份证和登机牌首字母都相同，可为何在运行的过程中会出现首字母不相同的情况，下面我们也通过图示的方式进行分析，如图 16-3 所示。

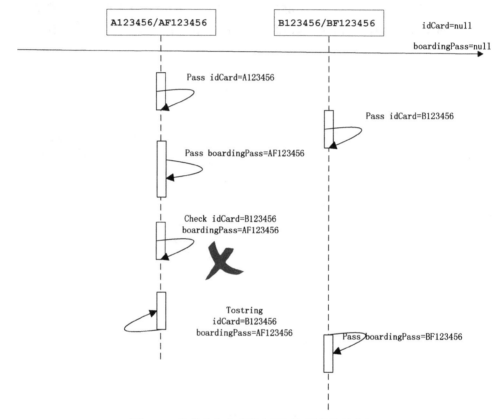

图 16-3　为什么会出现首字母不相同的情况分析

这种情况的执行步骤顺序如下。

1）线程 A 调用 pass 方法，传入"A123456""AF123456"并且对 idCard 赋值成功，由于 CPU 调度器时间片的轮转，CPU 的执行权归 B 线程所有。

2）线程 B 调用 pass 方法，传入"B123456""BF123456"并且对 idCard 赋值成功，覆盖 A 线程赋值的 idCard。

3）线程 A 重新获得 CPU 的执行权，将 boardingPass 赋予 AF123456，因此 check 无法通过。

4）线程 A 检查不通过，输出 idCard="A123456"和 boardingPass="BF123456"。

16.1.3 线程安全

16.1.1 节中出现的问题说到底就是数据同步的问题，虽然线程传递给 pass 方法的两个参数能够百分之百地保证首字母相同，可是在为 FlightSecurity 中的属性赋值的时候会出现多个线程交错的情况，结合我们在第一部分第 4 章的所讲内容可知，需要对共享资源增加同步保护，改进代码如下：

```
public synchronized void pass(String boardingPass, String idCard)
{
    this.boardingPass = boardingPass;
    this.idCard = idCard;
    this.count++;
    check();
}
```

修改后的 pass 方法，无论运行多久都不会再出现检查出错的情况了，为什么只在 pass 方法增加 synchronized 关键字，check 以及 toString 方法都有对共享资源的访问，难道它们不加同步就不会引起错误么？由于 check 方法是在 pass 方法中执行的，pass 方法加同步已经保证了 single thread execution，因此 check 方法不需要增加同步，toString 方法原因与此相同。

何时适合使用 single thread execution 模式呢？答案如下。
- 多线程访问资源的时候，被 synchronized 同步的方法总是排他性的。
- 多个线程对某个类的状态发生改变的时候，比如 FlightSecurity 的登机牌以及身份证。

在 Java 中经常会听到线程安全的类和线程非安全的类，所谓线程安全的类是指多个线程在对某个类的实例同时进行操作时，不会引起数据不一致的问题，反之则是线程非安全的类，在线程安全的类中经常会看到 synchronized 关键字的身影。

16.2 吃面问题

在本节中，我们将模拟两个人吃意大利面的场景，来演示交叉锁导致的程序死锁情况，如图 16-4 所示。

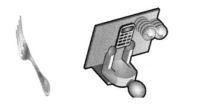

图 16-4　吃意大利面

16.2.1 吃面引起的死锁

虽然使用 synchronized 关键字可以保证 single thread execution，但是如果使用不得当则会导致死锁的情况发生，比如 A 手持刀等待 B 放下叉，而 B 手持叉等待 A 放下刀，示例代码如清单 16-3 所示。

代码清单 16-3　餐具类 Tableware.java

```java
package com.wangwenjun.concurrent.chapter16;

public class Tableware
{
    //餐具名称
    private final String toolName;

    public Tableware(String toolName)
    {
        this.toolName = toolName;
    }

    @Override
    public String toString()
    {
        return "Tool:" + toolName;
    }
}
```

Tableware 代表餐具的类，比较简单，其中 toolName 表示餐具的名称，我们再来创建吃面的线程，代码如清单 16-4 所示。

代码清单 16-4　吃面条的线程 EatNoodleThread.java

```java
package com.wangwenjun.concurrent.chapter16;

public class EatNoodleThread extends Thread
{
    private final String name;

    //左手边的餐具
    private final Tableware leftTool;

    //右手边的餐具
    private final Tableware rightTool;

    public EatNoodleThread(String name, Tableware leftTool, Tableware rightTool)
    {
        this.name = name;
        this.leftTool = leftTool;
        this.rightTool = rightTool;
```

```
        }

        @Override
        public void run()
        {
            while (true)
            {
                this.eat();
            }
        }
        // 吃面条的过程
        private void eat()
        {
            synchronized (leftTool)
            {
                System.out.println(name + " take up " + leftTool + "(left)");
                synchronized (rightTool)
                {
                    System.out.println(name + " take up " + rightTool + "(right)");
                    System.out.println(name + " is eating now.");
                    System.out.println(name + " put down " + rightTool + "(right)");
                }
                System.out.println(name + " put down " + leftTool + "(left)");
            }
        }
    }
```

在创建吃面线程时需要指定名称和吃面用的刀叉，在 eat 方法中先拿起左手的餐具，然后尝试获取右手的餐具，代码如下：

```
public static void main(String[] args)
{
    Tableware fork = new Tableware("fork");
    Tableware knife = new Tableware("knife");
    new EatNoodleThread("A", fork, knife).start();
    new EatNoodleThread("B", knife, fork).start();
}
```

运行上面的程序，吃不了几个回合便会陷入死锁，在程序最后卡住的地方不难发现 B 获取了刀企图获取叉，而 A 则相反手持叉企图获取刀，因此进入了阻塞，输出如下：

```
...
A is eating now.
A put down Tool:knife(right)
A put down Tool:fork(left)
B take up Tool:knife(left)
A take up Tool:fork(left)
```

16.2.2 解决吃面引起的死锁问题

虽然我们使用了 Single Thread Execution 对 eat 加以控制，但还是出现了死锁，其主要原因主要是交叉锁导致两个线程之间相互等待彼此释放持有的锁，关于这一点我们在 4.5 节中有过比较详细的介绍。

为了解决交叉锁的情况，我们需要将刀叉进行封装，使刀叉同属于一个类中，改进代码如清单 16-5 所示。

代码清单 16-5　TablewarePair.java

```java
package com.wangwenjun.concurrent.chapter16;
public class TablewarePair
{
    private final Tableware leftTool;

    private final Tableware rightTool;

    public TablewarePair(Tableware leftTool, Tableware rightTool)
    {
        this.leftTool = leftTool;
        this.rightTool = rightTool;
    }

    public Tableware getLeftTool()
    {
        return leftTool;
    }

    public Tableware getRightTool()
    {
        return rightTool;
    }
}
```

在 EatNoodleThread 中使用 TablewarePair 代替 leftTool 和 rightTool，这样就可以避免交叉锁的情况，代码如清单 16-6 所示。

代码清单 16-6　EatNoodleThread.java

```java
package com.wangwenjun.concurrent.chapter16;

public class EatNoodleThread extends Thread
{
    private final String name;

    private final TablewarePair tablewarePair;

    public EatNoodleThread(String name, TablewarePair tablewarePair)
```

```java
{
    this.name = name;
    this.tablewarePair = tablewarePair;
}

@Override
public void run()
{
    while (true)
    {
        this.eat();
    }
}

private void eat()
{
    synchronized (tablewarePair)
    {
        System.out.println(name + " take up " + tablewarePair.getLeftTool()
            + "(left)");
        System.out.println(name + " take up " + tablewarePair.getRightTool()
            + "(right)");
        System.out.println(name + " is eating now.");
        System.out.println(name + " put down " + tablewarePair.getRightTool()
            + "(right)");
        System.out.println(name + " put down " + tablewarePair.getLeftTool()
            + "(left)");
    }
}
}
```

更改后的程序无论运行多久都不会出现死锁的情况,因为在同一时间只能有一个线程获得刀和叉。

16.2.3 哲学家吃面

哲学家吃面是解释操作系统中多个进程竞争资源的经典问题,每个哲学家的左右手都有吃面用的刀叉,但是不足以同时去使用,比如 A 哲学家想要吃面,必须拿起左手边的叉和右手边的刀,但是有可能叉和刀都被其他哲学家拿走使用,或者是手持刀等待别人放下叉等容易引起死锁的问题,如图 16-5 所示。

关于哲学家吃面的描述可以参考资

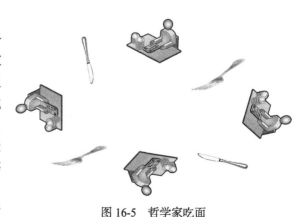

图 16-5 哲学家吃面

料 http://wiki.c2.com/?DiningPhilosophers。

16.3 本章总结

将某个类设计成线程安全的类，用 Single Thread Execution 控制是其中的方法之一，但是子类如果继承了线程安全的类并且打破了 Single Thread Execution 的方式，就会破坏方法的安全性，这种情况一般称为继承异常（inheritance anomaly）。

在 Single Thread Execution 中，synchronized 关键字起到了决定性的作用，但是 synchronized 的排他性是以性能的牺牲为代价的，因此在保证线程安全的前提下应尽量缩小 synchronized 的作用域。

CHAPTER17 · 第 17 章

读写锁分离设计模式

17.1 场景描述

在多线程的情况下访问共享资源，需要对资源进行同步操作以防止数据不一致的情况发生，通常我们可以使用 synchronized 关键字或者显式锁，比如我们在 5.4 节中定义的 BooleanLock 进行资源的同步操作，当然也可以使用 JDK1.5 以后的显式锁 Lock。

对资源的访问一般包括两种类型的动作——读和写（更新、删除、增加等资源会发生变化的动作），如果多个线程在某个时刻都在进行资源的读操作，虽然有资源的竞争，但是这种竞争不足以引起数据不一致的情况发生，那么这个时候直接采用排他的方式加锁，就显得有些简单粗暴了。表 17-1 将两个线程对资源的访问动作进行了枚举，除了多线程在同一时间都进行读操作时不会引起冲突之外，其余的情况都会导致访问的冲突，需要对资源进行同步处理。

表 17-1 共享资源在多个线程中同时进行读操作时不会引起冲突

线程	读	写
读	不冲突	冲突
写	冲突	冲突

如果对某个资源读的操作明显多过于写的操作，那么多线程读时并不加锁，很明显对程序性能的提升会有很大的帮助。在本章中，我们将使用之前掌握的知识点，实现一个读写分离的锁。

17.2 读写分离程序设计

17.2.1 接口定义

读写锁的类图如图 17-1 所示。

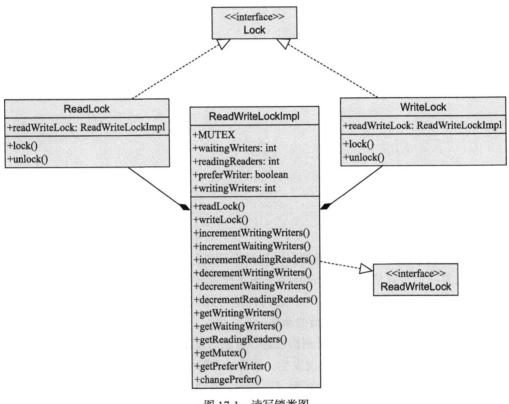

图 17-1　读写锁类图

1. Lock 接口定义

Lock 接口定义的代码如清单 17-1 所示。

代码清单 17-1　锁接口 Lock.java

```
package com.wangwenjun.concurrent.chapter17;

public interface Lock
{
    // 获取显式锁，没有获得锁的线程将被阻塞
    void lock() throws InterruptedException;
    // 释放获取的锁
    void unlock();
}
```

Lock 接口定义了锁的基本操作，加锁和解锁，显式锁的操作强烈建议与 try finally 语句块一起使用，加锁和解锁说明如下。

- lock()：当前线程尝试获得锁的拥有权，在此期间有可能进入阻塞。
- unlock()：释放锁，其主要目的就是为了减少 reader 或者 writer 的数量。

2. ReadWriteLock 接口定义

ReadWriteLock 接口定义的代码如清单 17-2 所示。

代码清单 17-2　ReadWriteLock.java

```java
package com.wangwenjun.concurrent.chapter17;

public interface ReadWriteLock
{
    //创建 reader 锁
    Lock readLock();

    //创建 write 锁
    Lock writeLock();

    //获取当前有多少线程正在执行写操作
    int getWritingWriters();

    //获取当前有多少线程正在等待获取写入锁
    int getWaitingWriters();

    //获取当前有多少线程正在等待获取 reader 锁
    int getReadingReaders();

    //工厂方法，创建 ReadWriteLock
    static ReadWriteLock readWriteLock()
    {
        return new ReadWriteLockImpl();
    }

    //工厂方法，创建 ReadWriteLock，并且传入 preferWriter
    static ReadWriteLock readWriteLock(boolean preferWriter)
    {
        return new ReadWriteLockImpl(preferWriter);
    }
}
```

ReadWriteLock 虽然名字中有 lock，但是它并不是 lock，它主要是用于创建 read lock 和 write lock 的，并且提供了查询功能用于查询当前有多少个 reader 和 writer 以及 waiting 中的 writer，根据我们在前文中的分析，如果 reader 的个数大于 0，那就意味着 writer 的个数等于 0，反之 writer 的个数大于 0（事实上 writer 最多只能为 1），则 reader 的个数等于 0，由于读和写，写和写之间都存在着冲突，因此这样的数字关系也就不奇怪了。

- readLock()：该方法主要用来获得一个 ReadLock。
- writeLock()：同 readLock 类似，该方法用来获得 WriteLock。
- getWritingWriters()：获取当前有多少个线程正在进行写的操作，最多是 1 个。
- getWaitingWriters()：获取当前有多少个线程由于获得写锁而导致阻塞。
- getReadingReaders()：获取当前有多少个线程正在进行读的操作。

17.2.2 程序实现

1. ReadWriteLockImpl

相对于 Lock，ReadWriteLockImpl 更像是一个工厂类，可以通过它创建不同类型的锁，我们将 ReadWriteLockImpl 设计为包可见的类，其主要目的是不想对外暴露更多的细节，在 ReadWriteLockImpl 中还定义了非常多的包可见方法，代码如清单 17-3 所示。

代码清单 17-3　ReadWriteLockImpl.java

```java
package com.wangwenjun.concurrent.chapter17;
//包可见，创建时使用 ReadWriterLock 的 create 方法
class ReadWriteLockImpl implements ReadWriteLock
{
    //定义对象锁
    private final Object MUTEX = new Object();

    //当前有多少个线程正在写入
    private int writingWriters = 0;
    //当前有多少个线程正在等待写入
    private int waitingWriters = 0;

    //当前有多少个线程正在 read
    private int readingReaders = 0;

    //read 和 write 的偏好设置
    private boolean preferWriter;

    //默认情况下 perferWriter 为 true
    public ReadWriteLockImpl()
    {
        this(true);
    }

    //构造 ReadWriteLockImpl 并且传入 preferWriter
    public ReadWriteLockImpl(boolean preferWriter)
    {
        this.preferWriter = preferWriter;
    }

    //创建 read lock
```

```java
public Lock readLock()
{
    return new ReadLock(this);
}

// 创建write lock
public Lock writeLock()
{
    return new WriteLock(this);
}

// 使写线程的数量增加
void incrementWritingWriters()
{
    this.writingWriters++;
}
// 使等待写入的线程数量增加
void incrementWaitingWriters()
{
    this.waitingWriters++;
}

// 使读线程的数量增加
void incrementReadingReaders()
{
    this.readingReaders++;
}

// 使写线程的数量减少
void decrementWritingWriters()
{
    this.writingWriters--;
}

// 使等待获取写入锁的数量减一
void decrementWaitingWriters()
{
    this.waitingWriters--;
}

// 使读取线程的数量减少
void decrementReadingReaders()
{
    this.readingReaders--;
}
// 获取当前有多少个线程正在进行写操作
public int getWritingWriters()
{
    return this.writingWriters;
}
```

```java
    // 获取当前有多少个线程正在等待获取写入锁
    public int getWaitingWriters()
    {
        return this.waitingWriters;
    }
    // 获取当前多少个线程正在进行读操作
    public int getReadingReaders()
    {
        return this.readingReaders;
    }

    // 获取对象锁
    Object getMutex()
    {
        return this.MUTEX;
    }
    // 获取当前是否偏向写锁
    boolean getPreferWriter()
    {
        return this.preferWriter;
    }

    // 设置写锁偏好
    void changePrefer(boolean preferWriter)
    {
        this.preferWriter = preferWriter;
    }
}
```

虽然我们在开发一个读写锁,但是在实现的内部也需要一个锁进行数据同步以及线程之间的通信,其中 MUTEX 的作用就在于此,而 preferWriter 的作用在于控制倾向性,一般来说读写锁非常适用于读多写少的场景,如果 preferWriter 为 false,很多读线程都在读数据,那么写线程将会很难得到写的机会。

2. ReadLock

读锁是 Lock 的实现,同样将其设计成包可见以透明其实现细节,让使用者只用专注于对接口的调用,代码如清单 17-4 所示。

代码清单 17-4 读取锁 ReadLock.java

```java
package com.wangwenjun.concurrent.chapter17;

//ReadLock 被设计为包可见
class ReadLock implements Lock
{
    private final ReadWriteLockImpl readWriteLock;

    ReadLock(ReadWriteLockImpl readWriteLock)
    {
```

```java
        this.readWriteLock = readWriteLock;
    }

    @Override
    public void lock() throws InterruptedException
    {
        // 使用 Mutex 作为锁
        synchronized (readWriteLock.getMutex())
        {
            // 若此时有线程在进行写操作，或者有写线程在等待并且偏向写锁的标识为 true 时，就会无
            // 法获得读锁，只能被挂起
            while (readWriteLock.getWritingWriters() > 0
                    || (readWriteLock.getPreferWriter()
                    && readWriteLock.getWaitingWriters() > 0))
            {
                readWriteLock.getMutex().wait();
            }
            // 成功获得读锁，并且使 readingReaders 的数量增加
            readWriteLock.incrementReadingReaders();
        }
    }

    @Override
    public void unlock()
    {
        // 使用 Mutex 作为锁，并且进行同步
        synchronized (readWriteLock.getMutex())
        {
            // 释放锁的过程就是使得当前 reading 的数量减一
            // 将 perferWriter 设置为 true，可以使得 writer 线程获得更多的机会
            // 通知唤醒与 Mutex 关联 monitor waitset 中的线程
            readWriteLock.decrementReadingReaders();
            readWriteLock.changePrefer(true);
            readWriteLock.getMutex().notifyAll();
        }
    }
}
```

- 当没有任何线程对数据进行写操作的时候，读线程才有可能获得锁的拥有权，当然除此之外，为了公平起见，如果当前有很多线程正在等待获得写锁的拥有权，同样读线程将会进入 Mutex 的 wait set 中，readingReader 的数量将增加。
- 读线程释放锁，这意味着 reader 的数量将减少一个，同时唤醒 wait 中的线程，reader 唤醒的基本上都是由于获取写锁而进入阻塞的线程，为了提高写锁获得锁的机会，需要将 preferWriter 修改为 true。

3. WriteLock

写锁是 Lock 的实现，同样将其设计成包可见以透明其实现细节，让使用者只用专注于

对接口的调用,由于写－写冲突的存在,同一时间只能由一个线程获得锁的拥有权,代码如清单17-5所示。

代码清单17-5　写入锁 WriteLock.java

```java
package com.wangwenjun.concurrent.chapter17;
//WriteLock被设计为包可见
class WriteLock implements Lock
{
    private final ReadWriteLockImpl readWriteLock;

    WriteLock(ReadWriteLockImpl readWriteLock)
    {
        this.readWriteLock = readWriteLock;
    }

    @Override
    public void lock() throws InterruptedException
    {
        synchronized (readWriteLock.getMutex())
        {
            try
            {
                // 首先使等待获取写入锁的数字加一
                readWriteLock.incrementWaitingWriters();
                // 如果此时有其他线程正在进行读操作,或者写操作,那么当前线程将被挂起
                while (readWriteLock.getReadingReaders() > 0
                        || readWriteLock.getWritingWriters() > 0)
                {
                    readWriteLock.getMutex().wait();
                }
            } finally
            {
                // 成功获取到了写入锁,使得等待获取写入锁的计数器减一
                this.readWriteLock.decrementWaitingWriters();
            }
            // 将正在写入的线程数量加一
            readWriteLock.incrementWritingWriters();
        }
    }

    @Override
    public void unlock()
    {
        synchronized (readWriteLock.getMutex())
        {
            // 减少正在写入锁的线程计数器
            readWriteLock.decrementWritingWriters();
            // 将偏好状态修改为false,可以使得读锁被最快速的获得
            readWriteLock.changePrefer(false);
```

```
            // 通知唤醒其他在 Mutext monitor waitset 中的线程
            readWriteLock.getMutex().notifyAll();
        }
    }
}
```

- 当有线程在进行读操作或者写操作的时候,若当前线程试图获得锁,则其将会进入 MUTEX 的 wait set 中而阻塞,同时增加 waitingWriter 和 writingWriter 的数量,但是当线程从 wait set 中被激活的时候 waitingWriter 将很快被减少。
- 写释放锁,意味着 writer 的数量减少,事实上变成了 0,同时唤醒 wait 中的线程,并将 preferWriter 修改为 false,以提高读线程获得锁的机会。

17.3 读写锁的使用

我们大致完成了读写锁的设计,现在是时候该去使用它了,其实使用读写锁也是一件非常简单的事情,用不同的锁来同步不同的读写动作,代码如清单 17-6 所示。

代码清单 17-6　ShareData.java

```java
package com.wangwenjun.concurrent.chapter17;

import java.util.ArrayList;
import java.util.List;
import java.util.concurrent.TimeUnit;

public class ShareData
{
    // 定义共享数据(资源)
    private final List<Character> container = new ArrayList<>();
    // 构造 ReadWriteLock
    private final ReadWriteLock readWriteLock = ReadWriteLock.readWriteLock();
    // 创建读取锁
    private final Lock readLock = readWriteLock.readLock();
    // 创建写入锁
    private final Lock writeLock = readWriteLock.writeLock();
    private final int length;

    public ShareData(int length)
    {
        this.length = length;
        for (int i = 0; i < length; i++)
        {
            container.add(i, 'c');
        }
    }

    public char[] read() throws InterruptedException
```

```java
    {
        try
        {
            // 首先使用读锁进行lock
            readLock.lock();
            char[] newBuffer = new char[length];
            for (int i = 0; i < length; i++)
            {
                newBuffer[i] = container.get(i);
            }
            slowly();
            return newBuffer;
        } finally
        {
            // 当操作结束之后,将锁释放
            readLock.unlock();
        }
    }

    public void write(char c) throws InterruptedException
    {
        try
        {
            // 使用写锁进行lock
            writeLock.lock();
            for (int i = 0; i < length; i++)
            {
                this.container.add(i, c);
            }
            slowly();
        } finally
        {
            // 当所有的操作都完成之后,对写锁进行释放
            writeLock.unlock();
        }
    }

    // 简单模拟操作的耗时
    private void slowly()
    {
        try
        {
            TimeUnit.SECONDS.sleep(1);
        } catch (InterruptedException e)
        {
            e.printStackTrace();
        }
    }
}
```

ShareData 中涉及了对数据的读写操作，因此它是需要进行线程同步控制的。首先，创建一个 ReadWriteLock 工厂类，然后用该工厂分别创建 ReadLock 和 WriteLock 的实例，在 read 方法中使用 ReadLock 对其进行加锁，而在 write 方法中则使用 WriteLock，清单 17-7 的程序则是关于对 ShareData 的使用。

代码清单 17-7　ReadWriteLockTest.java

```java
package com.wangwenjun.concurrent.chapter17;
import static java.lang.Thread.currentThread;
public class ReadWriteLockTest
{
    //This is the example for read write lock
    private final static String text = "Thisistheexampleforreadwritelock";

    public static void main(String[] args)
    {
        // 定义共享数据
        final ShareData shareData = new ShareData(50);
        // 创建两个线程进行数据写操作
        for (int i = 0; i < 2; i++)
        {
            new Thread(() ->
            {
                for (int index = 0; index < text.length(); index++)
                {
                    try
                    {
                        char c = text.charAt(index);
                        shareData.write(c);
                        System.out.println(currentThread() + " write " + c);
                    } catch (InterruptedException e)
                    {
                        e.printStackTrace();
                    }
                }
            }).start();
        }
        // 创建 10 个线程进行数据读操作
        for (int i = 0; i < 10; i++)
        {
            new Thread(() ->
            {
                while (true)
                    try
                    {
                        System.out.println(currentThread() + " read " + new
                            String(shareData.read()));
                    } catch (InterruptedException e)
                    {
```

```
                    e.printStackTrace();
                }
        }).start();
    }
}
```

为了在读多于写的场景中体现读写锁，我们创建了 10 个读的线程和两个写的线程，上面的程序需要你借助于 Ctrl+Break 进行停止，虽然我们比较完整地完成了一个读写锁的设计并且可以投入使用，但是还是存在着一定的缺陷和一些需要补强的地方，比如可以结合 BooleanLock 的方式增加超时的功能，提供用于查询哪些线程被陷入阻塞的方法，判断当前线程是否被某个 lock 锁定等，读者可以根据阅读本书所掌握的知识自行拓展。

17.4 本章总结

JDK 版本的几乎每一次升级都会看到对锁的优化，就连 synchronized 关键字的性能都得到了很大的提升，在某些场合下直接使用 synchronized 关键字甚至要比 java.util.concurrent.locks 包下提供的锁性能更佳，在 JDK 的并发包下包含了读写锁的实现，可见 Java 官方对读写锁分离方案的重视。

RW 锁（读写锁）允许多个线程在同一时间对共享资源进行读取操作，在读明显多于写的场合下，其对性能的提升是非常明显的，但是如果使用不当性能反倒会比较差，比如在写线程的数量和读线程的数量接近甚至多于读线程情况下，因此在 JDK1.8 中又增加了 StampedLock 的解决方案（关于 StampedLock 的解决方案不在本书讨论范畴之内）。

笔者在自己的电脑上做了三组（每组五轮）测试，分别对比了在不同数量的读写线程情况下的性能，测试中的 StampedLock 和 ReadWriteLock 都是 Java 并发包提供的显式锁，测试的主要目的是要读者掌握在不同的场景下该选择怎样的锁机制，测试对比结果分别如表 17-2、表 17-3 和表 17-4 所示。

- 5 个 reader vs 5 个 writer

表 17-2　读写操作各 5 个的三组测试结果

Synchronized	StampedLock	ReadWriteLock
220.1	68.6	90.7
189.3	72.6	84.7
207.9	59	86.8
228.3	83.7	76.6
169.2	59.4	75.7
202.96	68.66	82.9

最后一行为五轮测试的平均值，五个线程进行读操作，五个线程进行写操作的情况下

StampedLock 的表现是最优秀的,读写锁次之,性能最差的就当属 synchronized 关键字了。

- 10 个 reader vs 10 个 writer

表 17-3 读写操作各 10 个的三组测试结果

Synchronized	StampedLock	ReadWriteLock
177.4	168.1	1960.8
192.1	111.3	1473.6
173.3	216.8	2119.7
205.4	221.9	2772.2
181.2	189.3	2721.4
185.88	180.88	2209.54

synchronized 关键字的表现很稳定,几组测试数据抖动不大,性能仅次于 StampedLock,如果加大写的线程数量 synchronized 关键字的表现则会是最优的,但是读写锁的性能骤降,和其他两个对照组相比较简直就是数量级的降低。

- 16 个 reader vs 4 个 writer

表 17-4 16 个读操作 4 个写操作的三组测试结果

Synchronized	StampedLock	ReadWriteLock
546.6	532.7	447.3
446	386.9	469.9
597.7	541.4	458.8
638	438.4	462.8
543.8	386	467.4
554.42	455.28	461.24

在读的情况比较多的情况下,读写分离锁的性能优势也体现出来了,如果读者使用的是 JDK1.8 的开发环境,那么强烈建议直接使用 StampedLock,该锁提供了一种乐观的机制,性能在目前来说是最好的,因此被称为 lock 家族的 "宠儿"。

第 18 章 · CHAPTER18

不可变对象设计模式

18.1 线程安全性

所谓共享的资源，是指在多个线程同时对其进行访问的情况下，各线程都会使其发生变化，而线程安全性的主要目的就在于在受控的并发访问中防止数据发生变化。除了使用 synchronized 关键字同步对资源的写操作之外，还可以在线程之间不共享资源状态，甚至将资源的状态设置为不可变。在本章中，我们将讨论如何设计不可变对象，这样就可以不用依赖于 synchronized 关键字的约束。

18.2 不可变对象的设计

无论是 synchronized 关键字还是显式锁 Lock，都会牺牲系统的性能，不可变对象的设计理念在这几年变得越来越受宠，其中 Actor 模型（不可变对象在 Akka、jActor、Kilim 等 Actor 模型框架中得到了广泛的使用，关于 Actor 模型的更多资料请参考 http s://en.wikipedia.org/wiki/Actor_model）以及函数式编程语言 Clojure（Clojure 官网 https://www.clojure.org/）等都是依赖于不可变对象的设计达到 lock free（无锁）的。

Java 核心类库中提供了大量的不可变对象范例，其中 java.lang.String 的每一个方法都没有同步修饰，可是其在多线程访问的情况下是安全的，Java 8 中通过 Stream 修饰的 ArrayList 在函数式方法并行访问的情况下也是线程安全的，所谓不可变对象是没有机会去修改它，每一次的修改都会导致一个新的对象产生，比如 String s1 ="Hello"；s1=s1+"world"两者相加会产生新的字符串。

有些非线程安全可变对象被不可变机制加以处理之后，照样也具备不可变性，比如 ArrayList 生成的 stream 在多线程的情况下也是线程安全的，同样是因为其具备不可变性的结果，示例代码如清单 18-1 所示。

代码清单 18-1　ArrayListStream.java

```java
package com.wangwenjun.concurrent.chapter18;

import java.util.Arrays;
import java.util.List;

public class ArrayListStream
{
    public static void main(String[] args)
    {
        // 定义一个 list 并且使用 Arrays 的方式进行初始化
        List<String> list = Arrays.asList("Java", "Thread", "Concurrency", "Scala", "Clojure");

        // 获取并行的 stream，然后通过 map 函数对 list 中的数据进行加工，最后输出
        list.parallelStream().map(String::toUpperCase).forEach(System.out::println);
        list.forEach(System.out::println);
    }
}
```

list 虽然是在并行的环境下运行的，但是在 stream 的每一个操作中都是一个全新的 List，根本不会影响到最原始的 list，这样也是符合不可变对象的最基本思想。

18.2.1　非线程安全的累加器

不可变对象最核心的地方在于不给外部修改共享资源的机会，这样就会避免多线程情况下的数据冲突而导致的数据不一致的情况，又能避免因为对锁的依赖而带来的性能降低，好了，在本节中我们将模仿 java.lang.String 的方式实现一个不可变的 int 类型累加器，先来看看不加同步的累加器，代码如清单 18-2 所示。

代码清单 18-2　非线程安全的 IntegerAccumulator.java

```java
package com.wangwenjun.concurrent.chapter18;

import java.util.concurrent.TimeUnit;
import java.util.stream.IntStream;
public class IntegerAccumulator
{
    private int init;

    // 构造时传入初始值
    public IntegerAccumulator(int init)
    {
```

```java
        this.init = init;
    }
    // 对初始值增加i
    public int add(int i)
    {
        this.init += i;
        return this.init;
    }
    // 返回当前的初始值
    public int getValue()
    {
        return this.init;
    }

    public static void main(String[] args)
    {
        // 定义累加器,并且将设置初始值为0
        IntegerAccumulator accumulator = new IntegerAccumulator(0);
        // 定义三个线程,并且分别启动
        IntStream.range(0, 3).forEach(i -> new Thread(() ->
        {
            int inc = 0;
            while (true)
            {
                // 首先获得old value
                int oldValue = accumulator.getValue();
                // 然后调用add方法计算
                int result = accumulator.add(inc);
                System.out.println(oldValue + "+" + inc + "=" + result);
                // 经过验证,如果不合理,则输出错误信息
                if (inc + oldValue != result)
                {
                    System.err.println("ERROR:" + oldValue + "+" + inc + "=" + result);
                }
                inc++;
                // 模拟延迟
                slowly();
            }
        }).start());
    }

    private static void slowly()
    {
        try
        {
            TimeUnit.MILLISECONDS.sleep(1);
        } catch (InterruptedException e)
        {
            e.printStackTrace();
```

                }
            }
        }

这段程序既没有对共享资源进行共享锁的保护，也没有进行不可变的设计，在程序的运行过程中偶尔会出现错误的情况，输出如下：

```
... 省略
13348   7   05+2980=13351685
13351   6   85+2946=13354631
13354   6   31+2981=13357612
ERROR:13363669+3111=13369762
ERROR:13372874+2983=13378970
ERROR:13378970+2984=13385068
13357   6   12+2947=13360559
13360   5   59+3110=13363669
...
```

18.2.2 方法同步增加线程安全性

18.2.1 中的程序出现错误的原因是显而易见的，共享资源被多个线程操作未加任何同步控制，出现数据不一致的问题是情理之中的事情，下面我们对 init 变量的操作加以同步，情况就会变得不一样了，改进代码如下：

```java
public static void main(String[] args)
{
    IntegerAccumulator accumulator = new IntegerAccumulator(0);
    IntStream.range(0, 3).forEach(i -> new Thread(() ->
    {
        int inc = 0;
        while (true)
        {
            int oldValue;
            int result;
            // 使用 class 实例作为同步锁
            synchronized (IntegerAccumulator.class)
            {
                oldValue = accumulator.getValue();
                result = accumulator.add(inc);
            }
            System.out.println(oldValue + "+" + inc + "=" + result);
            if (inc + oldValue != result)
            {
                System.err.println("ERROR:" + oldValue + "+" + inc + "=" + result);
            }
            inc++;
            slowly();
        }
```

```java
        })).start());
    }
```

这里将数据同步的控制放在了线程的逻辑执行单元中，而在 IntegerAccumulator 中未增加任何同步的控制，如果单纯对 getValue 方法和 add 方法增加同步控制，虽然保证了单个方法的原子性，但是两个原子类型的操作在一起未必就是原子性的，因此在线程的逻辑执行单元中增加同步控制是最为合理的。

18.2.3 不可变的累加器对象设计

18.2.2 节中通过同步的方式解决了线程安全性的问题，正确的加锁方式固然能使得一个类变成线程安全的，比如 java.utils.Vector，但是我们需要的是设计出类似于 java.lang.String 的不可变类，示例代码如清单 18-3 所示。

代码清单 18-3　不可变累加器的设计 IntegerAccumulator.java

```java
package com.wangwenjun.concurrent.chapter18;

import java.util.concurrent.TimeUnit;
import java.util.stream.IntStream;

// 不可变对象不允许被继承
public final class IntegerAccumulator
{
    private final int init;

    // 构造时传入初始值
    public IntegerAccumulator(int init)
    {
        this.init = init;
    }

    // 构造新的累加器，需要用到另外一个 accumulator 和初始值
    public IntegerAccumulator(IntegerAccumulator accumulator, int init)
    {
        this.init = accumulator.getValue() + init;
    }

    // 每次相加都会产生一个新的 IntegerAccumulator
    public IntegerAccumulator add(int i)
    {
        return new IntegerAccumulator(this, i);
    }

    public int getValue()
    {
        return this.init;
    }
```

```java
public static void main(String[] args)
{
    //用同样的方式进行测试
    IntegerAccumulator accumulator = new IntegerAccumulator(0);
    IntStream.range(0, 3).forEach(i -> new Thread(() ->
    {
        int inc = 0;
        while (true)
        {
            int oldValue = accumulator.getValue();
            int result = accumulator.add(inc).getValue();
            System.out.println(oldValue + "+" + inc + "=" + result);
            if (inc + oldValue != result)
            {
                System.err.println("ERROR:" + oldValue + "+" + inc + "=" + result);
            }
            inc++;
            slowly();
        }
    }).start());
}
private static void slowly()
{
    try
    {
        TimeUnit.MILLISECONDS.sleep(1);
    } catch (InterruptedException e)
    {
        e.printStackTrace();
    }
}
```

重构后的 IntegerAccumulator，使用了 final 修饰其的目的是为了防止由于继承重写而导致失去线程安全性，另外 init 属性被 final 修饰不允许线程对其进行改变，在构造函数中赋值后将不会再改变。

add 方法并未在原有 init 的基础之上进行累加，而是创建了一个全新的 IntegerAccumulator，并未提供任何修改原始 IntegerAccumulator 的机会，运行上面的程序不会出现 ERROR 的情况。

18.3 本章总结

设计一个不可变的类共享资源需要具备不可破坏性，比如使用 final 修饰，另外针对共享资源操作的方法是不允许被重写的，以防止由于继承而带来的安全性问题，但是单凭这两点也不足以保证一个类是不可变的，比如下面的类用 final 修饰，并且其中的 list 也是

final 修饰的，只允许在构造时创建：

```
public final class Immutable
{
    private final List<String> list;

    public Immutable(List<String> list)
    {
        this.list = list;
    }

    public List<String> getList(){
        return this.list;
    }
}
```

Immutable 类被 final 修饰因此不允许更改，同样 list 只能在构造时被指定，但是该类同样是可变的（mutable），因为 getList 方法返回的 list 是可被其他线程修改的，如果想要使其真正的不可变，则需要在返回 list 的时候增加不可修改的约束 Collections.unmodifiableList(this.list) 或者克隆一个全新的 list 返回。

CHAPTER19 · 第 19 章

Future 设计模式

19.1 先给你一张凭据

假设有个任务需要执行比较长的的时间，通常需要等待任务执行结束或者出错才能返回结果，在此期间调用者只能陷入阻塞苦苦等待，对此，Future 设计模式提供了一种凭据式的解决方案。在我们日常生活中，关于凭据的使用非常多见，比如你去某西服手工作坊想订做一身合体修身的西服，西服的制作过程比较漫长，少则一个礼拜，多则一个月，你不可能一直待在原地等待，一般来说作坊会为你开一个凭据，此凭据就是 Future，在接下来的任意日子里你可以凭借此凭据到作坊获取西服。在本章中，我们将通过程序的方式实现 Future 设计模式，让读者体会这种设计的好处。

自 JDK1.5 起，Java 提供了比较强大的 Future 接口，在 JDK1.8 时更是引入了 CompletableFuture，其结合函数式接口可实现更强大的功能，由于本书不涉及讨论并发包的知识点，读者可自行查阅。

19.2 Future 设计模式实现

图 19-1 是 Future 设计模式所涉及的关键接口和它们之间的关系 UML 图，其中 FutureTest 用于测试。

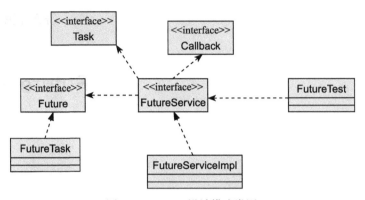

图 19-1 Future 设计模式类图

19.2.1 接口定义

1. Future 接口设计

Future 提供了获取计算结果和判断任务是否完成的两个接口，其中获取计算结果将会导致调用阻塞（在任务还未完成的情况下），相关代码如清单 19-1 所示。

代码清单 19-1 接口 Future.java

```java
package com.wangwenjun.concurrent.chapter19;

public interface Future<T>
{
    // 返回计算后的结果，该方法会陷入阻塞状态
    T get() throws InterruptedException;

    // 判断任务是否已经被执行完成
    boolean done();
}
```

2. FutureService 接口设计

FutureService 主要用于提交任务，提交的任务主要有两种，第一种不需要返回值，第二种则需要获得最终的计算结果。FutureService 接口中提供了对 FutureServiceImpl 构建的工厂方法，JDK 8 中不仅支持 default 方法还支持静态方法，JDK 9 甚至还支持接口私有方法。FutureService 接口的设计代码如清单 19-2 所示。

代码清单 19-2 FutureService.java

```java
package com.wangwenjun.concurrent.chapter19;

public interface FutureService<IN, OUT>
{
    // 提交不需要返回值的任务，Future.get 方法返回的将会是 null
    Future<?> submit(Runnable runnable);
```

```java
// 提交需要返回值的任务,其中Task接口代替了Runnable接口
Future<OUT> submit(Task<IN, OUT> task, IN input);

// 使用静态方法创建一个FutureService的实现
static <T, R> FutureService<T, R> newService()
{
    return new FutureServiceImpl<>();
}
}
```

3. Task 接口设计

Task 接口主要是提供给调用者实现计算逻辑之用的,可以接受一个参数并且返回最终的计算结果,这一点非常类似于 JDK1.5 中的 Callable 接口,Task 接口的设计代码如清单 19-3 所示。

代码清单 19-3　函数式接口 Task.java

```java
package com.wangwenjun.concurrent.chapter19;

@FunctionalInterface
public interface Task<IN, OUT>
{
    // 给定一个参数,经过计算返回结果
    OUT get(IN input);
}
```

19.2.2　程序实现

1. FutureTask

FutureTask 是 Future 的一个实现,除了实现 Future 中定义的 get() 以及 done() 方法,还额外增加了 protected 方法 finish,该方法主要用于接收任务被完成的通知,FutureTask 接口的设计代码如清单 19-4 所示。

代码清单 19-4　FutureTask.java

```java
package com.wangwenjun.concurrent.chapter19;

public class FutureTask<T> implements Future<T>
{
    // 计算结果
    private T result;
    // 任务是否完成
    private boolean isDone = false;
    // 定义对象锁
    private final Object LOCK = new Object();

    @Override
```

```java
    public T get() throws InterruptedException
    {
        synchronized (LOCK)
        {
            // 当任务还没完成时，调用 get 方法会被挂起而进入阻塞
            while (!isDone)
            {
                LOCK.wait();
            }
            // 返回最终计算结果
            return result;
        }
    }

    //finish 方法主要用于为 FutureTask 设置计算结果
    protected void finish(T result)
    {
        synchronized (LOCK)
        {
            //balking 设计模式
            if (isDone)
                return;
            // 计算完成，为 result 指定结果，并且将 isDone 设为 true，同时唤醒阻塞中的线程
            this.result = result;
            this.isDone = true;
            LOCK.notifyAll();
        }
    }

    // 返回当前任务是否已经完成
    @Override
    public boolean done()
    {
        return isDone;
    }
}
```

FutureTask 中充分利用了线程间的通信 wait 和 notifyAll，当任务没有被完成之前通过 get 方法获取结果，调用者会进入阻塞，直到任务完成并接收到其他线程的唤醒信号，finish 方法接收到了任务完成通知，唤醒了因调用 get 而进入阻塞的线程。

2. FutureServiceImpl

FutureServiceImpl 接口的设计代码如清单 19-5 所示。

代码清单 19-5　FutureServiceImpl.java

```java
package com.wangwenjun.concurrent.chapter19;

import java.util.concurrent.atomic.AtomicInteger;
```

```java
/**
 * FutureServiceImpl 的主要作用在于当提交任务时创建一个新的线程来受理该任务，进而达到任务异步执
 * 行的效果
 */
public class FutureServiceImpl<IN, OUT> implements FutureService<IN, OUT>
{
    // 为执行的线程指定名字前缀（再三强调，为线程起一个特殊的名字是一个非常好的编程习惯）
    private final static String FUTURE_THREAD_PREFIX = "FUTURE-";

    private final AtomicInteger nextCounter = new AtomicInteger(0);

    private String getNextName()
    {
        return FUTURE_THREAD_PREFIX + nextCounter.getAndIncrement();
    }

    @Override
    public Future<?> submit(Runnable runnable)
    {
        final FutureTask<Void> future = new FutureTask<>();
        new Thread(() ->
        {
            runnable.run();
            // 任务执行结束之后将 null 作为结果传给 future
            future.finish(null);
        }, getNextName()).start();

        return future;
    }

    @Override
    public Future<OUT> submit(Task<IN, OUT> task, IN input)
    {
        final FutureTask<OUT> future = new FutureTask<>();
        new Thread(() ->
        {
            OUT result = task.get(input);
            // 任务执行结束之后，将真实的结果通过 finish 方法传递给 future
            future.finish(result);
        }, getNextName()).start();

        return future;
    }
}
```

在 FutureServiceImpl 的 submit 方法中，分别启动了新的线程运行任务，起到了异步的作用，在任务最终运行成功之后，会通知 FutureTask 任务已完成。

19.3　Future 的使用以及技巧总结

Future 直译是"未来"的意思，主要是将一些耗时的操作交给一个线程去执行，从而达到异步的目的，提交线程在提交任务和获得计算结果的过程中可以进行其他的任务执行，而不至于傻傻等待结果的返回。

我们提供了两种任务的提交（无返回值和有返回值）方式，在这里分别对其进行测试。无返回值的任务提交测试如下：

```
// 定义不需要返回值的 FutureService
FutureService<Void, Void> service = FutureService.newService();
//submit 方法为立即返回的方法
    Future<?> future = service.submit(() ->
    {
        try
        {
            TimeUnit.SECONDS.sleep(10);
        } catch (InterruptedException e)
        {
            e.printStackTrace();
        }
        System.out.println("I am finish done.");
    });
//get 方法会使当前线程进入阻塞
    future.get();
```

上面的测试代码中提交了一个无返回值的任务，当调用了 submit 方法之后会立即返回不再进入阻塞。下面是有返回值的任务提交测试：

```
// 定义有返回值的 FutureService
FutureService<String, Integer> service = FutureService.newService();
//submit 方法会立即返回
Future<Integer> future = service.submit(input ->
{
    try
    {
        TimeUnit.SECONDS.sleep(10);
    } catch (InterruptedException e)
    {
        e.printStackTrace();
    }
    return input.length();
}, "Hello");
//get 方法使当前线程进入阻塞，最终会返回计算的结果
System.out.println(future.get());
```

上面的测试提交了一个有返回值类型的任务，用于计算字符串的长度，最后的计算结果将会返回输入字符串的长度。

至此，Future 模式设计已讲解完毕，虽然我们提交任务时不会进入任何阻塞，但是当调用者需要获取结果的时候，还是有可能陷入阻塞直到任务完成，其实这个问题不仅在我们设计的 Future 中有，在 JDK 1.5 时期也存在，直到 JDK 1.8 引入了 CompletableFuture 才得到了完美的增强，那么在此期间各种开源项目中都给出了各自的解决方案，比如 Google 的 Guava Toolkit 就提供了 ListenableFuture 用于支持任务完成时回调的方式。

19.4　增强 FutureService 使其支持回调

使用任务完成时回调的机制可以让调用者不再进行显式地通过 get 的方式获得数据而导致进入阻塞，可在提交任务的时候将回调接口一并注入，在这里对 FutureService 接口稍作修改，修改代码如清单 19-6 所示。

代码清单 19-6　增加回调机制，提高使用体验

```java
// 增加回调接口 Callback, 当任务执行结束之后，Callback 会得到执行
@Override
public Future<OUT> submit(Task<IN, OUT> task, IN input, Callback<OUT> callback)
{
    final FutureTask<OUT> future = new FutureTask<>();
    new Thread(() ->
    {
        OUT result = task.get(input);
        future.finish(result);
        // 执行回调接口
        if (null != callback)
            callback.call(result);
    }, getNextName()).start();
    return future;
}
```

修改后的 submit 方法，增加了一个 Callback 参数，主要用来接受并处理任务的计算结果，当提交的任务执行完成之后，会将结果传递给 Callback 接口进行进一步的执行，这样在提交任务之后不再会因为通过 get 方法获得结果而陷入阻塞。

Callback 接口非常简单，非常类似于 JDK 8 中的 Consumer 函数式接口，Callback 接口的代码如清单 19-7 所示。

代码清单 19-7　回调接口 Callback.java

```java
package com.wangwenjun.concurrent.chapter19;

@FunctionalInterface
public interface Callback<T>
{
    // 任务完成后会调用该方法，其中 T 为任务执行后的结果
    void call(T t);
}
```

好了，我们再测试一下增加了 Callback 之后的 Future 任务提交，代码如下：

```
public static void main(String[] args)
        throws InterruptedException
{
FutureService<String, Integer> service = FutureService.newService();
    service.submit(input ->
    {
        try
        {
            TimeUnit.SECONDS.sleep(10);
        } catch (InterruptedException e)
        {
            e.printStackTrace();
        }
        return input.length();
    }, "Hello", System.out::println);
}
```

System.out::println 是一个 Lambda 表达式的静态推导，其作用就是实现 call 方法，通过升级后的程序你会发现，我们再也不需要通过 get 的方式来获得结果了，当然你也可以继续使用 get 方法获得最终的计算结果。

19.5 本章总结

当某个任务运行需要较长的时间时，调用线程在提交任务之后的徒劳等待对 CPU 资源来说是一种浪费，在等待的这段时间里，完全可以进行其他任务的执行，这种场景完全符合 Future 设计模式的应用，虽然我们实现了一个简单的 Future 设计，但是仍旧存在诸多缺陷，读者在阅读完本章内容之后可以对其进行再次增强。

- 将提交的任务交给线程池运行，比如我们在第 8 章自定义的线程池。
- Get 方法没有超时功能，如果获取一个计算结果在规定的时间内没有返回，则可以抛出异常通知调用线程。
- Future 未提供 Cancel 功能，当任务提交之后还可以对其进行取消。
- 任务运行时出错未提供回调方式。
- 其他需要改进的地方请读者自行思考并解决。

Guarded Suspension 设计模式

20.1 什么是 Guarded Suspension 设计模式

Suspension 是 "挂起"、"暂停" 的意思, 而 Guarded 则是 "担保" 的意思, 连在一起就是确保挂起。当线程在访问某个对象时, 发现条件不满足, 就暂时挂起等待条件满足时再次访问, 这一点和 Balking 设计模式刚好相反 (Balking 在遇到条件不满足时会放弃)。

Guarded Suspension 设计模式是很多设计模式的基础, 比如生产者消费者模式, Worker Thread 设计模式, 等等, 同样在 Java 并发包中的 BlockingQueue 中也大量使用到了 Guarded Suspension 设计模式。

20.2 Guarded Suspension 的示例

我们来看一个比较简单的示例, 在学习生产者消费者模式以及 BooleanLock 时都曾写过类似的代码, 如代码清单 20-1 所示。

代码清单 20-1　GuardedSuspensionQueue.java

```java
package com.wangwenjun.concurrent.chapter20;

import java.util.LinkedList;

public class GuardedSuspensionQueue
{
    //定义存放 Integer 类型的 queue
    private final LinkedList<Integer> queue = new LinkedList<>();
```

```java
// 定义 queue 的最大容量为 100
private final int LIMIT = 100;

// 往 queue 中插入数据，如果 queue 中的元素超过了最大容量，则会陷入阻塞
public void offer(Integer data) throws InterruptedException
{
    synchronized (this)
    {
        // 判断 queue 的当前元素是否超过了 LIMIT
        while (queue.size() >= LIMIT)
        {
            // 挂起当前线程，使其陷入阻塞
            this.wait();
        }
        // 插入元素并且唤醒 take 线程
        queue.addLast(data);
        this.notifyAll();
    }
}

// 从队列中获取元素，如果队列此时为空，则会使当前线程阻塞
public Integer take() throws InterruptedException
{
    synchronized (this)
    {
        // 判断如果队列为空
        while (queue.isEmpty())
        {
            // 则挂起当前线程
            this.wait();
        }
        // 通知 offer 线程可以继续插入数据了
        this.notifyAll();
        return queue.removeFirst();
    }
}
```

在 GuardedSuspensionQueue 中，我们需要保证线程安全的是 queue，分别在 take 和 offer 方法中对应的临界值是 queue 为空和 queue 的数量 >=100，当 queue 中的数据已经满时，如果有线程调用 offer 方法则会被挂起（Suspension），同样，当 queue 没有数据的时候，调用 take 方法也会被挂起。

Guarded Suspension 模式是一个非常基础的设计模式，它主要关注的是当某个条件（临界值）不满足时将操作的线程正确地挂起，以防止出现数据不一致或者操作超过临界值的控制范围。

20.3 本章总结

Guarded Suspension 设计模式并不复杂，但是它是很多其他线程设计模式的基础，比如生产者消费者模式，后文中的 Thread Worker 设计模式、Balking 设计模式等，都可以看到 Guarded Suspension 模式的影子，Guarded Suspension 的关注点在于临界值的条件是否满足，当达到设置的临界值时相关线程则会被挂起。

第 21 章 · CHAPTER21

线程上下文设计模式

21.1 什么是上下文

关于上下文（context），我们在开发的过程中经常会遇到，比如开发struts2的ActionContext、Spring中的ApplicationContext，上下文是贯穿整个系统或阶段生命周期的对象，其中包含了系统全局的一些信息，比如登录之后的用户信息、账号信息，以及在程序每一个阶段运行时的数据。

在第14章中定义的单例对象实际上也是上下文，并且它是贯穿整个程序运行的生命周期的上下文，比如清单21-1中的代码就是典型的使用单例对象充当系统级别上下文的例子：

代码清单21-1　单例ApplicationContext.java

```java
package com.wangwenjun.concurrent.chapter21;

public final class ApplicationContext
{
    //在Context中保存configuration实例
    private ApplicationConfiguration configuration;
    //在Context中保存runtimeinfor实例
    private RuntimeInfo runtimeInfo;
    //...其他

    //采用Holder的方式实现单例
    private static class Holder
    {
        private static ApplicationContext instance = new ApplicationContext();
```

```java
    }

    public static ApplicationContext getContext()
    {
        return Holder.instance;
    }

    public void setConfiguration(ApplicationConfiguration configuration)
    {
        this.configuration = configuration;
    }

    public ApplicationConfiguration getConfiguration()
    {
        return this.configuration;
    }

    public void setRuntimeInfo(RuntimeInfo runtimeInfo)
    {
        this.runtimeInfo = runtimeInfo;
    }

    public RuntimeInfo getRuntimeInfo()
    {
        return this.runtimeInfo;
    }
}
```

在上面的代码中我们不难发现，如果 configuration 和 runtimeInfo 的生命周期会随着被创建一直到系统运行结束，我们就可以将 ApplicationContext 称为系统的上下文，诸如 configuration 和 runtimeInfo 等其他实例属性则称为系统上下文成员。

当然在设计系统上下文时，除了要考虑到它的全局唯一性（单例设计模式保证）之外，还要考虑到有些成员只能被初始化一次，比如配置信息的加载（在第 22 章所讲的 Balking 设计模式就可以保证这一点），以及在多线程环境下，上下文成员的线程安全性（第 16 章 "Single Thread Execution 设计模式"，第 18 章 "不可变对象设计模式" 等资源保护方法）。

21.2　线程上下文设计

在有些时候，单个线程执行的任务步骤会非常多，后一个步骤的输入有可能是前一个步骤的输出，比如在单个线程多步骤（阶段）执行时，为了使得功能单一，有时候我们会采用 GoF 职责链设计模式，如图 21-1 所示。

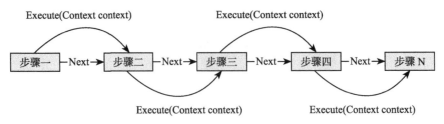

图 21-1　职责链设计模式使得每个步骤功能单一

虽然有些时候后一个步骤未必会需要前一个步骤的输出结果，但是都需要将 context 从头到尾进行传递，假如方法参数比较少还可以容忍，如果方法参数比较多，在七八次的调用甚至十几次的调用，都需要从头到尾地传递 context，很显然这是一种比较烦琐的设计，那么我们就可以尝试采用线程的上下文设计来解决这样的问题。

我们在 ApplicationContext 中增加 ActionContext（线程上下文）相关的内容，代码如下：

```
private ConcurrentHashMap<Thread, ActionContext> contexts =
        new ConcurrentHashMap<>();
public ActionContext getActionContext()
{
ActionContext actionContext = contexts.get(Thread.currentThread());
    if (actionContext == null)
    {
        actionContext = new ActionContext();
        contexts.put(Thread.currentThread(), actionContext);
    }
    return actionContext;
}
```

不同的线程访问 getActionContext() 方法，每一个线程都将会获得不一样的 ActionContext 实例，原因是我们采用 Thread.currentThread() 作为 contexts 的 key 值，这样就可以保证线程之间上下文的独立性，同时也不用考虑 ActionContext 的线程安全性（因为始终只有一个线程访问 ActionContext），因此线程上下文又被称为"线程级别的单例"。

> **注意**　通过这种方式定义线程上下文很可能会导致内存泄漏，contexts 是一个 Map 的数据结构，用当前线程做 key，当线程的生命周期结束后，contexts 中的 Thread 实例不会被释放，与之对应的 Value 也不会被释放，时间长了就会导致内存泄漏（Memory Leak），当然可以通过 soft reference 或者 weak reference 等引用类型，JVM 会主动尝试回收（关于 Java 中的四种引用类型，可以参考 25.3 节）。

21.3　ThreadLocal 详解

自 JDK1.2 版本起，Java 就提供了 java.lang.ThreadLocal，ThreadLocal 为每一个使用该

变量的线程都提供了独立的副本,可以做到线程间的数据隔离,每一个线程都可以访问各自内部的副本变量。

21.3.1 ThreadLocal 的使用场景及注意事项

ThreadLocal 在 Java 的开发中非常常见,一般在以下情况中会使用到 ThreadLocal。

- 在进行对象跨层传递的时候,可以考虑使用 ThreadLocal,避免方法多次传递,打破层次间的约束。
- 线程间数据隔离,比如 21.2 节中描述的线程上下文 ActionContext。
- 进行事务操作,用于存储线程事务信息。

ThreadLocal 并不是解决多线程下共享资源的技术,一般情况下,每一个线程的 ThreadLocal 存储的都是一个全新的对象(通过 new 关键字创建),如果多线程的 ThreadLocal 存储了一个对象的引用,那么其还将面临资源竞争,数据不一致等并发问题。

21.3.2 ThreadLocal 的方法详解及源码分析

我们先来看一个简单的 ThreadLocal 实例,让读者对使用 ThreadLocal 有一个快速的认识,代码如清单 21-2 所示。

代码清单 21-2　ThreadLocal 简单测试

```java
package com.wangwenjun.concurrent.chapter21;

import java.util.concurrent.TimeUnit;
import java.util.stream.IntStream;

import static java.lang.Thread.currentThread;

public class ThreadLocalExample
{
    public static void main(String[] args)
    {
        // 创建 ThreadLocal 实例
        ThreadLocal<Integer> tlocal = new ThreadLocal<>();
        // 创建十个线程,使用 tlocal
        IntStream.range(0, 10)
            .forEach(i -> new Thread(() ->
                {
                    try
                    {
                        // 每个线程都会设置 tlocal,但是彼此之间的数据是独立的
                        tlocal.set(i);
                        System.out.println(currentThread() + " set i " + tlocal.get());
                        TimeUnit.SECONDS.sleep(1);
                        System.out.println(currentThread() + " get i " + tlocal.
```

```
get());
                            } catch (InterruptedException e)
                            {
                                e.printStackTrace();
                            }
                        }).start()
                );
        }
    }
```

上面的代码中定义了一个全局唯一的 ThreadLocal<Integer>，然后启动了 10 个线程对 threadLocal 进行 set 和 get 操作，通过下面的输出可以发现，这 10 个线程之间彼此不会相互影响，每一个线程存入 threadLocal 中的 i 值也是完全不同彼此独立的。

```
Thread[Thread-0,5,main] set i 0
Thread[Thread-3,5,main] set i 3
Thread[Thread-8,5,main] set i 8
Thread[Thread-7,5,main] set i 7
Thread[Thread-4,5,main] set i 4
Thread[Thread-2,5,main] set i 2
Thread[Thread-6,5,main] set i 6
Thread[Thread-1,5,main] set i 1
Thread[Thread-5,5,main] set i 5
Thread[Thread-9,5,main] set i 9
Thread[Thread-0,5,main] get i 0
Thread[Thread-3,5,main] get i 3
Thread[Thread-7,5,main] get i 7
Thread[Thread-8,5,main] get i 8
Thread[Thread-4,5,main] get i 4
Thread[Thread-2,5,main] get i 2
Thread[Thread-6,5,main] get i 6
Thread[Thread-1,5,main] get i 1
Thread[Thread-5,5,main] get i 5
Thread[Thread-9,5,main] get i 9
```

在使用 ThreadLocal 的时候，最常用的方法就是 initialValue()、set(T t)、get()。

（1）initialValue() 方法

initialValue() 方法为 ThreadLocal 要保存的数据类型指定了一个初始化值，在 ThreadLocal 中默认返回值为 null，示例代码如下：

```
//ThreadLocal 中 initialValue() 方法源码
protected T initialValue() {
    return null;
}
```

但是我们可以通过重写 initialValue() 方法进行数据的初始化，如下面的代码所示，线程并未对 threadlocal 进行 set 操作，但是还可以通过 get 方法得到一个初始值，通过输出信

息也不难看出,每一个线程通过 get 方法获取的值都是不一样的(线程私有的数据拷贝):

```
ThreadLocal<Object> threadLocal = new ThreadLocal<Object>()
{
    @Override
    protected Object initialValue()
    {
        return new Object();
    }
};
new Thread(() ->
        System.out.println(threadLocal.get())
).start();
System.out.println(threadLocal.get());
```

在 get 和 get 方法源码分析的时候,我们会看到 initialValue() 方法何时被调用。

> **提醒** 上述重写 initialValue() 方法的方式若使用 Java 8 提供的 Supplier 函数接口会更加简化:
> `ThreadLocal<Object> threadLocal = ThreadLocal.withInitial(Object::new)`

(2) set(T t) 方法

set 方法主要是为 ThreadLocal 指定将要被存储的数据,如果重写了 initialValue() 方法,在不调用 set(T t) 方法的时候,数据的初始值是 initialValue() 方法的计算结果,示例代码如下:

```
//ThreadLocal 的 set 方法源码
public void set(T value) {
    Thread t = Thread.currentThread();
    ThreadLocalMap map = getMap(t);
    if (map != null)
        map.set(this, value);
    else
        createMap(t, value);
}

//ThreadLocal 的 createMap 方法源码
void createMap(Thread t, T firstValue) {
    t.threadLocals = new ThreadLocalMap(this, firstValue);
}

//ThreadLocalMap 的 set 方法源码
private void set(ThreadLocal<?> key, Object value) {
    Entry[] tab = table;
    int len = tab.length;
    int i = key.threadLocalHashCode & (len-1);
    for (Entry e = tab[i];
         e != null;
         e = tab[i = nextIndex(i, len)]) {
```

```
            ThreadLocal<?> k = e.get();
            if (k == key) {
                    e.value = value;
                    return;
            }
            if (k == null) {
                    replaceStaleEntry(key, value, i);
                    return;
            }
    }
    tab[i] = new Entry(key, value);
    int sz = ++size;
    if (!cleanSomeSlots(i, sz) && sz >= threshold)
            rehash();
}
```

上述代码的运行步骤具体如下。

1）获取当前线程 Thread.currentThread()。

2）根据当前线程获取与之关联的 ThreadLocalMap 数据结构。

3）如果 map 为 null 则进入第 4 步，否则进入第 5 步。

4）当 map 为 null 的时候创建一个 ThreadLocalMap，用当前 ThreadLocal 实例作为 key，将要存放的数据作为 Value，对应到 ThreadLocalMap 中则是创建了一个 Entry。

5）在 map 的 set 方法中遍历整个 map 的 Entry，如果发现 ThreadLocal 相同，则使用新的数据替换即可，set 过程结束。

6）在遍历 map 的 entry 过程中，如果发现有 Entry 的 Key 值为 null，则直接将其逐出并且使用新的数据占用被逐出数据的位置，这个过程主要是为了防止内存泄漏（关于 ThreadLocal 的内存泄漏在 21.3.3 节中有详细介绍）。

7）创建新的 entry，使用 ThreadLocal 作为 Key，将要存放的数据作为 Value。

8）最后再根据 ThreadLocalMap 的当前数据元素的大小和阀值做比较，再次进行 key 为 null 的数据项清理工作。

（3）get() 方法

get 用于返回当前线程在 ThreadLocal 中的数据备份，当前线程的数据都存放在一个称为 ThreadLocalMap 的数据结构中，我们稍后会介绍 ThreadLocalMap，get 方法示例代码如下：

```
//ThreadLocal 的 get 方法源码
public T get() {
    Thread t = Thread.currentThread();
    ThreadLocalMap map = getMap(t);
    if (map != null) {
        ThreadLocalMap.Entry e = map.getEntry(this);
        if (e != null) {
            @SuppressWarnings("unchecked")
```

```
            T result = (T)e.value;
            return result;
        }
    }
    return setInitialValue();
}

//ThreadLocal 的 setInitialValue 方法源码
private T setInitialValue() {
    T value = initialValue();
    Thread t = Thread.currentThread();
    ThreadLocalMap map = getMap(t);
    if (map != null)
        map.set(this, value);
    else
        createMap(t, value);
    return value;
}
```

通过上面的源码我们大致分析一下一个数据拷贝的 get 过程，运行步骤具体如下。

1）首先获取当前线程 Thread.currentThread() 方法。

2）根据 Thread 获取 ThreadLocalMap，其中 ThreadLocalMap 与 Thread 是关联的，而我们存入 ThreadLocal 中的数据事实上是存储在 ThreadLocalMap 的 Entry 中的。

3）如果 map 已经被创建过，则以当前的 ThreadLocal 作为 key 值获取对应的 Entry。

4）如果 Entry 不为 null，则直接返回 Entry 的 value 值，否则进入第 5 步。

5）如果在第 2 步获取不到对应的 ThreadLocalMap，则执行 setInitialValue() 方法。

6）在 setInitialValue() 方法中首先通过执行 initialValue() 方法获取初始值。

7）根据当前线程 Thread 获取对应的 ThreadLocalMap。

8）如果 ThreadLocalMap 不为 null，则为 map 指定 initialValue() 所获得的初始值，实际上是在 map.set(this，value) 方法中 new 了一个 Entry 对象。

9）如果 ThreadLocalMap 为 null（首次使用的时候），则创建一个 ThreadLocalMap，并且与 Thread 对象的 threadlocals 属性相关联（通过这里我们也可以发现 ThreadLocalMap 的构造过程是一个 Lazy 的方式）。

10）返回 initialValue() 方法的结果，当然这个结果在没有被重写的情况下结果为 null。

（4）ThreadLocalMap

无论是 get 方法还是 set 方法都不可避免地要与 ThreadLocalMap 和 Entry 打交道，ThreadLocalMap 是一个完全类似于 HashMap 的数据结构，仅仅用于存放线程存放在 ThreadLocal 中的数据备份，ThreadLocalMap 的所有方法对外部都完全不可见。

在 ThreadLocalMap 中用于存储数据的是 Entry，它是一个 WeakReference 类型的子类，之所以被设计成 WeakReference 是为了能够在 JVM 发生垃圾回收事件时，能够自动回收防止内存溢出的情况出现，通过 Entry 源码分析不难发现，在 Entry 中会存储 ThreadLocal 以及所需数据的备份。ThreadLocalMap 的 Entry 源码如下：

```
//ThreadLocalMap 的 Entry 源码
static class Entry extends WeakReference<ThreadLocal<?>> {
    /** The value associated with this ThreadLocal. */
    Object value;
    Entry(ThreadLocal<?> k, Object v) {
        super(k);
        value = v;
    }
}
```

21.3.3　ThreadLocal 的内存泄漏问题分析

在 21.2 节中实现的线程上下文和 ThreadLocal 非常类似，都是使用当前线程作为一个 map 的 Key 值用于线程间数据的隔离，但是 21.2 节中存在内存泄漏的隐患，比如某个线程结束了生命周期，但是 Thread 的实例和所要存储的数据还存在于 contexts 中，随着运行时间的不断增大（比如几个月、半年甚至更久的时间），在 contexts 中将会残留很多 thread 实例以及被保存的数据。

这个问题在 ThreadLocal 中也是存在的，尤其是在早些的 JDK 版本中，ThreadLocalMap 完全是由 HashMap 来充当的，在最新的 JDK 版本中，ThreadLocal 为解决内存泄漏做了很多工作，我们一起通过源码分析结合实战练习来研究下。

- WeakReference 在 JVM 中触发任意 GC（young gc、full gc）时都会导致 Entry 的回收，关于这一点我们在上文中已经说过了，如果想要了解更多关于 Reference 的知识，可以阅读 25.3 节的内容。
- 在 get 数据时增加检查，清除已经被垃圾回收器回收的 Entry（Weak Reference 可自动回收）。

ThreadLocalMap 的源代码片段如下：

```
//ThreadLocalMap 的源码片段
private Entry getEntryAfterMiss(ThreadLocal<?> key, int i, Entry e)
{
    Entry[] tab = table;
    int len = tab.length;
    //查找 Key 为 null 的 Entry
    while (e != null) {
        ThreadLocal<?> k = e.get();
        if (k == key)
            return e;
        if (k == null)
            expungeStaleEntry(i);// 将 key 为 null 的 Entry 删除
        else
            i = nextIndex(i, len);
        e = tab[i];
```

```java
        }
        return null;
    }
    private boolean cleanSomeSlots(int i, int n) {
        boolean removed = false;
        Entry[] tab = table;
        int len = tab.length;
        // 查找 Key 为 null 的 Entry
        do {
            i = nextIndex(i, len);
            Entry e = tab[i];
            if (e != null && e.get() == null) {
                n = len;
                removed = true;
                i = expungeStaleEntry(i);// 将 key 为 null 的 Entry 删除
            }
        } while ( (n >>>= 1) != 0);
        return removed;
    }

    // 执行 Entry 在 ThreadLocalMap 中的删除动作
    private int expungeStaleEntry(int staleSlot) {
        Entry[] tab = table;
        int len = tab.length;
        // expunge entry at staleSlot
        tab[staleSlot].value = null;
        tab[staleSlot] = null;
        size--;
        // Rehash until we encounter null
        Entry e;
        int i;
        for (i = nextIndex(staleSlot, len);
             (e = tab[i]) != null;
             i = nextIndex(i, len)) {
            ThreadLocal<?> k = e.get();
            if (k == null) {
                e.value = null;
                tab[i] = null;
                size--;
            } else {
                int h = k.threadLocalHashCode & (len - 1);
                if (h != i) {
                    tab[i] = null;
                    // Unlike Knuth 6.4 Algorithm R, we must scan until
                    // null because multiple entries could have been stale.
                    while (tab[h] != null)
                        h = nextIndex(h, len);
                    tab[h] = e;
                }
            }
```

```
        }
        return i;
}
```

- 在 set 数据时增加检查，删除已经被垃圾回收器清除的 Entry，并且将其移除，代码如下：

```
//ThreadLocalMap 的源码片段
private boolean cleanSomeSlots(int i, int n) {
    boolean removed = false;
    Entry[] tab = table;
    int len = tab.length;
    // 查找 Key 为 null 的 Entry
    do {
        i = nextIndex(i, len);
        Entry e = tab[i];
        if (e != null && e.get() == null) {
            n = len;
            removed = true;
            i = expungeStaleEntry(i);// 将 key 为 null 的 Entry 删除
        }
    } while ( (n >>>= 1) != 0);
    return removed;
}

// 执行 Entry 在 ThreadLocalMap 中的删除动作
private int expungeStaleEntry(int staleSlot) {
    // 省略
}
```

通过以上这三点的分析，ThreadLocal 可以在一定程度上保证不发生内存泄漏，我们来看看如下的代码：

```
ThreadLocal<byte[]> threadLocal = new ThreadLocal<>();
TimeUnit.SECONDS.sleep(30);
threadLocal.set(new byte[1024 * 1024 * 100]); //100Mb
threadLocal.set(new byte[1024 * 1024 * 100]); //100Mb
threadLocal.set(new byte[1024 * 1024 * 100]); //100Mb
threadLocal = null;
currentThread().join();
```

该代码首先定义了一个 ThreadLocal<byte[]> 分别设置了 100MB 的数据（最终存储于 threadLocal 中的数据以最后一次 set 为主），然后将 threadLocal 设置为 null，最后手动进行一次 full gc，查看内存的变化情况，如图 21-2 所示。

借助于 VisualVM 工具对 JVM 的进程进行监控我们发现，堆内存的大小维持在了 100MB 以上的水准，远远高于应该有的数值，无论进行多少次强制 GC，最后 100MB 的堆内存都不会得到释放，根据 21.3.2 节中对 ThreadLocal 源码的分析，我们来梳理一下

ThreadLocal 中对象的引用链（图 21-3）。

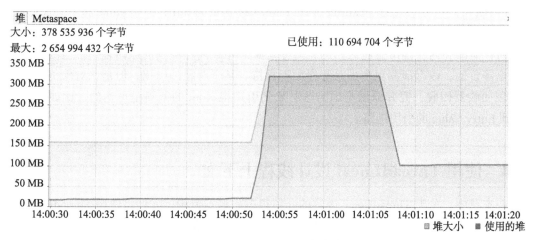

图 21-2　ThreadLocal 测试内存趋势图

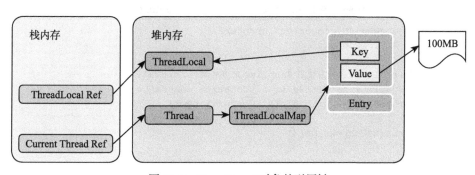

图 21-3　ThreadLocal 对象的引用链

当 Thread 和 ThreadLocal 发生绑定之后，关键对象引用链如图 21-3 所示，与我们在源码分析中的情况是一致的，将 ThreadLocal Ref 显式地指定为 null 时，引用关系链就变成了如图 21-4 所示的情况。

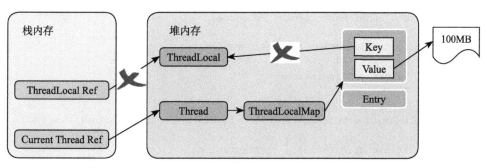

图 21-4　ThreadLocal 中的 Ref 对象为 null 时的引用链

当 ThreadLocal 被显式地指定为 null 之后，执行 GC 操作，此时堆内存中的 ThreadLocal 被回收，同时 ThreadLocalMap 中的 Entry.key 也成为 null，但是 value 将不会被释放，除非当前线程已经结束了生命周期的 Thread 引用被垃圾回收器回收。

内存泄漏和内存溢出是有区别的，内存泄漏是导致内存溢出的原因之一，但两者并不是完全等价的，内存泄漏更多的是程序中不再持有某个对象的引用，但是该对象仍然无法被垃圾回收器回收，究其原因是因为该对象到引用根 Root 的链路是可达的，比如 Thread Ref 到 Entry.Value 的引用链路。

21.4 使用 ThreadLocal 设计线程上下文

在本章中，关于 ThreadLocal 内容的介绍很多，在本节中，我们将使用 ThreadLocal 实现一个简单的线程上下文（一般某个动作或者任务都是交给一个线程去执行的，因此线程上下文的类可以命名为 ActionContext 或者 TaskContext 之类的），代码如清单 21-3 所示。

代码清单 21-3　线程上下文 ActionContext.java

```java
package com.wangwenjun.concurrent.chapter21;
public class ActionContext
{
    // 定义 ThreadLocal, 并且使用 Supplier 的方式重写 initValue
    private static final ThreadLocal<Context> context =
            ThreadLocal.withInitial(Context::new);

    public static Context get()
    {
        return context.get();
    }

    // 每一个线程都会有一个独立的 Context 实例
    static class Context
    {
        // 在 Context 中的其他成员
        private Configuration configuration;
        private OtherResource otherResource;

        public Configuration getConfiguration()
        {
            return configuration;
        }

        public void setConfiguration(Configuration configuration)
        {
            this.configuration = configuration;
        }

        public OtherResource getOtherResource()
```

```
        {
            return otherResource;
        }

        public void setOtherResource(OtherResource otherResource)
        {
            this.otherResource = otherResource;
        }
    }
}
```

为了确保线程间数据隔离的绝对性,重写 initialValue() 方法是一个比较不错的编程体验,ThreadLocal.withInitial(Context::new);在这个 Context 例子中,我们将所有需要被线程访问和操作的数据都封装在了 Context 中,每一个线程拥有不一样的上下文数据,当然你也可以选择为每一个数据定义一个 ThreadLocal,清单 21-4 是更改过后的 ActionContext。

代码清单 21-4　线程上下文第二个版本 ActionContext.java

```
package com.wangwenjun.concurrent.chapter21;

public class ActionContext
{
    // 为 Configuration 创建 ThreadLocal
    private static final ThreadLocal<Configuration> configuration =
            ThreadLocal.withInitial(Configuration::new);

    // 为 OtherResource 创建 ThreadLocal
    private static final ThreadLocal<OtherResource> otherResource =
            ThreadLocal.withInitial(OtherResource::new);

    public static void setConfiguration(Configuration conf)
    {
        configuration.set(conf);
    }

    public static Configuration getConfiguraiton()
    {
        return configuration.get();
    }

    public static void setOtherResource(OtherResource oResource)
    {
        otherResource.set(oResource);
    }

    public static OtherResource getOtherResource()
    {
        return otherResource.get();
    }
}
```

第一种方式只用一个 ThreadLocal，这也就意味着与之对应的 ThreadLocalMap 有一份 Entry，其中 Key 是 ThreadLocal，Value 是 Context，第二种方式使用了两个 ThreadLocal，那么与之对应的 ThreadLocalMap 中将会存在两个 Entry。

21.5 本章总结

线程上下文 ThreadLocal 又被称之为"线程保险箱"，在很多开源软件源码中都能看到 ThreadLocal 的应用，ThreadLocal 能够将指定的变量和当前线程进行绑定，线程之间彼此隔离，持有不同的对象实例，从而避免了数据资源的竞争。

当然，ThreadLocal 也存在着内存泄漏的问题，在本章中我们也进行了比较详细的解释，希望读者在使用 ThreadLocal 的时候引起注意。

第 22 章

Balking 设计模式

22.1 什么是 Balking 设计

多个线程监控某个共享变量，A 线程监控到共享变量发生变化后即将触发某个动作，但是此时发现有另外一个线程 B 已经针对该变量的变化开始了行动，因此 A 便放弃了准备开始的工作，我们把这样的线程间交互称为 Balking（犹豫）设计模式。其实这样的场景在生活中很常见，比如你去饭店吃饭，吃到途中想要再点一个小菜，于是你举起手示意服务员（见图 22-1），其中一个服务员看到了你举手正准备走过来的时候，发现距离你比较近的服务员已经准备要受理你的请求于是中途放弃了。

图 22-1　Balking 设计模式

再比如，我们在用 word 编写文档的时候，每次的文字编辑都代表着文档的状态发生了改变，除了我们可以使用 ctrl+s 快捷键手动保存以外，word 软件本身也会定期触发自动保存，如果 word 自动保存文档的线程在准备执行保存动作的时候，恰巧我们进行了主动保存，那么自动保存文档的线程将会放弃此次的保存动作。

看了以上两个例子的说明，想必大家已经清楚了 Balking 设计模式要解决的问题了吧，简短截说就是某个线程因为发现其他线程正在进行相同的工作而放弃即将开始的任务，在本章中，我们将通过模拟 word 文档自动保存与手动保存的功能讲解 Balking 模式的设计与应用。

22.2 Balking 模式之文档编辑

22.2.1 Document

在代码清单 22-1 中，设计了 Document 类代表文档本身，在 Document 中有两个主要方法 save 和 edit 分别用于保存文档和编辑文档。

代码清单 22-1　Document.java

```java
package com.wangwenjun.concurrent.chapter22;

import java.io.File;
import java.io.FileWriter;
import java.io.IOException;
import java.util.ArrayList;
import java.util.List;

import static java.lang.Thread.currentThread;

//代表正在编辑的文档类
public class Document
{
    //如果文档发生改变，changed 会被设置为 true
    private boolean changed = false;

    //一次需要保存的内容，可以将其理解为内容缓存
    private List<String> content = new ArrayList<>();

    private final FileWriter writer;

    //自动保存文档的线程
    private static AutoSaveThread autoSaveThread;

    //构造函数需要传入文档保存的路径和文档名称
    private Document(String documentPath, String documentName)
            throws IOException
    {
```

```java
        this.writer = new FileWriter(new File(documentPath, documentName), true);
    }

    // 静态方法，主要用于创建文档，顺便启动自动保存文档的线程
    public static Document create(String documentPath, String documentName)
            throws IOException
    {
        Document document = new Document(documentPath, documentName);
        autoSaveThread = new AutoSaveThread(document);
        autoSaveThread.start();
        return document;
    }

    // 文档的编辑，其实就是往 content 队列中提交字符串
    public void edit(String content)
    {
        synchronized (this)
        {
            this.content.add(content);
            // 文档改变，changed 会变为 true
            this.changed = true;
        }
    }

    // 文档关闭的时候首先中断自动保存线程，然后关闭 writer 释放资源
    public void close() throws IOException
    {
        autoSaveThread.interrupt();
        writer.close();
    }

    //save 方法用于为外部显式进行文档保存
    public void save() throws IOException
    {
        synchronized (this)
        {
            //balking, 如果文档已经被保存了，则直接返回
            if (!changed)
            {
                return;
            }

            System.out.println(currentThread() + " execute the save action");
            // 将内容写入文档中
            for (String cacheLine : content)
            {
                this.writer.write(cacheLine);
                this.writer.write("\r\n");
            }

            this.writer.flush();
```

```
            // 将 changed 修改为 false，表明此刻再没有新的内容编辑
            this.changed = false;
            this.content.clear();
        }
    }
}
```

在上述代码中：

- edit 方法和 save 方法进行方法同步，其目的在于防止当文档在保存的过程中如果遇到新的内容被编辑时引起的共享资源冲突问题。
- changed 在默认情况下为 false，当有新的内容被编辑的时候将被修改为 true。
- 在进行文档保存的时候，首先查看 changed 是否为 true，如果文档发生过编辑则在文档中保存新的内容，否则就会放弃此次保存动作，changed 是 balking pattern 关注的状态，当 changed 为 true 的时候就像远处的服务员看到客户的请求被另外一个服务员接管了一样，于是放弃了任务的执行。
- 在创建 Document 的时候，顺便还会启动自动保存文档的线程，该线程的主要目的在于在固定时间里执行一次文档保存动作。

22.2.2 AutoSaveThread

与平日里编写 word 文档一样，word 会定期自动保存我们编辑的文档，如果在电脑出现故障重启之时，没有来得及对文档保存，也不至于损失太多劳动成果，它甚至能够百分之百的恢复，AutoSaveThread 类扮演的角色便在于此，见代码清单 22-2。

代码清单 22-2　自动保存文档的线程 AutoSaveThread.java

```
package com.wangwenjun.concurrent.chapter22;

import java.io.IOException;
import java.util.concurrent.TimeUnit;

public class AutoSaveThread extends Thread
{
    private final Document document;

    public AutoSaveThread(Document document)
    {
        super("DocumentAutoSaveThread");
        this.document = document;
    }

    @Override
    public void run()
    {
        while (true)
```

```
        {
            try
            {
                //每隔一秒自动保存一次文档
                document.save();
                TimeUnit.SECONDS.sleep(1);
            } catch (IOException | InterruptedException e)
            {
                break;
            }
        }
    }
}
```

AutoSaveThread 比较简单，其主要的工作就是每隔一秒的时间调用一次 document 的 save 方法。

22.2.3　DocumentEditThread

AutoSaveThread 线程用于文档自动保存，那么 DocumentEditThread 线程则类似于主动编辑文档的作者，在 DocumentEditThread 中除了对文档进行修改编辑之外，还会同时按下 Ctrl+S 组合键（调用 save 方法）主动保存，见代码清单 22-3。

代码清单 22-3　DocumentEditThread.java

```java
package com.wangwenjun.concurrent.chapter22;

import java.io.IOException;
import java.util.Scanner;

//该线程代表的是主动进行文档编辑的线程，为了增加交互性，我们使用了 Scanner
public class DocumentEditThread extends Thread
{
    private final String documentPath;

    private final String documentName;

    private final Scanner scanner = new Scanner(System.in);

    public DocumentEditThread(String documentPath, String documentName)
    {
        super("DocumentEditThread");
        this.documentPath = documentPath;
        this.documentName = documentName;
    }

    @Override
    public void run()
```

```java
    {
        int times = 0;
        try
        {
            Document document = Document.create(documentPath, documentName);
            while (true)
            {
                // 获取用户的键盘输入
                String text = scanner.next();
                if ("quit".equals(text))
                {
                    document.close();
                    break;
                }
                // 将内容编辑到document中
                document.edit(text);
                if (times == 5)
                {
                    // 用户在输入了5次之后进行文档保存
                    document.save();
                    times = 0;
                }
                times++;
            }
        } catch (IOException e)
        {
            throw new RuntimeException(e);
        }
    }
}
```

DocumentEditThread 类代表了主动编辑文档的线程，在该线程中，我们使用 Scanner 交互的方式，每一次对文档的修改都不可能直接保存（Ctrl+S），因此在程序中约定了五次以后主动执行保存动作，当输入 quit 时，表示要退出此次文档编辑，代码如下：

```java
package com.wangwenjun.concurrent.chapter22;

public class BalkingTest
{
    public static void main(String[] args)
    {
        new DocumentEditThread("G:\\", "balking.txt").start();
    }
}
```

运行上面的程序，然后与系统交互提交内容，交互截图如图 22-2 所示。

```
hello
Thread[DocumentAutoSaveThread,5,main] execute the save action
xxxx
Thread[DocumentAutoSaveThread,5,main] execute the save action
aa
Thread[DocumentAutoSaveThread,5,main] execute the save action
sdsd
Thread[DocumentAutoSaveThread,5,main] execute the save action
sdsdsd
Thread[DocumentAutoSaveThread,5,main] execute the save action
wwwww
Thread[DocumentEditThread,5,main] execute the save action
sdsd
Thread[DocumentAutoSaveThread,5,main] execute the save action
```

图 22-2　文档交互截图

可以看到，自动保存线程和主动保存线程在分别进行文档的保存，它们并不会在不改变内容的情况下执行 save 动作。

22.3　本章总结

Balking 模式在日常的开发中很常见，比如在系统资源的加载或者某些数据的初始化时，在整个系统中声明周期中资源可能只被加载一次，我们就可以采用 balking 模式加以解决，代码如下：

```java
public synchronized Map<String, Resource> load()
{
    //balking
    if (loaded)
    {
        return resourceMap;
    } else
    {
        //do the resource load task.
        //...
        this.loaded = true;
        return resourceMap;
    }
}
```

第 23 章 · CHAPTER23

Latch 设计模式

23.1 什么是 Latch

做程序员是个苦差事,并不像外人看到的那么光鲜,加班熬夜是家常便饭的事,Alex 在结束了连续一个月的加班之后决定叫上几个同为程序员的好朋友(Jack、Gavin、Dillon)打算在周末的时间放松一下,他们相约周六的早上十点在城市广场北口见面,然后一同前往郊区的"同心湖"垂钓、烧烤,最后登上本市的最高建筑"汇聚塔",计划了出游路线之后,每个人或乘坐公交,或搭乘地铁,总之他们都会在城市广场北口相会,然后统一去做同一件事情,那就是一起前往"同心湖"。

诸如此类的情形在我们的日常工作中也是屡见不鲜,比如若干线程并发执行某个特定的任务,然后等到所有的子任务都执行结束之后再统一汇总,比如用户想要查询自己三年以来银行账号的流水,为了保证运行数据库的数据量在一个恒定的范围之内,通常数据只会保存一年的记录,其他的历史记录或被备份到磁盘,或者被存储于 hive 数据仓库,或者被转存至备份数据库之中,总之想要三年的流水记录,需要若干个渠道的查询才可以汇齐。如果一个线程负责执行这样的任务,则需要经历若干次的查询最后汇总返回给用户,很明显这样的操作性能低下,用户体验差,如果我们将每一个渠道的查询交给一个线程或者若干线程去查询,然后统一汇总,那么性能会提高很多,响应时间也会缩短不少。

再回到程序员出游的例子,每个程序员到达城市广场北口的时间不一定,同样到达广场的方式也不尽相同,但是他们都会陆续到达汇集的地点,假设 Jack 由于堵车迟迟未到,其他三个人只能一直等待,直到 Jack 到达城市广场北口才能一同前往。

在本章中,我们将介绍 Latch(门阀)设计模式(见图 23-1),该模式指定了一个屏障,只有所有的条件都达到满足的时候，门阀才能打开。

图 23-1　CountDown Latch 设计模式

23.2　CountDownLatch 程序实现

23.2.1　无限等待的 Latch

在代码清单 23-1 中，首先定义了一个无限等待的抽象类 Latch，在 Latch 抽象类中定义了 await 方法、countDown 方法以及 getUnarrived 方法，这些方法的用途在代码注释中都有详细介绍，当然在 Latch 中的 limit 属性至关重要，当 limit 降低到 0 时门阀将会被打开。

代码清单 23-1　无限等待的 Latch.java

```
package com.wangwenjun.concurrent.chapter23;

public abstract class Latch
{
    //用于控制多少个线程完成任务时才能打开阀门
    protected int limit;

    //通过构造函数传入 limit
    public Latch(int limit)
    {
        this.limit = limit;
    }

    //该方法会使得当前线程一直等待，直到所有的线程都完成工作，被阻塞的线程是允许被中断的
    public abstract void await()
        throws InterruptedException;
```

```java
// 当任务线程完成工作之后调用该方法使得计数器减一
public abstract void countDown();

// 获取当前还有多少个线程没有完成任务
public abstract int getUnarrived();
}
```

子任务数量达到 limit 的时候，门阀才能打开，await() 方法用于等待所有的子任务完成，如果到达数量未达到 limit 的时候，将会无限等待下去，当子任务完成的时候调用 countDown() 方法使计数器减少一个，表明我已经完成任务了，getUnarrived() 方法主要用于查询当前有多少个子任务还未结束。

1. 无限等待 CountDownLatch 实现

下面来实现一个无限制等待门阀打开的 Latch 实现，当 limit >0 时调用 await 方法的线程将会进入无限的等待，见代码清单 23-2。

代码清单 23-2　无限等待 CountDownLatch.java

```java
package com.wangwenjun.concurrent.chapter23;

public class CountDownLatch extends Latch
{
    public CountDownLatch(int limit)
    {
        super(limit);
    }

    @Override
    public void await()
            throws InterruptedException
    {
        synchronized (this)
        {
            // 当 limit>0 时，当前线程进入阻塞状态
            while (limit > 0)
            {
                this.wait();
            }
        }
    }

    @Override
    public void countDown()
    {
        synchronized (this)
        {
            if (limit <=0)
                throw new IllegalStateException("all of task already arrived");
            // 使 limit 减一，并且通知阻塞线程
```

```
            limit--;
            this.notifyAll();
        }
    }

    @Override
    public int getUnarrived()
    {
        // 返回有多少线程还未完成任务
        return limit;
    }
}
```

在上述代码中，await() 方法不断判断 limit 的数量，大于 0 时门阀将不能打开，需要持续等待直到 limit 数量为 0 为止；countDown() 方法调用之后会导致 limit-- 操作，并且通知 wait 中的线程再次判断 limit 的值是否等于 0，当 limit 被减少到了 0 以下，则抛出状态非法的异常；getUnarrived() 获取当前还有多少个子任务未完成，这个返回值并不一定就是准确的，在多线程的情况下，某个线程在获得 Unarrived 任务数量并且返回之后，有可能 limit 又被减少，因此 getUnarrived() 是一个评估值。

2. 程序测试齐心协力打开门阀

连续的加班让这几位程序员游玩的心愿非常迫切，他们力争能够在礼拜六的早上十点到达城市广场，因此每个人都非常积极，清单 23-3 的程序模拟了程序员乘坐不同的交通工具到达城市广场的场景。

代码清单 23-3　ProgrammerTravel.java

```
package com.wangwenjun.concurrent.chapter23;

import java.util.concurrent.ThreadLocalRandom;
import java.util.concurrent.TimeUnit;
/**
 * 程序员旅游线程
 */
public class ProgrammerTravel extends Thread
{
    // 门阀
    private final Latch latch;
    // 程序员
    private final String programmer;
    // 交通工具
    private final String transportation;

    // 通过构造函数传入 latch, programmer, transportation
    public ProgrammerTravel(Latch latch, String programmer, String transportation)
    {
```

```java
        this.latch = latch;
        this.programmer = programmer;
        this.transportation = transportation;
    }

    @Override
    public void run()
    {
        System.out.println(programmer + " start take the transportation [" + transportation + "]");
        try
        {
            // 程序员乘坐交通工具花费在路上的时间（使用随机数字模拟）
            TimeUnit.SECONDS.sleep(ThreadLocalRandom.current().nextInt(10));
        } catch (InterruptedException e)
        {
            e.printStackTrace();
        }
        System.out.println(programmer + " arrived by " + transportation);
        // 完成任务时使计数器减一
        latch.countDown();
    }
}
```

ProgrammerTravel 继承自 Thread 代表程序员，需要三个构造参数，第一个是前文中设计的 latch（门阀），第二个是程序员的名称，比如前文中的 Jack、Gavin 等，第三个参数则表示他们所搭乘的交通工具。

TimeUnit.SECONDS.sleep(ThreadLocalRandom.current().nextInt(10)) 子句在 run 方法中模拟每个人到达目的地所花费的时间，当他们分别到达目的地的时候，需要执行 latch.countDown()，使计数器减少一个以标明自己已到达，代码如下：

```java
public static void main(String[] args)
        throws InterruptedException
{
    // 定义 Latch, limit 为 4
    Latch latch = new CountDownLatch(4);
    new ProgrammerTravel(latch, "Alex", "Bus").start();
    new ProgrammerTravel(latch, "Gavin", "Walking").start();
    new ProgrammerTravel(latch, "Jack", "Subway").start();
    new ProgrammerTravel(latch, "Dillon", "Bicycle").start();
    // 当前线程（main 线程会进入阻塞，直到四个程序员全部都到达目的地）
    latch.await();
    System.out.println("== all of programmer arrived ==");
}
```

在测试代码中，我们定义了四个程序员实例并启动，latch.await() 等待所有的程序员都到达，如果其中一个没有到达则会无限地等待下去，程序输出代码如下：

```
Gavin start take the transportation [Walking]
Dillon start take the transportation [Bicycle]
Jack start take the transportation [Subway]
Alex start take the transportation [Bus]
Jack arrived by Subway
Alex arrived by Bus
Gavin arrived by Walking
Dillon arrived by Bicycle
== all of programmer arrived ==
```

23.2.2 有超时设置的 Latch

在23.2.1节中，我们虽然实现了latch的设计，但是该latch有一个问题，假设Jack在搭乘地铁的途中突然接到公司领导的来电说临时有急事需要处理而不能参加此次旅游，或者是Gavin睡了个懒觉忘记了关于旅游的约定，到达广场的其他人不可能一直等待下去。在本节中，我们将为Latch增加可超时的功能，如果在等待了指定时间之后，还有人未完成任务，则会收到超时通知。

1. 可超时的等待

在Latch中增加可超时的抽象方法await(TimeUnit unit, long time)的示例代码如下：

```java
public abstract void await(TimeUnit unit, long time)
    throws InterruptedException, WaitTimeoutException;
```

其中TimeUnit代表wait的时间单位，而time则是指定数量的时间单位，在该方法中又增加了WaitTimeoutException用于通知当前的等待已经超时，与之相关的代码如清单23-4所示。

代码清单23-4　WaitTimeoutException.java

```java
package com.wangwenjun.concurrent.chapter23;
//当子任务线程执行超时的时候将会抛出该异常
public class WaitTimeoutException extends Exception
{
    public WaitTimeoutException(String message)
    {
        super(message);
    }
}
```

好了，我们看看在CountDownLatch中应该如何实现超时功能，代码如下：

```java
@Override
public void await(TimeUnit unit, long time)
        throws InterruptedException, WaitTimeoutException
{
```

```java
        if (time <= 0)
            throw new IllegalArgumentException("The time is invalid.");
        long remainingNanos = unit.toNanos(time);   // 将time转换为纳秒
        // 等待任务将在endNanos纳秒后超时
        final long endNanos = System.nanoTime() + remainingNanos;
        synchronized (this)
        {
            while (limit > 0)
            {
                // 如果超时则抛出WaitTimeoutException异常
                if (TimeUnit.NANOSECONDS.toMillis(remainingNanos) <= 0)
                    throw new WaitTimeoutException("The wait time over specify time.");
                // 等待remainingNanos，在等待的过程中有可能会被中断，需要重新计算remainingNanos
                this.wait(TimeUnit.NANOSECONDS.toMillis(remainingNanos));
                remainingNanos = endNanos - System.nanoTime();
            }
        }
    }
```

为了方便计算，我们将所有的时间单位都换算成了纳秒，但是Object的wait方法只能够接受毫秒，因此该方法还涉及了时间的换算，另外如果等待剩余时间不足1毫秒，那么将会抛出WaitTimeoutException异常通知等待者。

2. 收到超时通知

世事无常，由于Jack的临时任务，四人旅游的约定就此作罢，当然等到超时之后，其他人可就此终结此次旅行，或者再次商议接下来的休闲活动。收到超时通知后的代码处理如下：

```java
public static void main(String[] args)
        throws InterruptedException
{
    Latch latch = new CountDownLatch(4);
    new ProgrammerTravel(latch, "Alex", "Bus").start();
    new ProgrammerTravel(latch, "Gavin", "Walking").start();
    new ProgrammerTravel(latch, "Jack", "Subway").start();
    new ProgrammerTravel(latch, "Dillon", "Bicycle").start();
    try
    {
        latch.await(TimeUnit.SECONDS, 5);
        System.out.println("== all of programmer arrived ==");
    } catch (WaitTimeoutException e)
    {
        e.printStackTrace();
    }
```

}

修改后的测试代码如上所示，在等待 5 秒之后如果还有未完成的任务，那么将会收到超时通知。

23.3 本章总结

Latch 设计模式提供了等待所有子任务完成，然后继续接下来工作的一种设计方法，自 JDK1.5 起也提供了 CountDownLatch 的工具类，其作用与我们创建的并无两样，无论是我们开发的 CountDownLatch 还是 JDK 所提供的，当 await 超时的时候，已完成任务的线程自然正常结束，但是未完成的则不会被中断还会继续执行下去，也就是说 CountDownLatch 只提供了门阀的功能，并不负责对线程的管理控制，对线程的控制还需要程序员自己控制。

Latch 的作用是为了等待所有子任务完成后再执行其他任务，因此可以对 Latch 进行再次的扩展，增加回调接口用于运行所有子任务完成后的其他任务，增加了回调功能的 CountDownLatch 代码如下：

```java
public CountDownLatch(int limit, Runnable runnable)
{
    super(limit);
    this.runnable = runnable;
}
@Override
public void await()
        throws InterruptedException
{
    synchronized (this)
    {
        while (limit > 0)
        {
            this.wait();
        }
    }
    if (null != runnable)
    {
        runnable.run();
    }
}
@Override
public void await(TimeUnit unit, long time)
        throws InterruptedException, WaitTimeoutException
{
    if (time <= 0)
        throw new IllegalArgumentException("The time is invalid.");
    long remainingNanos = unit.toNanos(time);
    final long endNanos = System.nanoTime() + remainingNanos;
```

```java
        synchronized (this)
        {
            while (limit > 0)
            {
                if (TimeUnit.NANOSECONDS.toMillis(remainingNanos) <= 0)
                    throw new WaitTimeoutException("The wait time over specify time.");
                this.wait(TimeUnit.NANOSECONDS.toMillis(remainingNanos));
                remainingNanos = endNanos - System.nanoTime();
            }
        }
        if (null != runnable)
        {
            runnable.run();
        }
    }
```

CHAPTER24 · 第 24 章

Thread-Per-Message 设计模式

24.1 什么是 Thread-Per-Message 模式

Thread-Per-Message 的意思是为每一个消息的处理开辟一个线程使得消息能够以并发的方式进行处理，从而提高系统整体的吞吐能力。这就好比电话接线员一样，收到的每一个电话投诉或者业务处理请求，都会提交对应的工单，然后交由对应的工作人员来处理，如图 24-1 所示。

图 24-1　每个电话接线员处理一个电话请求

24.2 每个任务一个线程

在本节中，我们首先实现一个简单的 Thread-Per-Message，但是在开发中不建议采用这种方式，在后文中笔者会对此进行详细解释。Request 的代码如清单 24-1 所示。

代码清单 24-1　Request.java

```
package com.wangwenjun.concurrent.chapter24;

public class Request
{
    private final String business;
```

```java
    public Request(String business)
    {
        this.business = business;
    }

    @Override
    public String toString()
    {
        return business;
    }
}
```

客户提交的任何业务受理请求都会被封装成 Request 对象。

TaskHandler 代码如清单 24-2 所示。

代码清单 24-2　TaskHandler.java

```java
package com.wangwenjun.concurrent.chapter24;

import java.util.concurrent.TimeUnit;

import static java.util.concurrent.ThreadLocalRandom.current;

/**
 * TaskHandler 用于处理每一个提交的 Request 请求,由于 TaskHandler 将被
 *Thread 执行,因此需要实现 Runnable 接口
 */
public class TaskHandler implements Runnable
{
    // 需要处理的 Request 请求
    private final Request request;

    public TaskHandler(Request request)
    {
        this.request = request;
    }

    @Override
    public void run()
    {
        System.out.println("Begin handle " + request);
        slowly();
        System.out.println("End handle " + request);
    }

    // 模拟请求处理比较耗时,使线程进入短暂的休眠阶段
    private void slowly()
    {
        try
        {
```

```java
            TimeUnit.SECONDS.sleep(current().nextInt(10));
        } catch (InterruptedException e)
        {
            e.printStackTrace();
        }
    }
}
```

TaskHandler 代表了每一个工作人员接收到任务后的处理逻辑。

Operator 代码如清单 24-3 所示。

代码清单 24-3　Operator.java

```java
package com.wangwenjun.concurrent.chapter24;

public class Operator
{
    public void call(String business)
    {
        // 为每一个请求创建一个线程去处理
        TaskHandler taskHandler = new TaskHandler(new Request(business));
        new Thread(taskHandler).start();
    }
}
```

Operator 代表了接线员，当有电话打进来时，话务员会将客户的请求封装成一个工单 Request，然后开辟一个线程（工作人员）去处理。

截至目前，我们完成了关于 Thread-Per-Message 的设计，但是这种设计方式存在着很严重的问题，经过第 2 章的学习，我们知道每一个 JVM 中可创建的线程数量是有限的，针对每一个任务都创建一个新的线程，假如每一个线程执行的时间比较长，那么在某个时刻 JVM 会由于无法再创建新的线程而导致栈内存的溢出；再假如每一个任务的执行时间都比较短，频繁地创建销毁线程对系统性能的开销也一个不小的影响。

这种处理方式虽然有很多问题，但不代表其就一无是处了，其实它也有自己的使用场景，比如在基于 Event 的编程模型中，当系统初始化事件发生时，需要进行若干资源的后台加载，由于系统初始化时的任务数量并不多，可以考虑使用该模式响应初始化 Event，或者系统在关闭时，进行资源回收也可以考虑将销毁事件触发的动作交给该模式。

我们可以将 call 方法中的创建新线程的方式交给线程池去处理，这样可以避免线程频繁创建和销毁带来的系统开销，还能将线程数量控制在一个可控的范围之内。下面使用我们在第 8 章中开发的线程池重构 Operator，代码如清单 24-4 所示。

代码清单 24-4　使用线程池重构 Operator.java

```java
package com.wangwenjun.concurrent.chapter24;

import com.wangwenjun.concurrent.chapter08.BasicThreadPool;
```

```java
import com.wangwenjun.concurrent.chapter08.ThreadPool;

public class Operator
{
    // 使用线程池替代为每一个请求创建线程
    private final ThreadPool threadPool = new BasicThreadPool(2, 6, 4, 1000);

    public void call(String business)
    {
        TaskHandler taskHandler = new TaskHandler(new Request(business));
        threadPool.execute(taskHandler);
    }
}
```

当然，读者也可以使用 Java 并发包中的 ExecutorService 替代我们自定义的线程，效果将会更好。

24.3　多用户的网络聊天

Thread-Per-Message 模式在网络通信中的使用也是非常广泛的，比如在本节中介绍的网络聊天程序，在服务端每一个连接到服务端的连接都将创建一个独立的线程进行处理，当客户端的连接数超过了服务端的最大受理能力时，客户端将被存放至排队队列中。

24.3.1　服务端程序

下面编写服务端程序 ChatServer 用于接收来自客户端的链接，并且与之进行 TCP 通信交互，当服务器端接收到了每一次的客户端连接后便会给线程池提交一个任务用于与客户端进行交互，进而提高并发响应能力，见代码清单 24-5。

代码清单 24-5　ChatServer.java

```java
package com.wangwenjun.concurrent.chapter24;

import com.wangwenjun.concurrent.chapter08.BasicThreadPool;
import com.wangwenjun.concurrent.chapter08.ThreadPool;

import java.io.IOException;
import java.net.ServerSocket;
import java.net.Socket;

public class ChatServer
{
    //服务端端口
    private final int port;

    //定义线程池，该线程池是我们在第 8 章中定义的
```

```java
    private ThreadPool threadPool;

    // 服务端 Socket
    private ServerSocket serverSocket;

    // 通过构造函数传入端口
    public ChatServer(int port)
    {
        this.port = port;
    }

    // 默认使用 13312 端口
    public ChatServer()
    {
        this(13312);
    }

    public void startServer() throws IOException
    {
        // 创建线程池，初始化一个线程，核心线程数量为 2，最大线程数为 4，阻塞队列中最大可加入
        1000 个任务
        this.threadPool = new BasicThreadPool(1, 4, 2, 1000);
        this.serverSocket = new ServerSocket(port);
        this.serverSocket.setReuseAddress(true);
        System.out.println("Chat server is started and listen at port: " + port);
        this.listen();
    }

    private void listen() throws IOException
    {
        for (; ; )
        {
//accept 方法是阻塞方法，当有新的链接进入时才会返回，并且返回的是客户端的连接
            Socket client = serverSocket.accept();
// 将客户端连接作为一个 Request 封装成对应的 Handler 然后提交给线程池
            this.threadPool.execute(new ClientHandler(client));
        }
    }
}
```

在上面的程序中，当接收到了新的客户端连接时，会为每一个客户端连接创建一个线程 ClientHandler 与客户端进行交互，当客户端的连接个数超过线程池的最大数量时，客户端虽然可以成功接入服务端，但是会进入阻塞队列。

24.3.2　响应客户端连接的 Handler

待服务器端接收到客户端的连接之后，便会创建一个新的 ChatHandler 任务提交给线程池，ChatHandler 任务是 Runnable 接口的实现，主要负责和客户端进行你来我往的简单通

信交互，见代码清单 24-6。

代码清单 24-6　ChatHandler.java

```java
package com.wangwenjun.concurrent.chapter24;

import java.io.*;
import java.net.Socket;

//ChatHandler 同样是一个 Runnable 接口的实现
public class ClientHandler implements Runnable
{
    // 客户端的 socket 连接
    private final Socket socket;

    // 客户端的 identity
    private final String clientIdentify;

    // 通过构造函数传入客户端连接
    public ClientHandler(final Socket socket)
    {
        this.socket = socket;
        this.clientIdentify = socket.getInetAddress().getHostAddress() + ":" + socket.getPort();
    }

    @Override
    public void run()
    {
        try
        {
            this.chat();
        } catch (IOException e)
        {
            e.printStackTrace();
        }
    }

    private void chat() throws IOException
    {
        BufferedReader bufferedReader = wrap2Reader(this.socket.getInputStream());
        PrintStream printStream = wrap2Print(this.socket.getOutputStream());
        String received;
        while ((received = bufferedReader.readLine()) != null)
        {
            // 将客户端发送的消息输出到控制台
            System.out.printf("client:%s-message:%s\n", clientIdentify, received);
            if (received.equals("quit"))
            {
                // 如果客户端发送了 quit 指令，则断开与客户端的连接
```

```
                write2Client(printStream, "client will close");
                socket.close();
                break;
            }
            // 向客户端发送消息
            write2Client(printStream, "Server:" + received);
        }
    }

    // 将输入字节流封装成 BufferedReader 缓冲字符流
    private BufferedReader wrap2Reader(InputStream inputStream)
    {
        return new BufferedReader(new InputStreamReader(inputStream));
    }
    // 将输出字节流封装成 PrintStream
    private PrintStream wrap2Print(OutputStream outputStream)
    {
        return new PrintStream(outputStream);
    }

    // 该方法主要用于向客户端发送消息
    private void write2Client(PrintStream print, String message)
    {
        print.println(message);
        print.flush();
    }
}
```

上面的通信方式是一种典型的一问一答的聊天方式，客户端连接后发送消息给服务端，服务端回复消息给客户端，每一个线程负责处理一个来自客户端的连接。

24.3.3 聊天程序测试

启动服务端程序，为了简单起见，我们直接使用 telnet 命令进行客户端测试即可，测试结果分别如图 24-2 至图 24-5 所示。

```
public static void main(String[] args) throws IOException
{
    new ChatServer().startServer();
}
```

图 24-2　第一个客户端连接成功且正常交互

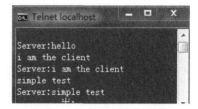

图 24-3　第二个客户端连接成功且正常交互

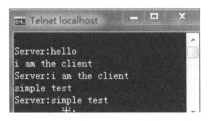

图 24-4　第三个客户端连接成功且正常交互　　图 24-5　第四个客户端连接成功且正常交互

但是当创建第五个客户端连接的时候，将会被加入到线程池的任务队列中而无法与服务器进行交互，其原因是因为当前线程池中再没有任何线程可以负责处理与该客户端的交互了。

24.4　本章总结

Thread-Per-Message 设计模式在日常的开发中非常常见，但是也要灵活使用，比如为了避免频繁创建线程带来的系统开销，可以用线程池来代替，在本书的第 28 章"Event Bus 设计"中也有对 Thread-Per-Message 设计模式的使用。

CHAPTER25 · 第 25 章

Two Phase Termination 设计模式

25.1 什么是 Two Phase Termination 模式

当一个线程正常结束，或者因被打断而结束，或者因出现异常而结束时，我们需要考虑如何同时释放线程中资源，比如文件句柄、socket 套接字句柄、数据库连接等比较稀缺的资源。Two Phase Termination 设计模式如图 25-1 所示。

如图 25-1 所示，我们使用"作业中"表示线程的执行状态，当希望结束这个线程时，发出线程结束请求（关于结束线程的内容，在本书的 3.9 节中有详细的介绍），接下来线程不会立即结束，而是会执行相应的资源释放动作直到真正的结束，在终止处理状态时，线程虽然还在运行，但是进行的是终止处理工作，因此终止处理又称为线程结束的第二个阶段，而受理终止要求则被称为线程结束的第一个阶段。

在进行线程两阶段终结的时候需要考虑如下几个问题。

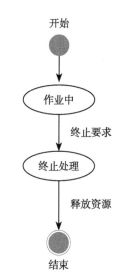

图 25-1　Two Phase Termination 模式

- ❑ 第二阶段的终止保证安全性，比如涉及对共享资源的操作。
- ❑ 要百分之百地确保线程结束，假设在第二个阶段出现了死循环、阻塞等异常导致无法结束。

❏ 对资源的释放时间要控制在一个可控的范围之内。

25.2　Two Phase Termination 的示例

25.2.1　线程停止的 Two Phase Termination

Two Phase Termination 与其说是一个模式，还不如说是线程使用的一个技巧（best practice）。其主要针对的是当线程结束生命周期时，能有机会做一些资源释放的动作，我们曾在第 24 章中定义了用于处理客户端连接的 Handler（ClientHandler），在本节中我们继续使用它作为我们的示例：

```
@Override
public void run()
{
    try
    {
        this.chat();
    } catch (IOException e)
    {
        e.printStackTrace();
    }
}
private void chat() throws IOException
{
    BufferedReader bufferedReader = wrap2Reader(this.socket.getInputStream());
    PrintStream printStream = wrap2Print(this.socket.getOutputStream());
    String received;
    while ((received = bufferedReader.readLine()) != null)
    {
        System.out.printf("client:%s-message:%s\n", clientIdentify, received);
        if (received.equals("quit"))
        {
            write2Client(printStream, "client will close");
            socket.close();
            break;
        }
        write2Client(printStream, "Server:" + received);
    }
}
```

当客户端发送 quit 指令时，服务端会断开与客户端的连接"socket.close();"，如果客户端发送正常信息后发生异常，chat 方法会抛出错误，那么此时该线程的任务执行也将结束，socket 和 thread 一样都属于严重依赖操作系统资源的对象，各个操作系统中可供创建的线程数量有限（第 2 章中做了详细的分析），为了能够确保客户端即使异常关闭，我们也能够尽快地释放 socket 资源，two phase Termination 设计模式将是一种比较好的解决方案，示例

代码如下:

```java
@Override
public void run()
{
    try
    {
        this.chat();
    } catch (IOException e)
    {
        e.printStackTrace();
    } finally
    {
        // 任务执行结束时,执行释放资源的工作
        this.release();
    }
}
private void release()
{
    try
    {
        if (socket != null)
        {
            socket.close();
        }
    } catch (Throwable e)
    {
        //ignore
    }
}
```

当 chat 方法出现异常会被 run 方法捕获,在 run 方法中加入 finally 子句,用于执行客户端 socket 的主动关闭,为了使客户端的关闭不影响线程任务的结束,我们捕获了 Throwable 异常(在关闭客户端连接时出现的异常,视为不可恢复的异常,基本上没有针对该异常进行处理的必要)。

上面的这个例子是针对线程(任务)关闭时进行资源释放的举例,那么当一个进程主动关闭或者异常关闭的时候,可不可以也进行两阶段的 termination 呢?

25.2.2 进程关闭的 Two Phase Termination

如同线程关闭需要进行资源释放一样,不论是进程的主动关闭还是被动关闭(出现异常)都需要对持有的资源进行释放,我们在本书的第 7 章的 7.2 节中做了比较详细的讲解,通过为进程注入一个或者多个 hook 线程的方式进行两阶段的结束,在这里就不再赘述了,读者可以返回前文自行查阅。

25.3　知识扩展

无论是 File 还是 Socket 等重量级的资源（严重依赖操作系统资源），在进行释放时并不能百分之百的保证成功（可能是操作系统的原因），如 25.2.1 节中对 socket 在第二阶段的关闭有可能会失败，然后 socket 的实例会被垃圾回收器回收，但是 socket 实例对应的底层系统资源或许并未释放。

那么我们有什么办法可以再次尝试对 socket 进行资源释放操作呢？这看起来似乎是不可能的事情，但是 JDK 为我们提供了对象被垃圾回收时的可跟踪机制——PhantomReference。我们借助于 PhantomReference 就可以很好地获取到是哪个对象即将被垃圾回收器清除，在被清除之前还可以尝试一次资源回收，尽最大的努力回收重量级的资源是一种非常好的编程体验。

我们在知识拓展中不会单纯地介绍 PhantomReference，会连同 SoftReference、WeakReference 等一起介绍，如图 25-2 所示，Reference 有 3 个子类，分别是我们即将要介绍到的三种 Reference 类型，除了这三种引用类型以外，还有一种称为 Strong Reference，我们平时使用最多的就是 Strong Reference（强引用），当一个对象被关键字 new 实例化出来的时候，它就是强引用。

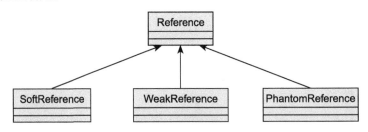

图 25-2　Reference 继承关系图

25.3.1　Strong Reference 及 LRUCache

强引用（Strong Reference）是我们平时使用最多的一种对象引用，当一个对象被关键字 new 实例化出来的时候，JVM 会在堆（heap）内存中开辟一片内存区域，用于存放与该实例对应的数据结构。JVM 垃圾回收器线程会在达到 GC 条件的时候尝试回收（Full GC，Young GC）堆栈内存中的数据，强引用的特点是只要引用到 ROOT 根的路径可达，无论怎样的 GC 都不会将其释放，而是宁可出现 JVM 内存溢出。

Cache 是一种用于提高系统性能，提高数据检索效率的机制，而 LRU（Least recently used，最近最少使用）算法和 Cache 的结合是最常见的一种 Cache 实现，在本节以及接下来的几节中，我们将通过 LRUCache 的方式来对比（Strong，Soft）这三种引用类型的不同并且总结他们各自的特点。

首先我们定义一个占用内存 1MB+ 的类 Reference，代码如清单 25-1 所示。

代码清单 25-1　Reference.java

```java
package com.wangwenjun.concurrent.chapter25;

public class Reference
{
    //1M
    private final byte[] data = new byte[2 << 19];

    @Override
    protected void finalize() throws Throwable
    {
        System.out.println("the reference will be GC.");
    }
}
```

当 Reference 对象被实例化之后，会在堆内存中创建 1M+（+代表的是除了 byte[] data 之外 Reference 对象自身占有的少许内存，如果要求不那么精确，+可以被忽略）的内存空间，finalize 方法会在垃圾回收的标记阶段被调用（垃圾回收器在回收一个对象之前，首先会进行标记，标记过程则会调用该对象的 finalize 方法，所以千万不要认为该方法被调用之后，就代表对象已被垃圾回收器回收，对象在 finalize 方法中是可以"自我救赎"的）。

LRU 其实是数据冷热治理的一种思想，不经常使用的数据被称为冷数据，经常使用的则被称为热数据，对冷数据分配很少的资源或者提前释放，可以帮助我们节省更多的内存资源，LRUCache 的实现方式有很多种，很多人喜欢借助于 LinkedHashMap 去实现，我们在这里使用双向链表+hash 表的方式来实现（其实 LinkedHashMap 自身也是双向链表和 hash 表的方式实现的），如代码清单 25-2 所示。

代码清单 25-2　LRUCache.java

```java
package com.wangwenjun.concurrent.chapter25;

import java.util.HashMap;
import java.util.LinkedList;
import java.util.Map;

public class LRUCache<K, V>
{
    //用于记录key值的顺序
    private final LinkedList<K> keyList = new LinkedList<>();

    //用于存放数据
    private final Map<K, V> cache = new HashMap<>();

    //cache的最大容量
    private final int capacity;

    //cacheLoader接口提供了一种加载数据的方式
```

```java
    private final CacheLoader<K, V> cacheLoader;

    public LRUCache(int capacity, CacheLoader<K, V> cacheLoader)
    {
        this.capacity = capacity;
        this.cacheLoader = cacheLoader;
    }

    public void put(K key, V value)
    {
        // 当元素数量超过容量时,将最老的数据清除
        if (keyList.size() >= capacity)
        {
            K eldestKey = keyList.removeFirst();//eldest data
            cache.remove(eldestKey);
        }
        // 如果数据已经存在,则从 key 的队列中删除
        if (keyList.contains(key))
            keyList.remove(key);
        // 将 key 存放至队尾
        keyList.addLast(key);
        cache.put(key, value);
    }

    public V get(K key)
    {
        V value;
        // 先将 key 从 key list 中删除
        boolean success = keyList.remove(key);
        if (!success)// 如果删除失败则表明该数据不存在
        {
            // 通过 cacheloader 对数据进行加载
            value = cacheLoader.load(key);
            // 调用 put 方法 cache 数据
            this.put(key, value);
        } else
        {
            // 如果删除成功,则从 cache 中返回数据,并且将 key 再次放到队尾
            value = cache.get(key);
            keyList.addLast(key);
        }
        return value;
    }

    @Override
    public String toString()
    {
        return this.keyList.toString();
    }
}
```

在上述代码中：

- 在 LRUCache 中，keyList 主要负责对 key 的顺序进行管理，而 Cache 则主要用于存储真正的数据（K-V）。
- CacheLoader 主要用于进行数据的获取。
- put 方法可将 Value 缓存至 Cache 中，如果当前 Cache 的容量超过了指定容量的大小，则会将最先保存至 Cache 中的数据丢弃掉。
- get 方法根据 key 从 Cache 中获取数据，如果数据存在则先从 keyList 中删除，然后再插入到队尾，否则调用 CacheLoader 的 load 方法进行加载，如清单 25-3 所示。

代码清单 25-3　接口 CacheLoader.java

```java
package com.wangwenjun.concurrent.chapter25;

@FunctionalInterface
public interface CacheLoader<K, V>
{
    // 定义加载数据的方法
    V load(K k);
}
```

CacheLoader 比较简单，是一种标准的函数式接口，下面的程序是对 LRUCache 的基本测试：

```java
public static void main(String[] args)
{
    LRUCache<String, Reference> cache = new LRUCache<>(5, key -> new Reference());
    cache.get("Alex");
    cache.get("Jack");
    cache.get("Gavin");
    cache.get("Dillon");
    cache.get("Leo");
    // 上面的数据在缓存中的新旧程度为 Leo>Dillon>Gavin>Jack>Alex

    cache.get("Jenny"); //Alex 将会被踢出
    System.out.println(cache.toString());
}
```

通过测试，我们发现 LRU 算法的基本功能都能够满足，好了，该说我们的重点了。我们在前面定义了 Reference 类，创建一个 Reference 的实例会产生 1MB 的内存开销，当不断地往 Cache 中存放数据或者存放固定数量大小（capacity）的数据时，由于是 Strong Reference 的缘故，可能会引起内存溢出的情况，代码如下：

```java
LRUCache<Integer, Reference> cache = new LRUCache<>(200, key -> new Reference());
for (int i = 0; i < Integer.MAX_VALUE; i++)
{
```

```
        cache.get(i);
        TimeUnit.SECONDS.sleep(1);
        System.out.println("The " + i + " reference stored at cache.");
    }
```

为了方便测试，在程序启动时我们增加了如下几个 JVM 参数。

- -Xmx128M：指定最大的堆内存大小。
- -Xms64M：指定初始化的堆内存大小。
- -XX:+PrintGCDetails：在控制台输出 GC 的详细信息。

运行程序大约在插入了 98 个 Reference 左右的时候，JVM 出现了堆内存溢出，输出如下：

```
The 98 reference stored at cache.
    [Full GC (Ergonomics) [PSYoungGen: 15679K->15362K(18944K)]
[ParOldGen: 86963K->86962K(87552K)] 102642K->102325K(106496K), [Metaspace:
4249K->4249K(1056768K)], 0.0441757 secs] [Times: user=0.11 sys=0.00, real=0.04 secs]
    [Full GC (Allocation Failure) Exception in thread "main"
[PSYoungGen: 15362K->15362K(18944K)] [ParOldGen: 86962K->86962K(87552K)]
102325K->102325K(106496K), [Metaspace: 4249K->4249K(1056768K)], 0.0108377 secs]
[Times: user=0.00 sys=0.00, real=0.01 secs]
    java.lang.OutOfMemoryError: Java heap space
```

98 个 Reference 大约占用了 100MB 左右的堆内存大小，JVM 自身启动时也需要加载和初始化很多对象实例。

既然数据是被 Cache 的，那么能不能在 JVM 进行垃圾回收的时候帮我们进行数据清除呢？当需要的时候再次加载就可以了，这就是我们接下来需要介绍的内容 Soft Reference。

25.3.2 Soft Reference 及 SoftLRUCache

当 JVM Detect（探测）到内存即将溢出，它会尝试 GC soft 类型的 reference，我们参考 LRUCache 编写了一个 SoftLRUCache，代码如清单 25-4 所示。

代码清单 25-4　SoftLRUCache.java

```java
package com.wangwenjun.concurrent.chapter25;

import java.lang.ref.SoftReference;
import java.util.HashMap;
import java.util.LinkedList;
import java.util.Map;
public class SoftLRUCache<K, V>
{
    private final LinkedList<K> keyList = new LinkedList<>();

    //Value 采用 SoftReference 进行修饰
    private final Map<K, SoftReference<V>> cache = new HashMap<>();
```

```java
    private final int capacity;

    private final CacheLoader<K, V> cacheLoader;

    public SoftLRUCache(int capacity, CacheLoader<K, V> cacheLoader)
    {
        this.capacity = capacity;
        this.cacheLoader = cacheLoader;
    }

    public void put(K key, V value)
    {
        if (keyList.size() >= capacity)
        {
            K eldestKey = keyList.removeFirst();//eldest data
            cache.remove(eldestKey);
        }
        if (keyList.contains(key))
        {
            keyList.remove(key);
        }

        keyList.addLast(key);

        // 保存SoftReference
        cache.put(key, new SoftReference<>(value));
    }

    public V get(K key)
    {
        V value;
        boolean success = keyList.remove(key);
        if (!success)
        {
            value = cacheLoader.load(key);
            this.put(key, value);
        } else
        {
            value = cache.get(key).get();
            keyList.addLast(key);
        }
        return value;
    }

    @Override
    public String toString()
    {
        return keyList.toString();
    }
}
```

在 SoftLRUCache 中，cache 的 Value 被声明成 SoftReference<V>，在 put 一个键值对的时候，需要将 V 实例封装成 new SoftReference<>(V)，这也是 SoftReference 引用的构造方式；在 get 数据的时候首先从 cache 中获取 SoftReference，然后再通过 get 方法得到 put 进去的 Value 值，示例代码如下：

```
public static void main(String[] args) throws InterruptedException
{
    SoftLRUCache<Integer, Reference> cache = new SoftLRUCache<>(1000, key -> new Reference());
    System.out.println(cache);
    for (int i = 0; i < Integer.MAX_VALUE; i++)
    {
        cache.get(i);
        TimeUnit.SECONDS.sleep(1);
        System.out.println("The " + i + " reference stored at cache.");
    }
}
```

使用同 25.3.1 节一样的 JVM 参数，上面的程序不论运行多久都不会出现 JVM 溢出的问题（但是不代表 SoftReference 引用不会引起内存溢出，如果 cache 中插入的速度太快，那么 GC 线程没有来得及回收对象，很有可能也会引起溢出），输出结果如下：

```
The 1642 reference stored at cache.
[GC (Allocation Failure) [PSYoungGen: 26211K->13408K(29184K)] 89557K->88019K(116736K), 0.0125298 secs] [Times: user=0.03 sys=0.00, real=0.01 secs]
[Full GC (Ergonomics) [PSYoungGen: 13408K->1024K(29184K)] [ParOldGen: 74610K->86900K(87552K)] 88019K->87924K(116736K), [Metaspace: 8975K->8975K(1058816K)], 0.0758078 secs] [Times: user=0.12 sys=0.00, real=0.08 secs]
the reference will be GC.
the reference will be GC.
the reference will be GC.
```

通过 GC 输出信息和 Reference 的 finalize 方法输出，我们可以断定 Reference 对象在不断地被回收，图 25-3 是 JVM 进程的内存监控图（锯齿状是最理想的 JVM 内存状态）。

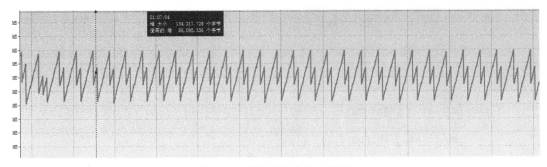

图 25-3　锯齿状的内存走势图

25.3.3　Weak Reference

无论是 young GC 还是 full GC Weak Reference 的引用都会被垃圾回收器回收，Weak Reference（弱引用）可以用来做 Cache，但是一般很少使用。

（1）任何类型的 GC 都可导致 Weak Reference 对象被回收

```
Reference ref = new Reference();
WeakReference<Reference> reference = new WeakReference<>(ref);
ref = null;
// 执行GC操作
System.gc();
```

在上面的代码中我们定义了一个 WeakReference，并且显式地将 ref 设置为 null，当调用 gc 方法时 Weak Reference 对象将被立即回收，输出如下：

```
[GC (System.gc()) [PSYoungGen: 5019K->2040K(18944K)] 5019K->2048K(62976K),
0.1362097 secs] [Times: user=0.11 sys=0.00, real=0.14 secs]
    [Full GC (System.gc()) [PSYoungGen: 2040K->0K(18944K)] [ParOldGen:
8K->1968K(44032K)] 2048K->1968K(62976K), [Metaspace: 4029K->4029K(1056768K)],
0.0589787 secs] [Times: user=0.06 sys=0.00, real=0.06 secs]
    the reference will be GC.
```

（2）获取被垃圾回收器回收的对象

无论是 SoftReference 还是 WeakReference 引用，被垃圾回收器回收后，都会被存放到与之关联的 ReferenceQueue 中，代码如下：

```
// 被垃圾回收的 Reference 会被加入与之关联的 Queue 中
ReferenceQueue<Reference> queue = new ReferenceQueue<>();
Reference ref = new Reference();
// 定义 WeakReference 并且指定关联的 Queue
WeakReference<Reference> reference = new WeakReference<>(ref, queue);
ref = null;
System.out.println(reference.get());

// 手动执行 gc 操作
System.gc();
TimeUnit.SECONDS.sleep(1); //make sure GC thread triggered.
//remove 方法是阻塞方法
java.lang.ref.Reference<? extends Reference> gcedRef = queue.remove();
// 被垃圾回收之后，会从队列中获得
System.out.println(gcedRef);
```

在上面的代码中，我们定义了 WeakReference，同时传入了 ReferenceQueue 队列，当有对象被回收的时候，WeakReference 实例会被加入到 ReferenceQueue 中，输出如下：

```
com.wangwenjun.concurrent.chapter25.Reference@3a71f4dd
    [GC (System.gc()) [PSYoungGen: 5019K->2072K(18944K)] 5019K->2080K(62976K),
0.0045337 secs] [Times: user=0.00 sys=0.00, real=0.01 secs]
```

```
[Full GC (System.gc()) [PSYoungGen: 2072K->0K(18944K)] [ParOldGen: 
8K->1969K(44032K)] 2080K->1969K(62976K), [Metaspace: 4021K->4021K(1056768K)], 
0.0222895 secs] [Times: user=0.08 sys=0.00, real=0.02 secs]
java.lang.ref.WeakReference@7adf9f5f
the reference will be GC.
```

25.3.4 Phantom Reference

Phantom reference objects, which are enqueued after the collector determines that their referents may otherwise be reclaimed. Phantom references are most often used for scheduling pre-mortem cleanup actions in a more flexible way than is possible with the Java finalization mechanism.

这是 JDK 官网对 Phantom reference（幻影引用）的说明，与 SoftReference 和 WeakReference 相比较 Phantom Reference 有如下几点不同之处。

- Phantom Reference 必须和 ReferenceQueue 配合使用。
- Phantom Reference 的 get 方法返回的始终是 null。
- 当垃圾回收器决定回收 Phantom Reference 对象的时候会将其插入关联的 ReferenceQueue 中。
- 使用 Phantom Reference 进行清理动作要比 Object 的 finalize 方法更加灵活。

```
ReferenceQueue<Reference> queue = new ReferenceQueue<>();
PhantomReference<Reference> reference = new PhantomReference<>(new Reference(), 
queue);
System.out.println(reference.get());// 始终返回 null
System.gc();
java.lang.ref.Reference<? extends Reference> gcedRef = queue.remove();
System.out.println(gcedRef);
```

在 25.2.1 节中，我们对 ClientHandler 增加了 Two Phase Termination 的机制，目的是为了在客户端线程结束的时候确保 socket 资源能够被回收，但是 socket.close() 方法并不能百分之百的保证一定能够成功关闭 socket 资源，我们可以借助 PhantomReference 的特性，在垃圾回收器对 socket 对象进行回收的时候再次尝试一次清理，虽然也不能百分之百地保证资源能够彻底释放，但是这样做能够提高资源释放的概率，示例代码如下：

```
private void release()
{
    try
    {
        if (socket != null)
        {
            socket.close();
        }
    } catch (Throwable e)
    {
```

第25章 Two Phase Termination设计模式

```
            if (socket != null)
            {
                // 将socket实例加入Tracker中
                SocketCleaningTracker.tracker(socket);
            }
        }
    }
```

再次对ClientHandler进行重构，关闭socket时失败，将其交给SocketCleaningTracker进行跟踪，当JVM对socket对象进行垃圾回收时可尝试再次进行资源释放，示例代码如清单25-5所示。

代码清单25-5　SocketCleaningTracker.java

```java
class SocketCleaningTracker
{
    // 定义ReferenceQueue
    private static final ReferenceQueue<Object> queue = new ReferenceQueue<>();

    static
    {
        // 启动Cleaner线程
        new Cleaner().start();
    }

    private static void track(Socket socket)
    {
        new Tracker(socket, queue);
    }

    private static class Cleaner extends Thread
    {
        private Cleaner()
        {
            super("SocketCleaningTracker");
            setDaemon(true);

        }

        @Override
        public void run()
        {
            for (; ; )
            {
                try
                {
                    // 当Tracker被垃圾回收器回收时会加入Queue中
                    Tracker tracker = (Tracker) queue.remove();
                    tracker.close();
                } catch (InterruptedException e)
```

```
                {
                }
            }
        }
    }
    //Tracker 是一个 PhantomReference 的子类
    private static class Tracker extends PhantomReference<Object>
    {
        private final Socket socket;

        Tracker(Socket socket, ReferenceQueue<? super Object> queue)
        {
            super(socket, queue);
            this.socket = socket;
        }

        public void close()
        {
            try
            {
                socket.close();
            } catch (IOException e)
            {
                e.printStackTrace();
            }
        }
    }
}
```

在 SocketCleaningTracker 中启动 Cleaner 线程（Cleaner 线程被设置为守护线程，清理动作一般是系统的清理工作，用于防止 JVM 无法正常关闭，如同 JVM 的 GC 线程的作用一样），不断地从 ReferenceQueue 中 remove Tracker 实例（Tracker 是一个 PhantomReference 的子类），然后尝试进行最后的清理动作。

25.4 本章总结

Two Phase Termination 是一个比较好的编程技巧，在线程正常 / 异常结束生命周期的时候，尽最大的努力释放资源是程序设计人员的职责。在本章中，我们通过对聊天程序增加 Two Phase Termination，当线程结束的时候尽最大的努力进行 socket 资源的释放，PhantomReference 是一种比 finalize() 方法更好的跟踪引用被释放的机制，为了使我们的知识点尽可能的系统化，我们顺便介绍了 SoftReference、WeakReference，并且实现了一个基于 LRU 算法的 Cache。

CHAPTER26 · 第 26 章

Worker-Thread 设计模式

26.1 什么是 Worker-Thread 模式

　　Worker-Thread 模式有时也称为流水线设计模式，这种设计模式类似于工厂流水线，上游工作人员完成了某个电子产品的组装之后，将半成品放到流水线传送带上，接下来的加工工作则会交给下游的工人，如图 26-1 所示。

　　线程池在某种意义上也算是 Worker-Thread 模式的一种实现，线程池初始化时所创建的线程类似于在流水线等待工作的工人，提交给线程池的 Runnable 接口类似于需要加工的产品，而 Runnable 的 run 方法则相当于组装该产品的说明书。

图 26-1　工厂流水线

26.2　Worker-Thread 模式实现

　　根据我们前面的描述可以看出 Worker-Thread 模式需要有如下几个角色。

❑ 流水线工人：流水线工人主要用来对传送带上的产品进行加工。
❑ 流水线传送带：用于传送来自上游的产品。
❑ 产品组装说明书：用来说明该产品如何组装。

26.2.1 产品及组装说明书

抽象类 InstructionBook 的代码如清单 26-1 所示。

代码清单 26-1　InstructionBook.java

```java
package com.wangwenjun.concurrent.chapter26;

// 在流水线上需要被加工的产品，create 作为一个模板方法，提供了加工产品的说明书
public abstract class InstructionBook
{
    public final void create()
    {
        this.firstProcess();
        this.secondProcess();
    }
    protected abstract void firstProcess();
    protected abstract void secondProcess();
}
```

抽象类 InstructionBook，代表着组装产品的说明书，其中经过流水线传送带的产品将通过 create() 方法进行加工，而 firstProcess() 和 secondProcess() 则代表着加工每个产品的步骤，这就是说明书的作用。

传送带上的产品除了说明书以外还需要有产品自身，产品继承了说明书，每个产品都有产品编号，通用 Production 的代码如清单 26-2 所示。

代码清单 26-2　通用 Production.java

```java
package com.wangwenjun.concurrent.chapter26;

public class Production extends InstructionBook
{
    // 产品编号
    private final int prodID;
    public Production(int prodID)
    {
        this.prodID = prodID;
    }
    @Override
    protected void firstProcess()
    {
        System.out.println("execute the " + prodID + " first process");
    }
    @Override
```

```java
    protected void secondProcess()
    {
        System.out.println("execute the " + prodID + " second process");
    }
}
```

26.2.2 流水线传送带

流水线的传送带主要用于传送待加工的产品，上游的工作人员将完成的半成品放到传送带上，工作人员从传送带上取下产品进行再次加工，传送带 ProductionChannel 的代码如清单 26-3 所示。

代码清单 26-3　传送带 ProductionChannel.java

```java
package com.wangwenjun.concurrent.chapter26;

// 产品传送带，在传送带上除了负责产品加工的工人之外，还有在传送带上等待加工的产品
public class ProductionChannel
{
    // 传送带上最多可以有多少个待加工的产品
    private final static int MAX_PROD = 100;
    // 主要用来存放待加工的产品，也就是传送带
    private final Production[] productionQueue;
    // 队列尾
    private int tail;
    // 队列头
    private int head;
    // 当前在流水线上有多少个待加工的产品
    private int total;

    // 在流水线上工作的工人
    private final Worker[] workers;
    // 创建 ProductionChannel 时应指定需要多少个流水线工人
    public ProductionChannel(int workerSize)
    {
        this.workers = new Worker[workerSize];
        this.productionQueue = new Production[MAX_PROD];
        // 实例化每一个工人（Worker 线程）并且启动
        for (int i = 0; i < workerSize; i++)
        {
            workers[i] = new Worker("Worker-" + i, this);
            workers[i].start();
        }
    }

    // 接受来自上游的半成品（待加工的产品）
    public void offerProduction(Production production)
    {
        synchronized (this)
```

```java
        {
            // 当传送带上待加工的产品超过了最大值时需要阻塞上游再次传送产品
            while (total >= productionQueue.length)
            {
                try
                {
                    this.wait();
                } catch (InterruptedException e)
                {
                }
            }
            // 将产品放到传送带,并且通知工人线程工作
            productionQueue[tail] = production;
            tail = (tail + 1) % productionQueue.length;
            total++;
            this.notifyAll();
        }
    }
    // 工人线程(Worker)从传送带上获取产品,并且进行加工
    public Production takeProduction()
    {
        synchronized (this)
        {
            // 当传送带上没有产品时,工人等待着产品从上游输送到传送带上
            while (total <= 0)
            {
                try
                {
                    this.wait();
                } catch (InterruptedException e)
                {
                }
            }
            // 获取产品
            Production prod = productionQueue[head];
            head = (head + 1) % productionQueue.length;
            total--;
            this.notifyAll();
            return prod;
        }
    }
}
```

传送带是搁置产品的地方,如果工人们处理比较慢则会导致无限制的产品积压,因此我们需要做的是让上游的流水线阻塞并等待,直至流水线有位置可以用于放置新的产品为止,MAX_PROD 的作用就在于此,其用于控制传送带的最大容量,传送带被创建的同时,流水线上的工人们也已经就绪到位,等待着流水线上产品的到来。

26.2.3 流水线工人

流水线工人是 Thread 的子类，不断地从流水线上提取产品然后进行再次加工，加工的方法是 create()（对该产品的加工方法说明书），流水线工人示例代码如清单 26-4 所示。

代码清单 26-4　流水线工人 Worker.java

```java
package com.wangwenjun.concurrent.chapter26;

import java.util.Random;
import java.util.concurrent.TimeUnit;

public class Worker extends Thread
{
    private final ProductionChannel channel;
    // 主要用于获取一个随机值，模拟加工一个产品需要耗费一定的时间，当然每个工人操作时所花费的时间可也能不一样
    private final static Random random =
            new Random(System.currentTimeMillis());

    public Worker(String workerName, ProductionChannel channel)
    {
        super(workerName);
        this.channel = channel;
    }

    @Override
    public void run()
    {
        while (true)
        {
            try
            {
                // 从传送带上获取产品
                Production production = channel.takeProduction();
                System.out.println(getName() + " process the " + production);
                // 对产品进行加工
                production.create();
                TimeUnit.SECONDS.sleep(random.nextInt(10));
            } catch (InterruptedException e)
            {
                e.printStackTrace();
            }
        }
    }
}
```

26.3 本章总结

26.3.1 产品流水线测试

关于 Worker-Thread 模式我们已经实现完成，下面写一个简单的程序进行测试，代码如清单 26-5 所示。

代码清单 26-5　测试 Worker Thread 模式

```java
package com.wangwenjun.concurrent.chapter26;

import java.util.concurrent.ThreadLocalRandom;
import java.util.concurrent.TimeUnit;
import java.util.concurrent.atomic.AtomicInteger;
import java.util.stream.IntStream;
import static java.util.concurrent.ThreadLocalRandom.current;
public class Test
{
    public static void main(String[] args)
    {
        //流水线上有5个工人
        final ProductionChannel channel = new ProductionChannel(5);
        AtomicInteger productionNo = new AtomicInteger();
        //流水线上有8个工作人员往传送带上不断地放置等待加工的半成品
        IntStream.range(1, 8).forEach(i ->
                new Thread(() ->
                {
                    while (true)
                    {
                        channel.offerProduction(new Production(productionNo.getAndIncrement()));
                        try
                        {
                            TimeUnit.SECONDS.sleep(current().nextInt(10));
                        } catch (InterruptedException e)
                        {
                            e.printStackTrace();
                        }
                    }
                }).start()
        );
    }
}
```

在测试中，假设上游的流水线上有 8 个工人将产品放到传送带上，我们的传送带上定义了 5 个工人，运行上面的程序，Worker 将根据产品的使用说明书对产品进行再次加工，程序输出如下：

```
... 省略
execute the 3 second process
Worker-0 process the PROD:5
execute the 5 first process
execute the 5 second process
Worker-4 process the PROD:7
Worker-0 process the PROD:6
execute the 7 first process
execute the 6 first process
execute the 7 second process
execute the 6 second process
... 省略
```

26.3.2　Worker-Thread 和 Producer-Consumer

笔者在进行互联网授课的过程中，很多人觉得无法区分 Worker-Thread 和 Producer-Consumer 模式，其实这也可以理解，毕竟多线程的架构设计模式是很多程序员在日常的开发过程中累积沉淀下来的优秀编程模式，并不像 GoF23 中设计模式那样具有官方的认可，但是两者之间的区别还是很明显的。

（1）Producer-Consumer 模式

图 26-2 是生产者消费者模式关键角色之间的关系图。

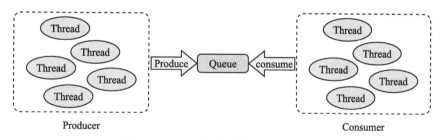

图 26-2　生产者消费者模式角色关系图

首先 Producer、Consumer 对 Queue 都是依赖关系，其次 Producer 要做的就是不断地往 Queue 中生产数据，而 Consumer 则是不断地从 Queue 中获取数据，Queue 既不知道 Producer 的存在也不知道 Consumer 的存在，最后 Consumer 对 Queue 中数据的消费并不依赖于数据本身的方法（使用说明书）。

（2）Worker-Thread 模式

图 26-3 是 Worker-Thread 模式关键角色之间的关系图。

左侧的线程，也就是传送带上游的线程，同样在不断地往传送带（Queue）中生产数据，而当 Channel 被启动的时候，就会同时创建并启动若干数量的 Worker 线程，因此我们可以看出 Worker 于 Channel 来说并不是单纯的依赖关系，而是聚合关系，Channel 必须知道 Worker 的存在。

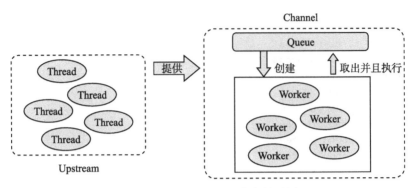

图 26-3　Worker-Thread 角色关系图

CHAPTER27 · 第 27 章

Active Objects 设计模式

27.1 接受异步消息的主动对象

Active 是"主动"的意思，Active Object 是"主动对象"的意思，所谓主动对象就是指其拥有自己的独立线程，比如 java.lang.Thread 实例就是一个主动对象，不过 Active Object Pattern 不仅仅是拥有独立的线程，它还可以接受异步消息，并且能够返回处理的结果。

我们在本书中频繁使用的 System.gc() 方法就是一个"接受异步消息的主动对象"，调用 gc 方法的线程和 gc 自身的执行线程并不是同一个线程，在本章中，我们将实现一个类似于 System.gc 的可接受异步消息的主动对象。接受异步消息的主动对象工作原理如图 27-1 所示。

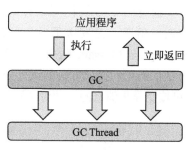

图 27-1 接受异步消息的主动对象的工作原理

27.2 标准 Active Objects 模式设计

在本节中，我们首先从标准的 Active Objects 设计入手，将一个接口的方法调用转换成可接受异步消息的主动对象，也就是说方法的执行和方法的调用是在不同的线程中进行的，那么如何使得执行线程知道应该如何正确执行接口方法呢？我们需要将接口方法的参数以及具体实现封装成特定的 Message 告知执行线程。如果该接口方法需要返回值，则必须得设计成 Future 的返回形式，图 27-2 为标准 Active Objects 模型设计的结构类图。

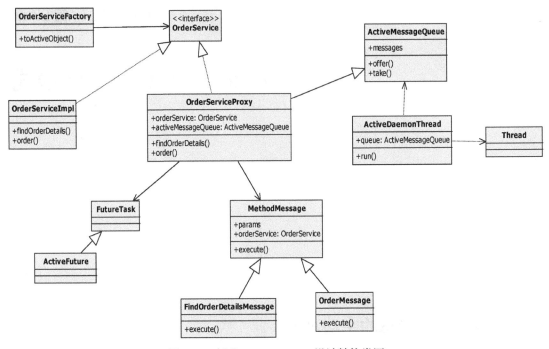

图 27-2　标准 Active Objects 设计结构类图

通过图 27-3 我们可以看出，当某个线程调用 OrderService 接口的 findOrderDetails 方法时，事实上是发送了一个包含 findOrderDetails 方法参数以及 OrderService 具体实现的 Message 至 Message 队列，执行线程通过从队列中获取 Message 来调用具体的实现，接口方法的调用和接口方法的执行分别处于不同的线程中，因此我们称该接口为 Active Objects（可接受异步消息的主动对象）。

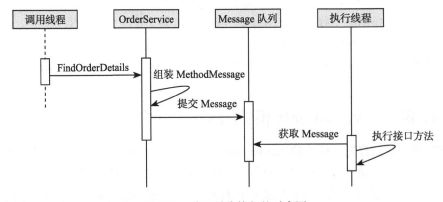

图 27-3　接口异步执行的时序图

27.2.1　OrderService 接口设计

OrderService 是一个比较简单的接口，包含两个方法，其中第一个为有返回值的方法，第二个方法则没有返回值，代码如清单 27-1 所示：

代码清单 27-1　OrderService.java

```java
package com.wangwenjun.concurrent.chapter27;

import com.wangwenjun.concurrent.chapter19.Future;

public interface OrderService
{
    /**
     * 根据订单编号查询订单明细，有入参也有返回值，但是返回类型必须是Future
     */
    Future<String> findOrderDetails(long orderId);

    /**
     * 提交订单，没有返回值
     */
    void order(String account, long orderId);
}
```

- findOrderDetails（long orderId）：通过订单编号获取订单详情，有返回值的方法必须是 Future 类型的，因为方法的执行是在其他线程中进行的，势必不会立即得到正确的最终结果，通过 Future 可以立即得到返回。
- Order（String account,long orderId）：提交用户的订单信息，是一种无返回值的方法。

27.2.2　OrderServiceImpl 详解

OrderServiceImpl 的代码如清单 27-2 所示。

代码清单 27-2　OrderServiceImpl.java

```java
package com.wangwenjun.concurrent.chapter27;

import com.wangwenjun.concurrent.chapter19.Future;
import com.wangwenjun.concurrent.chapter19.FutureService;
import java.util.concurrent.TimeUnit;

public class OrderServiceImpl implements OrderService
{
    @Override
    public Future<String> findOrderDetails(long orderId)
    {
        // 使用在第19章中实现的Future返回结果
        return FutureService.<Long, String>newService().submit(input ->
            {
```

```java
            try
            {
                // 通过休眠来模拟该方法的执行比较耗时
                TimeUnit.SECONDS.sleep(10);
                System.out.println("process the orderID->" + orderId);
            } catch (InterruptedException e)
            {
                e.printStackTrace();
            }
            return "The order Details Information";
        }, orderId, null);
    }

    @Override
    public void order(String account, long orderId)
    {
        try
        {
            TimeUnit.SECONDS.sleep(10);
            System.out.println("process the order for account " + account + ",orderId " + orderId);
        } catch (InterruptedException e)
        {
            e.printStackTrace();
        }
    }
}
```

OrderServiceImpl 类是 OrderService 的一个具体实现,该类是在执行线程中将要被使用的类,其中 findOrderDetails 方法通过第 19 章中我们开发的 Future 立即返回一个结果,order 方法则通过休眠来模拟该方法的执行比较耗时。

27.2.3 OrderServiceProxy 详解

OrderServiceProxy 是 OrderService 的子类,它的作用是将 OrderService 的每一个方法都封装成 MethodMessage,然后提交给 ActiveMessage 队列,在使用 OrderService 接口方法的时候,实际上是在调用 OrderServiceProxy 中的方法,代码如清单 27-3 所示。

代码清单 27-3　OrderServiceProxy.java

```java
package com.wangwenjun.concurrent.chapter27;

import com.wangwenjun.concurrent.chapter19.Future;

import java.util.HashMap;
import java.util.Map;

public class OrderServiceProxy implements OrderService
{
```

```java
        private final OrderService orderService;
        private final ActiveMessageQueue activeMessageQueue;

        public OrderServiceProxy(OrderService orderService, ActiveMessageQueue activeMessageQueue)
        {
            this.orderService = orderService;
            this.activeMessageQueue = activeMessageQueue;
        }

        @Override
        public Future<String> findOrderDetails(long orderId)
        {
            //定义一个ActiveFuture,并且可支持立即返回
            final ActiveFuture<String> activeFuture = new ActiveFuture<>();
            //收集方法入参以及返回的ActiveFuture封装成MethodMessage
            Map<String, Object> params = new HashMap<>();
            params.put("orderId", orderId);
            params.put("activeFuture", activeFuture);
            MethodMessage message = new FindOrderDetailsMessage(params, orderService);
            //将MethodMessage保存至activeMessageQueue中
            activeMessageQueue.offer(message);
            return activeFuture;
        }

        @Override
        public void order(String account, long orderId)
        {
            //收集方法参数,并且封装成MethodMessage,然后offer至队列中
            Map<String, Object> params = new HashMap<>();
            params.put("account", account);
            params.put("orderId", orderId);
            MethodMessage message = new OrderMessage(params, orderService);
            activeMessageQueue.offer(message);
        }
    }
```

OrderServiceProxy 作为 OrderService 的一个实现,看上去与 OrderService 没多大关系,其主要作用是将 OrderService 接口定义的方法封装成 MethodMessage,然后 offer 给 ActiveMessageQueue。若是无返回值的方法,则只需要提交 Message 到 ActiveMessageQueue 中即可,但若是有返回值的方法,findOrderDetails 是比较特殊的,它需要返回一个 ActiveFuture,该 Future 的作用是可以立即返回,当调用线程获取结果时将进入阻塞状态,代码如清单 27-4 所示。

代码清单 27-4　ActiveFuture.java

```java
package com.wangwenjun.concurrent.chapter27;
```

```java
import com.wangwenjun.concurrent.chapter19.FutureTask;
public class ActiveFuture<T> extends FutureTask<T>
{
    @Override
    public void finish(T result)
    {
        super.finish(result);
    }
}
```

ActiveFuture 非常简单，是 FutureTask 的直接子类，其主要作用是重写 finish 方法，并且将 protected 的权限换成 public，可以使得执行线程完成任务之后传递最终结果。

27.2.4 MethodMessage

MethodMessage 的主要作用是收集每一个接口的方法参数，并且提供 execute 方法供 ActiveDaemonThread 直接调用，该对象就是典型的 Worker Thread 模型中的 Product（附有使用说明书的半成品，等待流水线工人的加工），execute 方法则是加工该产品的说明书。MethodMessage 的代码如清单 27-5 所示。

代码清单 27-5　MethodMessage.java

```java
package com.wangwenjun.concurrent.chapter27;

import java.util.Map;

public abstract class MethodMessage
{
    // 用于收集方法参数，如果又返回 Future 类型则一并收集
    protected final Map<String, Object> params;

    protected final OrderService orderService;

    public MethodMessage(Map<String, Object> params, OrderService orderService)
    {
        this.params = params;
        this.orderService = orderService;
    }

    // 抽象方法，扮演 work thread 的说明书
    public abstract void execute();
}
```

其中，params 主要用来收集方法参数，orderService 是具体的接口实现，每一个方法都会被拆分成不同的 Message。在 OrderService 中，我们定义了两个方法，因此需要实现两个 MethodMessage。

(1) FindOrderDetailsMessage

FindOrderDetailsMessage 的代码如清单 27-6 所示。

代码清单 27-6　FindOrderDetailsMessage.java

```java
package com.wangwenjun.concurrent.chapter27;

import com.wangwenjun.concurrent.chapter19.Future;

import java.util.Map;

public class FindOrderDetailsMessage extends MethodMessage
{
    public FindOrderDetailsMessage(Map<String, Object> params
            , OrderService orderService)
    {
        super(params, orderService);
    }

    @Override
    public void execute()
    {
        //①
        Future<String> realFuture = orderService.findOrderDetails((Long) params.get("orderId"));
        ActiveFuture<String> activeFuture = (ActiveFuture<String>) params.get("activeFuture");
        try
        {
            //②
            String result = realFuture.get();
            //③
            activeFuture.finish(result);
        } catch (InterruptedException e)
        {
            activeFuture.finish(null);
        }
    }
}
```

在上述代码中：

①执行 orderService 的 findOrderDetails 方法。

②调用 orderServiceImpl 返回的 Future.get()，此方法会导致阻塞直到 findOrderDetails 方法完全执行结束。

③当 findOrderDetails 执行结束时，将结果通过 finish 的方法传递给 activeFuture。

(2) OrderMessage

OrderMessage 的代码如清单 27-7 所示。

代码清单 27-7　OrderMessage.java

```java
package com.wangwenjun.concurrent.chapter27;

import java.util.Map;

public class OrderMessage extends MethodMessage
{
    public OrderMessage(Map<String, Object> params,
                        OrderService orderService)
    {
        super(params, orderService);
    }

    @Override
    public void execute()
    {
        // 获取参数
        String account = (String) params.get("account");
        long orderId = (long) params.get("orderId");

        // 执行真正的 order 方法

        orderService.order(account, orderId);
    }
}
```

OrderMessage 主要处理 order 方法，从 param 中获取接口参数，然后执行真正的 OrderService 的 order 方法。

27.2.5　ActiveMessageQueue

ActiveMessageQueue 对应于 Worker-Thread 模式中的传送带，主要用于传送调用线程通过 Proxy 提交过来的 MethodMessage，但是这个传送带允许存放无限的 MethodMessage（没有 limit 的约束，理论上可以放无限多个 MethodMessage 直到发生堆内存溢出的异常），代码如清单 27-8 所示。

代码清单 27-8　ActiveMessageQueue.java

```java
package com.wangwenjun.concurrent.chapter27;

import java.util.LinkedList;

public class ActiveMessageQueue
{
    // 用于存放提交的 MethodMessage 消息
    private final LinkedList<MethodMessage> messages = new LinkedList<>();

    public ActiveMessageQueue()
```

```java
{
    // 启动 Worker 线程
    new ActiveDaemonThread(this).start();
}

public void offer(MethodMessage methodMessage)
{
    synchronized (this)
    {
        messages.addLast(methodMessage);
        // 因为只有一个线程负责 take 数据，因此没有必要使用 notifyAll 方法
        this.notify();
    }
}

protected MethodMessage take()
{
    synchronized (this)
    {
        // 当 MethodMessage 队列中没有 Message 的时候，执行线程进入阻塞
        while (messages.isEmpty())
        {
            try
            {
                this.wait();
            } catch (InterruptedException e)
            {
                e.printStackTrace();
            }
        }
        // 获取其中一个 MethodMessage 并且从队列中移除
        return messages.removeFirst();
    }
}
}
```

上述代码中：

- 在创建 ActiveMessageQueue 的同时启动 ActiveDaemonThread 线程，ActiveDaemon-Thread 主要用来进行异步的方法执行，后面我们会介绍。
- 执行 offer 方法没有进行 limit 的判断，允许提交无限个 MethodMessage（直到发生堆内存溢出），并且当有新的 Message 加入时会通知 ActiveDaemonThread 线程。
- take 方法主要是被 ActiveDaemonThread 线程使用，当 message 队列为空时 Active-DaemonThread 线程将会被挂起（Guarded Suspension），如代码清单 27-9 所示。

代码清单 27-9　ActiveDaemonThread.java

```java
package com.wangwenjun.concurrent.chapter27;

class ActiveDaemonThread extends Thread
```

```java
{
    private final ActiveMessageQueue queue;

    public ActiveDaemonThread(ActiveMessageQueue queue)
    {
        super("ActiveDaemonThread");
        this.queue = queue;
        //ActiveDaemonThread 为守护线程
        setDaemon(true);
    }

    @Override
    public void run()
    {
        for (; ; )
        {
            /* 从 MethodMessage 队列中获取一个 MethodMessage，然后执行 execute 方法
            */
            MethodMessage methodMessage = this.queue.take();
            methodMessage.execute();
        }
    }
}
```

ActiveDaemonThread 是一个守护线程，主要是从 queue 中获取 Message 然后执行 execute 方法（注意：保持为线程命名的习惯是一个比较好的编程习惯）。

27.2.6　OrderServiceFactory 及测试

我们基本上已经完成了一个标准 Active Objects 的设计，接口方法的每一次调用实际上都是向 Queue 中提交一个对应的 Message 信息，当然这个工作主要是由 Proxy 完成的，但是为了让 Proxy 的构造透明化，我们需要设计一个 Factory 工具类，代码如清单 27-10 所示。

代码清单 27-10　OrderServiceFactory.java

```java
package com.wangwenjun.concurrent.chapter27;

public final class OrderServiceFactory
{
    // 将 ActiveMessageQueue 定义成 static 的目的是，保持其在整个 JVM 进程中是唯一的，并且
ActiveDaemonThread 会在此刻启动
    private final static ActiveMessageQueue activeMessageQueue = new ActiveMessageQueue();

    // 不允许外部通过 new 的方式构建
    private OrderServiceFactory()
    {
    }

    // 返回 OrderServiceProxy
```

```java
    public static OrderService toActiveObject(OrderService orderService)
    {
        return new OrderServiceProxy(orderService, activeMessageQueue);
    }
}
```

toActiveObject 方法主要用于创建 OrderServiceProxy，下面就来简单地测试一下我们的 Active Object：

```java
public static void main(String[] args)
        throws InterruptedException
{
    // 在创建 OrderService 时需要传递 OrderService 接口的具体实现
    OrderService orderService = OrderServiceFactory
            .toActiveObject(new OrderServiceImpl());
    orderService.order("hello", 453453);
    // 立即返回
    System.out.println("Return immediately");

    currentThread().join();
}
```

运行上面的测试代码会立即得到返回，10 秒之后，order 方法执行结束，调用 order 方法的线程是主线程，但是执行该方法的线程却是其他线程（ActiveDaemonThread），这也正是 Active Objects 可接受异步消息的意思。

27.3　通用 Active Objects 框架设计

标准的 Active Objects 要将每一个方法都封装成 Message（比如 27.2 节中定义的 FindOrderDetailsMessage，OrderMessage），然后提交至 Message 队列中，这样的做法有点类似于远程方法调用（RPC：Remote Process Call）。如果某个接口的方法很多，那么需要封装很多的 Message 类；同样如果有很多接口需要成为 Active Object，则需要封装成非常多的 Message 类，这样显然不是很友好。在本节中，我们将设计一个更加通用的 Active Object 框架，可以将任意的接口转换成 Active Object。

在本节中，我们将使用 JDK 动态代理的方式实现一个更为通用的 Active Objects，可以将任意接口方法转换为 Active Objects，当然如果接口方法有返回值，则必须要求返回 Future 类型才可以，否则将会抛出 IllegalActiveMethod 异常，代码如清单 27-11 所示。

代码清单 27-11　IllegalActiveMethod.java

```java
package com.wangwenjun.concurrent.chapter27;
// 若方法不符合则其被转换为 Active 方法时会抛出该异常
public class IllegalActiveMethod extends Exception
{
```

```
    public IllegalActiveMethod(String message)
    {
        super(message);
    }
}
```

通用的 Active Objects 设计消除了为每一个接口方法定义 MethodMessage 的过程，同时也摒弃掉了为每一个接口创建定义 Proxy 的实现，所有的操作都会被支持动态代理的工厂类 ActiveServiceFactory 所替代，通用 Active Objects 框架详细类图如图 27-4 所示。

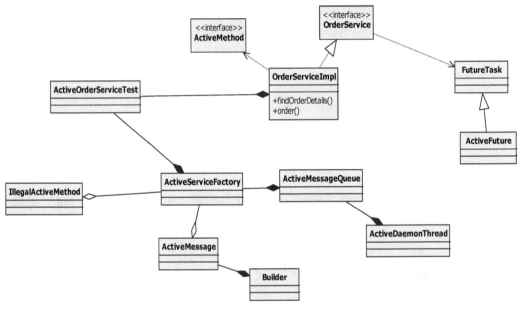

图 27-4　通用 Active Objects 架构类图

27.3.1　ActiveMessage 详解

相比较于 MethodMessage，ActiveMessage 更加通用，其可以满足所有 Active Objects 接口方法的要求，与 MethodMessage 类似，ActiveMessage 也是用于收集接口方法信息和具体的调用方法的。ActiveMessage 的代码如清单 27-12 所示。

代码清单 27-12　ActiveMessage.java

```
package com.wangwenjun.concurrent.chapter27;

import com.wangwenjun.concurrent.chapter19.Future;

import java.lang.reflect.Method;

// 包可见，ActiveMessage 只在框架内部使用，不会对外暴露
```

```java
class ActiveMessage
{
    // 接口方法的参数
    private final Object[] objects;

    // 接口方法
    private final Method method;

    // 有返回值的方法，会返回ActiveFuture<?>类型
    private final ActiveFuture<Object> future;

    // 具体的Service接口
    private final Object service;

    // 构造ActiveMessage是由Builder来完成的
    private ActiveMessage(Builder builder)
    {
        this.objects = builder.objects;
        this.method = builder.method;
        this.future = builder.future;
        this.service = builder.service;
    }

    //ActiveMessage的方法通过反射的方式调用执行的具体实现
    public void execute()
    {
        try
        {
            // 执行接口的方法
            Object result = method.invoke(service, objects);
            if (future != null)
            {
                // 如果是有返回值的接口方法，则需要通过get方法获得最终的结果
                Future<?> realFuture = (Future<?>) result;
                Object realResult = realFuture.get();
                // 将结果交给ActiveFuture，接口方法的线程会得到返回
                future.finish(realResult);
            }
        } catch (Exception e)
        {
            // 如果发生异常，那么有返回值的方法将会显式地指定结果为null，无返回值的接口方法则
            // 会忽略该异常
            if (future != null)
            {
                future.finish(null);
            }
        }
    }

//Builder主要负责对ActiveMessage的构建，是一种典型的Gof Builder设计模式
```

```java
static class Builder
{
    private Object[] objects;

    private Method method;

    private ActiveFuture<Object> future;

    private Object service;
    public Builder useMethod(Method method)
    {
        this.method = method;
        return this;
    }

    public Builder returnFuture(ActiveFuture<Object> future)
    {
        this.future = future;
        return this;
    }

    public Builder withObjects(Object[] objects)
    {
        this.objects = objects;
        return this;
    }

    public Builder forService(Object service)
    {
        this.service = service;
        return this;
    }
    // 构建ActiveMessage实例
    public ActiveMessage build()
    {
        return new ActiveMessage(this);
    }
}
}
```

构造 ActiveMessage 必须使用 Builder 方式进行 build，其中包含了调用某个方法必需的入参（objects），代表该方法的 java.lang.reflect.Method 实例，将要执行的 ActiveService 实例（service），以及如果该接口方法有返回值，需要返回的 Future 实例（future）。

27.3.2　@ActiveMethod

通用的 Active Objects 更加灵活，它允许你将某个接口的任意方法转换为 Active-

Method，如果不需要转换，则需要按照普通方法来执行，而不会被单独的线程执行，要做到这一点，就需要使用 @ActiveMethod 注解来进行标记，代码如清单 27-13 所示。

代码清单 27-13　ActiveMethod.java

```
package com.wangwenjun.concurrent.chapter27;

import java.lang.annotation.ElementType;
import java.lang.annotation.Retention;
import java.lang.annotation.RetentionPolicy;
import java.lang.annotation.Target;
@Retention(RetentionPolicy.RUNTIME)
@Target(ElementType.METHOD)
public @interface ActiveMethod
{
}
```

比如，我们用 @ActiveMethod 标记前文中的 OrderServiceImpl 类，代码如下：

```
...
public class OrderServiceImpl implements OrderService
{
    @ActiveMethod
    @Override
    public Future<String> findOrderDetails(long orderId)
    {
        ...
    }

    @ActiveMethod
    @Override
    public void order(String account, long orderId)
    {
        ...
    }
}
```

27.3.3　ActiveServiceFactory 详解

ActiveServiceFactory 是通用 Active Objects 的核心类，其负责生成 Service 的代理以及构建 ActiveMessage，代码如清单 27-14 所示。

代码清单 27-14　通用 ActiveServiceFactory.java

```
package com.wangwenjun.concurrent.chapter27;

import com.wangwenjun.concurrent.chapter19.Future;

import java.lang.reflect.InvocationHandler;
import java.lang.reflect.Method;
```

```java
import java.lang.reflect.Proxy;
public class ActiveServiceFactory
{
    // 定义 ActiveMessageQueue, 用于存放 ActiveMessage
    private final static ActiveMessageQueue queue = new ActiveMessageQueue();

    public static <T> T active(T instance)
    {
        // 生成 Service 的代理类
        Object proxy = Proxy.newProxyInstance(instance.getClass().getClassLoader(),
                instance.getClass().getInterfaces(), new ActiveInvocationHandler
                <>(instance));
        return (T) proxy;
    }

    //ActiveInvocationHandler 是 InvocationHandler 的子类, 生成 Proxy 时需要使用到
    private static class ActiveInvocationHandler<T> implements InvocationHandler
    {
        private final T instance;

        ActiveInvocationHandler(T instance)
        {
            this.instance = instance;
        }

        @Override
        public Object invoke(Object proxy, Method method, Object[] args)
                throws Throwable
        {
            // 如果接口方法被 @ActiveMethod 标记, 则会转换为 ActiveMessage
            if (method.isAnnotationPresent(ActiveMethod.class))
            {
                // 检查该方法是否符合规范
                this.checkMethod(method);
                ActiveMessage.Builder builder = new ActiveMessage.Builder();
                builder.useMethod(method).withObjects(args).forService(instance);
                Object result = null;
                if (this.isReturnFutureType(method))
                {
                    result = new ActiveFuture<>();
                    builder.returnFuture((ActiveFuture) result);
                }
                // 将 ActiveMessage 加入至队列中
                queue.offer(builder.build());
                return result;
            } else
            {
                // 如果是普通方法 ( 没有使用 @ActiveMethod 标记 ), 则会正常执行
                return method.invoke(instance, args);
            }
```

```
            }

            // 检查有返回值的方法是否为 Future,否则将会抛出
            //IllegalActiveMethod 异常
            private void checkMethod(Method method) throws IllegalActiveMethod
            {
                // 有返回值,必须是 ActiveFuture 类型的返回值
                if (!isReturnVoidType(method) && !isReturnFutureType(method))
                {
                    throw new IllegalActiveMethod("the method [" + method.getName()
                        + " return type must be void/Future");
                }
            }

            // 判断方法是否为 Future 返回类型
            private boolean isReturnFutureType(Method method)
            {
                return method.getReturnType().isAssignableFrom(Future.class);
            }

            // 判断方法是否无返回类型
            private boolean isReturnVoidType(Method method)
            {
                return method.getReturnType().equals(Void.TYPE);
            }
        }
    }
```

在上述代码中:

- 静态方法 active() 会根据 Active Service 实例生成一个动态代理实例,其中会用到 ActiveInvocationHandler 作为 newProxyInstance 的 InvocationHandler。
- 在 ActiveInvocationHandler 的 invoke 方法中,首先会判断该方法是否被 @ActiveMethod 标记,如果没有则被当作正常方法来使用。
- 如果接口方法被 @ActiveMethod 标记,则需要判断方法是否符合规范:有返回类型,必须是 Future 类型。
- 定义 ActiveMessage.Builder 分别使用 method、方法参数数组以及 Active Service 实例,如果该方法是 Future 的返回类型,则还需要定义 ActiveFuture。
- 最后将 ActiveMessage 插入 ActiveMessageQueue 中,并且返回 method 方法 invoke 结果。

27.3.4 ActiveMessageQueue 及其他

在 27.2.5 节中,插入到 ActiveMessageQueue 中的数据为 MethodMessage,由于我们定义了更加通用的 ActiveMessage,因此需要修改 Queue 中的数据类型,相关代码如清单

27-15 所示。

代码清单 27-15　通用 ActiveMessageQueue.java

```java
public class ActiveMessageQueue
{
    // 与27.2节中的标准Active Objects不一样的是，通用的ActiveMessageQueue只需要提交
ActiveMessage
    private final LinkedList<ActiveMessage> messages = new LinkedList<>();

    public ActiveMessageQueue()
    {
        // 同样启动ActiveDaemonThread
        new ActiveDaemonThread(this).start();
    }

    public void offer(ActiveMessage activeMessage)
    {
        synchronized (this)
        {
            messages.addLast(activeMessage);
            this.notify();
        }
    }

    public ActiveMessage take()
    {
        synchronized (this)
        {
            while (messages.isEmpty())
            {
                ...
            }
            return messages.removeFirst();
        }
    }
}
```

与 ActiveMessageQueue 紧密关联的 ActiveDaemonThread 也得进行简单修改，修改代码如下：

```java
...
class ActiveDaemonThread extends Thread
{
...
    @Override
    public void run()
    {
        for (; ; )
        {
```

```
            ActiveMessage activeMessage = this.queue.take();
            activeMessage.execute();
        }
    }
}
```

关于这个版本的 Active Objects,其用法与之前的版本差别不大,但是它能够将满足规范的任意接口方法都转换成 Active Objects,代码如下:

```
public static void main(String[] args) throws InterruptedException
{
    OrderService orderService = active(new OrderServiceImpl());
    Future<String> future = orderService.findOrderDetails(23423);
    System.out.println("i will be returned immediately");
    System.out.println(future.get());
}
```

运行上面的测试代码,future 将会立即返回,但是 get 方法会进入阻塞,10 秒钟以后订单的详细信息将会返回,同样,OrderService 接口的调用线程和具体的执行线程不是同一个,OrderServiceImpl 通过 active 方法具备了可接受异步消息的能力。

27.4 本章总结

在本章中,我们通过 System.gc() 方法的原理分析,分别设计了两种不同的 Active Objects 模式实现,第二种方式更加通用一些,因为它摒弃了第一种方式需要手动定义方法的 Message 以及 Proxy 等缺陷,通过动态代理的方式动态生成代理类,当然读者可以通过开源的第三方 Proxy 来实现动态代理的功能,比如 cglib 以及 asm 等。

Active Objects 模式既能够完整地保留接口方法的调用形式,又能让方法的执行异步化,这也是其他接口异步调用模式(Future 模式:只提供了任务的异步执行方案,但是无法保留接口原有的调用形式)无法同时做到的。

Active Objects 模式中使用了很多其他设计模式,代理类的生成(代理设计模式)、ActiveMessageQueue(Guarded Suspension Pattern 以 及 Worker Thread Pattern)、findOrderDetails 方法(Future 设计模式),希望读者能够熟练掌握在 Active Objects 设计模式中用到的其他设计模式。

第 28 章 · CHAPTER28

Event Bus 设计模式

相信每一位读者都有使用过消息中间件的经历，比如 Apache ActiveMQ 和 Apache Kafka 等，某 subscriber 在消息中间件上注册了某个 topic（主题），当有消息发送到了该 topic 上之后，注册在该 topic 上的所有 subscriber 都将会收到消息，如图 28-1 所示。

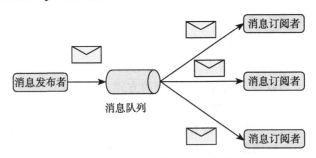

图 28-1　消息中间件的消息订阅与发布

消息中间件提供了系统之间的异步处理机制，比如在某电商网站上支付订单之后，会触发库存计算、物流调度计算，甚至是营销人员绩效计算、报表统计等的，诸如此类的操作一般会耗费比订单购买商品本身更多的时间，加之这样的操作没有即时的时效性要求，用户在下单之后完全没有必要等待电商后端做完所有的操作才算成功，那么此时消息中间件便是一种非常好的解决方案，用户下单成功支付之后即可向用户返回购买成功的通知，然后提交各种消息至消息中间件，这样注册在消息中间件的其他系统就可以顺利地接收订单通知了，然后执行各自的业务逻辑。消息中间件主要用于解决进程之间消息异步处理的解决方案，在本章中，我们使用消息中间件的思想设计一个 Java 进程内部的消息中间件——Event Bus。

28.1　Event Bus 设计

Event Bus 的设计稍微复杂一些，所涉及的类比较多（10 个左右），图 28-2 是类 / 接口之间的类关系图。

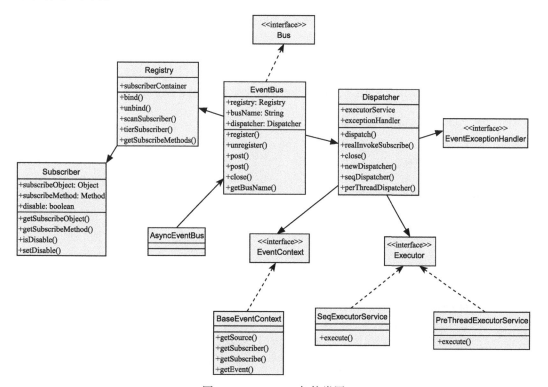

图 28-2　EventBus 架构类图

- Bus 接口对外提供了几种主要的使用方式，比如 post 方法用来发送 Event，register 方法用来注册 Event 接收者（Subscriber）接受响应事件，EventBus 采用同步的方式推送 Event，AsyncEventBus 采用异步的方式（Thread-Per-Message）推送 Event。
- Registry 注册表，主要用来记录对应的 Subscriber 以及受理消息的回调方法，回调方法我们用注解 @Subscribe 来标识。
- Dispatcher 主要用来将 event 广播给注册表中监听了 topic 的 Subscriber。

28.1.1　Bus 接口详解

Bus 接口的相关代码如清单 28-1 所示。

代码清单 28-1　接口方法 Bus.java

```
package com.wangwenjun.concurrent.chapter28;
```

```java
/**
 * Bus 接口定义了 EventBus 的所有使用方法
 */
public interface Bus
{
    /**
     * 将某个对象注册到 Bus 上，从此之后该类就成为 Subscriber 了
     */
    void register(Object subscriber);

    /**
     * 将某个对象从 Bus 上取消注册，取消注册之后就不会再接收到来自 Bus 的任何消息
     */
    void unregister(Object subscriber);

    /**
     * 提交 Event 到默认的 topic
     */
    void post(Object event);

    /**
     * 提交 Event 到指定的 topic
     */
    void post(Object Event, String topic);

    /**
     * 关闭该 bus
     */
    void close();

    /**
     * 返回 Bus 的名称标识
     */
    String getBusName();
}
```

Bus 接口中定义了注册 topic 的方法和 Event 发送的方法，具体如下。

❑ register（Object subscriber）：将某个对象实例注册给 Event Bus。

❑ unregister（Object subscriber）：取消对该对象实例的注册，会在 Event Bus 的注册表（Registry）中将其移除。

❑ post（Object event）：提交 Event 到 Event Bus 中，如果未指定 topic 则会将 event 广播给 Event Bus 默认的 topic。

❑ post（Object Event，String topic）：提交 Event 的同时指定了 topic。

❑ close()：销毁该 Event Bus。

❑ getBusName()：返回该 Event Bus 的名称。

注册对象给 Event Bus 的时候需要指定接收消息时的回调方法，我们采用注解的方式进

行 Event 回调，代码如清单 28-2 所示。

代码清单 28-2　Subscribe.java

```java
package com.wangwenjun.concurrent.chapter28;

import java.lang.annotation.ElementType;
import java.lang.annotation.Retention;
import java.lang.annotation.RetentionPolicy;
import java.lang.annotation.Target;

@Retention(RetentionPolicy.RUNTIME)
@Target(ElementType.METHOD)
public @interface Subscribe
{
    String topic() default "default-topic";
}
```

@Subscribe 要求注解在类中的方法，注解时可指定 topic，不指定的情况下为默认的 topic（default-topic）。

28.1.2　同步 EventBus 详解

同步 EventBus 是最核心的一个类，它实现了 Bus 的所有功能，但是该类对 Event 的广播推送采用的是同步的方式，如果想要使用异步的方式进行推送，可使用 EventBus 的子类 AsyncEventBus，代码如清单 28-3 所示。

代码清单 28-3　EventBus.java

```java
package com.wangwenjun.concurrent.chapter28;

import java.util.concurrent.Executor;
public class EventBus implements Bus
{
    //用于维护 Subscriber 的注册表
    private final Registry registry = new Registry();
    //Event Bus 的名字
    private String busName;

    //默认的 Event Bus 的名字
    private final static String DEFAULT_BUS_NAME = "default";

    //默认的 topic 的名字
    private final static String DEFAULT_TOPIC = "default-topic";

    //用于分发广播消息到各个 Subscriber 的类
    private final Dispatcher dispatcher;

    public EventBus()
    {
```

```java
            this(DEFAULT_BUS_NAME, null, Dispatcher.SEQ_EXECUTOR_SERVICE);
        }

        public EventBus(String busName)
        {
            this(busName, null, Dispatcher.SEQ_EXECUTOR_SERVICE);
        }

        EventBus(String busName, EventExceptionHandler exceptionHandler, Executor executor)
        {
            this.busName = busName;
            this.dispatcher = Dispatcher.newDispatcher(exceptionHandler, executor);
        }

        public EventBus(EventExceptionHandler exceptionHandler)
        {
            this(DEFAULT_BUS_NAME, exceptionHandler, Dispatcher.SEQ_EXECUTOR_SERVICE);
        }

        // 将注册 Subscriber 的动作直接委托给 Registry
        @Override
        public void register(Object subscriber)
        {
            this.registry.bind(subscriber);
        }

        // 接触注册同样委托给 Registry
        @Override
        public void unregister(Object subscriber)
        {
            this.registry.unbind(subscriber);
        }

        // 提交 Event 到默认的 topic
        @Override
        public void post(Object event)
        {
            this.post(event, DEFAULT_TOPIC);
        }

        // 提交 Event 到指定的 topic，具体的动作是由 Dispatcher 来完成的
        @Override
        public void post(Object event, String topic)
        {
            this.dispatcher.dispatch(this, registry, event, topic);
        }

        // 关闭销毁 Bus
        @Override
```

```java
    public void close()
    {
        this.dispatcher.close();
    }

    // 返回 Bus 的名称
    @Override
    public String getBusName()
    {
        return this.busName;
    }
}
```

在上述代码中:
- EventBus 的构造除了名称之外,还需要有 ExceptionHandler 和 Executor,后两个主要是给 Dispatcher 使用的。
- registry 和 unregister 都是通过 Subscriber 注册表来完成的。
- Event 的提交则是由 Dispatcher 来完成的。
- Executor 并没有使用我们在第 8 章中开发的线程池,而是使用 JDK 中的 Executor 接口,我们自己开发的 ThreadPool 天生就是多线程并发执行任务的线程池,自带异步处理能力,但是无法做到同步任务处理,因此我们使用 Executor 可以任意扩展同步、异步的任务处理方式。

28.1.3 异步 EventBus 详解

异步的 EventBus 比较简单,继承自同步 Bus,然后将 Thread-Per-Message 用异步处理任务的 Executor 替换 EventBus 中的同步 Executor 即可,代码如清单 28-4 所示。

代码清单 28-4　异步 EventBus AsyncEventBus.java

```java
package com.wangwenjun.concurrent.chapter28;

import java.util.concurrent.ThreadPoolExecutor;

public class AsyncEventBus extends EventBus
{
    AsyncEventBus(String busName, EventExceptionHandler exceptionHandler,
ThreadPoolExecutor executor)
    {
        super(busName, exceptionHandler, executor);
    }

    public AsyncEventBus(String busName, ThreadPoolExecutor executor)
    {
        this(busName, null, executor);
    }
```

```java
    public AsyncEventBus(ThreadPoolExecutor executor)
    {
        this("default-async", null, executor);
    }

    public AsyncEventBus(EventExceptionHandler exceptionHandler,
ThreadPoolExecutor executor)
    {
        this("default-async", exceptionHandler, executor);
    }
}
```

上述代码重写了父类 EventBus 的构造函数，使用 ThreadPoolExecutor 替代 Executor。

28.1.4　Subscriber 注册表 Registry 详解

注册表维护了 topic 和 subscriber 之间的关系，当有 Event 被 post 之后，Dispatcher 需要知道该消息应该发送给哪个 Subscriber 的实例和对应的方法，Subscriber 对象没有任何特殊要求，就是普通的类不需要继承任何父类或者实现任何接口，注册表 Registry 的代码如清单 28-5 所示。

代码清单 28-5　注册表 Registry.java

```java
package com.wangwenjun.concurrent.chapter28;
import java.lang.reflect.Method;
import java.lang.reflect.Modifier;
import java.util.ArrayList;
import java.util.Arrays;
import java.util.List;
import java.util.concurrent.ConcurrentHashMap;
import java.util.concurrent.ConcurrentLinkedQueue;

class Registry
{
    // 存储 Subscriber 集合和 topic 之间关系的 map
    private final ConcurrentHashMap<String, ConcurrentLinkedQueue<Subscriber>> subscriberContainer
            = new ConcurrentHashMap<>();
    public void bind(Object subscriber)
    {
        // 获取 Subscriber Object 的方法集合然后进行绑定
        List<Method> subscribeMethods = getSubscribeMethods(subscriber);
        subscribeMethods.forEach(m -> tierSubscriber(subscriber, m));
    }

    public void unbind(Object subscriber)
    {
        //unbind 为了提高速度，只对 Subscriber 进行失效操作
```

```java
        subscriberContainer.forEach((key, queue) ->
            queue.forEach(s ->
            {
                if (s.getSubscribeObject() == subscriber)
                {
                    s.setDisable(true);
                }
            }));
    }

    public ConcurrentLinkedQueue<Subscriber> scanSubscriber(final String topic)
    {
        return subscriberContainer.get(topic);
    }

    private void tierSubscriber(Object subscriber, Method method)
    {
        final Subscribe subscribe = method.getDeclaredAnnotation(Subscribe.class);
        String topic = subscribe.topic();
        // 当某topic没有Subscriber Queue的时候创建一个
        subscriberContainer.computeIfAbsent(topic, key -> new ConcurrentLinkedQueue<>());

        // 创建一个Subscriber并且加入Subscriber列表中
        subscriberContainer.get(topic).add(new Subscriber(subscriber, method));
    }
    private List<Method> getSubscribeMethods(Object subscriber)
    {
        final List<Method> methods = new ArrayList<>();
        Class<?> temp = subscriber.getClass();
        // 不断获取当前类和父类的所有@Subscribe方法
        while (temp != null)
        {
            //获取所有的方法
            Method[] declaredMethods = temp.getDeclaredMethods();
            // 只有public方法 && 有一个入参 && 最重要的是被@Subscribe标记的方法才符合回调方法
            Arrays.stream(declaredMethods)
                    .filter(m -> m.isAnnotationPresent(Subscribe.class)
                            && m.getParameterCount() == 1
                            && m.getModifiers() == Modifier.PUBLIC)
                    .forEach(methods::add);
            temp = temp.getSuperclass();
        }
        return methods;
    }
}
```

由于Registry是在Bus中使用的，不能暴露给外部，因此Registry被设计成了包可见的类，我们所设计的EventBus对Subscriber没有做任何限制，但是要接受event的回调则需要将方法使用注解 @Subscribe 进行标记（可指定 topic），同一个 Subscriber 的不同方法通过 @Subscribe 注解之后可接受来自两个不同的 topic 消息，代码如下所示：

```java
/**
 * 非常普通的对象
 */
public class SimpleObject
{
    /**
     *subscribe 方法，比如使用 @Subscribe 标记，并且是 void 类型且有一个参数
     */
    @Subscribe(topic = "alex-topic")
    public void test2(Integer x)
    {

    }
    @Subscribe(topic = "test-topic")
    public void test3(Integer x)
    {
    }
}
```

SimpleObject 的实例被注册到了 Event Bus 之后，test2 和 test3 这两个方法将会被加入到注册表中，分别用来接受来自 alex-topic 和 test-topic 的 event。

28.1.5　Event 广播 Dispatcher 详解

前文中已经说过，Dispatcher 的主要作用是将 EventBus post 的 event 推送给每一个注册到 topic 上的 subscriber 上，具体的推送其实就是执行被 @Subscribe 注解的方法，示例代码如清单 28-6 所示。

代码清单 28-6　Dispatcher.java

```java
package com.wangwenjun.concurrent.chapter28;
import java.lang.reflect.Method;
import java.util.concurrent.ConcurrentLinkedQueue;
import java.util.concurrent.Executor;
import java.util.concurrent.ExecutorService;
public class Dispatcher
{
    private final Executor executorService;
    private final EventExceptionHandler exceptionHandler;

    public static final Executor SEQ_EXECUTOR_SERVICE = SeqExecutorService.INSTANCE;

    public static final Executor PRE_THREAD_EXECUTOR_SERVICE = PreThreadExecutorService.INSTANCE;

    private Dispatcher(Executor executorService, EventExceptionHandler
```

```java
exceptionHandler)
{
    this.executorService = executorService;
    this.exceptionHandler = exceptionHandler;
}

public void dispatch(Bus bus, Registry registry, Object event, String topic)
{
    // 根据topic获取所有的Subscriber列表
    ConcurrentLinkedQueue<Subscriber> subscribers = registry.scanSubscriber(topic);
    if (null == subscribers)
    {
        if (exceptionHandler != null)
        {
            exceptionHandler.handle(new IllegalArgumentException("The topic " + topic + " not bind yet"),
                    new BaseEventContext(bus.getBusName(), null, event));
        }
        return;
    }

    // 遍历所有的方法，并且通过反射的方式进行方法调用
    subscribers.stream()
            .filter(subscriber -> !subscriber.isDisable())
            .filter(subscriber ->
            {
                Method subscribeMethod = subscriber.getSubscribeMethod();
                Class<?> aClass = subscribeMethod.getParameterTypes()[0];
                return (aClass.isAssignableFrom(event.getClass()));
            }).forEach(subscriber -> realInvokeSubscribe(subscriber, event, bus));
}

private void realInvokeSubscribe(Subscriber subscriber, Object event, Bus bus)
{
    Method subscribeMethod = subscriber.getSubscribeMethod();
    Object subscribeObject = subscriber.getSubscribeObject();
    executorService.execute(() ->
    {
        try
        {
            subscribeMethod.invoke(subscribeObject, event);
        } catch (Exception e)
        {
            if (null != exceptionHandler)
            {
                exceptionHandler.handle(e, new BaseEventContext(bus.
                        getBusName(), subscriber, event));
            }
        }
```

```java
            });
        }

        public void close()
        {
            if (executorService instanceof ExecutorService)
                ((ExecutorService) executorService).shutdown();
        }

        static Dispatcher newDispatcher(EventExceptionHandler exceptionHandler,
Executor executor)
        {
            return new Dispatcher(executor, exceptionHandler);
        }

        static Dispatcher seqDispatcher(EventExceptionHandler exceptionHandler)
        {
            return new Dispatcher(SEQ_EXECUTOR_SERVICE, exceptionHandler);
        }

        static Dispatcher perThreadDispatcher(EventExceptionHandler exceptionHandler)
        {
            return new Dispatcher(PRE_THREAD_EXECUTOR_SERVICE, exceptionHandler);
        }

        // 顺序执行的 ExecutorService
        private static class SeqExecutorService implements Executor
        {
            private final static SeqExecutorService INSTANCE = new SeqExecutorService();

            @Override
            public void execute(Runnable command)
            {
                command.run();
            }
        }

        // 每个线程负责一次消息推送
        private static class PreThreadExecutorService implements Executor
        {
            private final static PreThreadExecutorService INSTANCE = new PreThreadExecutorService();

            @Override
            public void execute(Runnable command)
            {
                new Thread(command).start();
            }
        }

        // 默认的 EventContext 实现
```

```java
        private static class BaseEventContext implements EventContext
        {
            private final String eventBusName;

            private final Subscriber subscriber;

            private final Object event;

            private BaseEventContext(String eventBusName, Subscriber subscriber, Object event)
            {
                this.eventBusName = eventBusName;
                this.subscriber = subscriber;
                this.event = event;
            }

            @Override
            public String getSource()
            {
                return this.eventBusName;
            }

            @Override
            public Object getSubscriber()
            {
                return subscriber != null ? subscriber.getSubscribeObject() : null;
            }

            @Override
            public Method getSubscribe()
            {
                return subscriber != null ? subscriber.getSubscribeMethod() : null;
            }

            @Override
            public Object getEvent()
            {
                return this.event;
            }
        }
    }
```

在 Dispatcher 中，除了从 Registry 中获取对应的 Subscriber 执行之外，我们还定义了几个静态内部类，其主要是实现了 JDK1.5 以后的 Executor 接口和 EventContent。

28.1.6　其他类接口设计

除了上面一些比较核心的类之外，还需要 Subscriber 封装类以及 EventContext、Event-

ExceptionHandler 接口。

(1) Subscriber 类

Subscriber 类的代码如清单 28-7 所示。

代码清单 28-7　Subscriber.java

```java
package com.wangwenjun.concurrent.chapter28;

import java.lang.reflect.Method;
public class Subscriber
{
    private final Object subscribeObject;

    private final Method subscribeMethod;

    private boolean disable = false;

    public Subscriber(Object subscribeObject, Method subscribeMethod)
    {
        this.subscribeObject = subscribeObject;
        this.subscribeMethod = subscribeMethod;
    }

    public Object getSubscribeObject()
    {
        return subscribeObject;
    }

    public Method getSubscribeMethod()
    {
        return subscribeMethod;
    }

    public boolean isDisable()
    {
        return disable;
    }

    public void setDisable(boolean disable)
    {
        this.disable = disable;
    }
}
```

Subscriber 类封装了对象实例和被 @Subscribe 标记的方法，也就是说一个对象实例有可能会被封装成若干个 Subscriber。

(2) EventExceptionHandler 接口

EventBus 会将方法的调用交给 Runnable 接口去执行，我们都知道 Runnable 接口不能抛出 checked 异常信息，并且在每一个 subscribe 方法中，也不允许将异常抛出从而影响 EventBus 对后续 Subscriber 进行消息推送，但是异常信息又不能被忽略掉，因此注册一个异常回调接口就可以知道在进行消息广播推送时都发生了什么，代码如清单 28-8 所示。

代码清单 28-8　EventExceptionHandler.java

```java
package com.wangwenjun.concurrent.chapter28;
public interface EventExceptionHandler
{
    void handle(Throwable cause, EventContext context);
}
```

（3）EventContext 接口

Event 接口提供了获取消息源、消息体，以及该消息是由哪一个 Subscriber 的哪个 subscribe 方法所接受，主要用于消息推送出错时被回调接口 EventExceptionHandler 使用，代码如清单 28-9 所示。

代码清单 28-9　EventContext.java

```java
package com.wangwenjun.concurrent.chapter28;

import java.lang.reflect.Method;

public interface EventContext
{
    String getSource();

    Object getSubscriber();

    Method getSubscribe();

    Object getEvent();
}
```

28.1.7　Event Bus 测试

关于 Event Bus 的设计已经完成，虽然代码比较多，但是原理其实并不复杂，在本节中，我们将分别对同步的 Event Bus 和异步的 Event Bus 进行简单的测试。

（1）简单的 Subscriber

我们简单地定义两个普通对象 SimpleSubscriber1 和 SimpleSubscriber2，由于两者比较类似，因此省略了 SimpleSubscriber2，代码如下：

```java
package com.wangwenjun.concurrent.chapter28;
public class SimpleSubscriber1
{
    @Subscribe
```

```java
    public void method1(String message)
    {
        System.out.println("==SimpleSubscriber1==method1==" + message);
    }

    @Subscribe(topic = "test")
    public void method2(String message)
    {
        System.out.println("==SimpleSubscriber1==method2==" + message);
    }
}
```

(2）同步 Event Bus

```java
public static void main(String[] args)
{
    Bus bus = new EventBus("TestBus");
    bus.register(new SimpleSubscriber1());
    bus.register(new SimpleSubscriber2());
    bus.post("Hello");
    System.out.println("------------");
    bus.post("Hello","test");
}
```

上面的这段程序定义了同步的 EventBus，然后将两个普通的对象注册给了 bus，当 bus 发送 Event 的时候 topic 相同，Event 类型相同的 subscribe 方法将被执行，程序输出如下所示：

```
==SimpleSubscriber1==method1==Hello
==SimpleSubscriber2==method1==Hello
------------
==SimpleSubscriber1==method2==Hello
==SimpleSubscriber2==method2==Hello
```

(3）异步 Event Bus

同步的 Event Bus 有个缺点，若其中的一个 subscribe 方法运行时间比较长，则会影响下一个 subscribe 方法的执行，因此采用 AsyncEventBus 是另外一个比较好的选择，示例代码如下：

```java
public static void main(String[] args)
{
    Bus bus = new AsyncEventBus("TestBus", (ThreadPoolExecutor) Executors.newFixedThreadPool(10));
    bus.register(new SimpleSubscriber1());
    bus.register(new SimpleSubscriber2());
    bus.post("Hello");
    System.out.println("------------");
    bus.post("Hello", "test");
}
```

28.2 Event Bus 实战——监控目录变化

记得笔者刚参加工作的时候，第一个开发任务就是监控某个硬件设备的运行时数据，然后记录在数据库中，该硬件设备在运行的过程中，会将一些性能信息等写入特殊的数据文件中，我要做的就是监控到该文件的变化，读取最后一行数据，然后根据格式将其解析出来插入数据库，实现的思路大致是：在程序首次启动时获取该文件的最后修改时间并且做文件的首次解析，然后每隔一段指定的时间检查一次文件最后被修改的时间，如果与记录的时间相等则等待下次的采集（Balking Pattern），否则进行新一轮的采集并且更新时间。

虽然这个程序足够简单，但是上述的实现方式还是存在着诸多问题，比如在采集时间间隔内，如果文件发生了 N 次变化，我只能获取到最后一次，其根本原因是文件的变化不会通知到应用程序，所以只能比较笨的主动去轮询。

JDK 自 1.7 版本后提供了 WatchService 类，该类可以基于事件通知的方式监控文件或者目录的任何变化，文件的改变相当于每一个事件（Event）的发生，针对不同的时间执行不同的动作，本节将结合 NIO2.0 中提供的 WatchService 和我们实现的 Event Bus 实现文件目录的监控的功能。

28.2.1 WatchService 遇到 EventBus

代码清单 28-10　DirectoryTargetMonitor.java

```java
package com.wangwenjun.concurrent.chapter28;

import java.nio.file.*;
public class DirectoryTargetMonitor
{
    private WatchService watchService;

    private final EventBus eventBus;

    private final Path path;

    private volatile boolean start = false;

    public DirectoryTargetMonitor(final EventBus eventBus,
        final String targetPath)
    {
            this(eventBus, targetPath, "");
    }

    // 构造 Monitor 的时候需要传入 EventBus 以及需要监控的目录
    public DirectoryTargetMonitor(final EventBus eventBus,
final String targetPath,final String... morePaths)
    {
        this.eventBus = eventBus;
```

```java
            this.path = Paths.get(targetPath, morePaths);
    }

    public void startMonitor() throws Exception
    {
        this.watchService = FileSystems.getDefault().newWatchService();
        // 为路径注册感兴趣的事件
        this.path.register(watchService, StandardWatchEventKinds.ENTRY_MODIFY,
                StandardWatchEventKinds.ENTRY_DELETE, StandardWatchEventKinds.
                ENTRY_CREATE);
        System.out.printf("The directory [%s] is monitoring... \n", path);
        this.start = true;
        while (start)
        {
            WatchKey watchKey = null;
            try
            {
                // 当有事件发生时会返回对应的WatchKey
                watchKey = watchService.take();
                watchKey.pollEvents().forEach(event ->
                {
                    WatchEvent.Kind<?> kind = event.kind();
                    Path path = (Path) event.context();
                    Path child = DirectoryTargetMonitor.this.path.resolve(path);
                    // 提交FileChangeEvent到EventBus
                    eventBus.post(new FileChangeEvent(child, kind));
                });
            } catch (Exception e)
            {
                this.start = false;
            } finally
            {
                if (watchKey != null)
                    watchKey.reset();
            }
        }
    }

    public void stopMonitor() throws Exception
    {
        System.out.printf("The directory [%s] monitor will be stop...\n", path);
        Thread.currentThread().interrupt();
        this.start = false;
        this.watchService.close();
        System.out.printf("The directory [%s] monitor will be stop done.\n", path);
    }
}
```

在创建 WatchService 之后将文件的修改、删除、创建等注册给了 WatchService，在指定目录下发生诸如此类的事件之后便会收到通知，我们将事件类型和发生变化的文件 Path 封装成 FileChangeEvent 提交给 Event Bus。

28.2.2 FileChangeEvent

FileChangeEvent 比较简单，就是对 WatchEvent.Kind 和 Path 的包装，一旦目录发生任何改变，都会提交 FileChangeEvent 事件，代码如清单 28-11 所示。

代码清单 28-11　FileChangeEvent.java

```java
package com.wangwenjun.concurrent.chapter28;

import java.nio.file.Path;
import java.nio.file.WatchEvent;
public class FileChangeEvent
{
    private final Path path;
    private final WatchEvent.Kind<?> kind;
    public FileChangeEvent(Path path, WatchEvent.Kind<?> kind)
    {
        this.path = path;
        this.kind = kind;
    }
    public Path getPath()
    {
        return path;
    }

    public WatchEvent.Kind<?> getKind()
    {
        return kind;
    }
}
```

28.2.3 监控目录变化

好了，目录监控的程序我们已经实现了，下面就来写一个接受文件目录变化的 Subscriber，也就是当目录发生变化时用来接受事件的方法，代码如清单 28-12 所示。

代码清单 28-12　FileChangeListener.java

```java
package com.wangwenjun.concurrent.chapter28;
public class FileChangeListener
{
    @Subscribe
    public void onChange(FileChangeEvent event)
    {
    System.out.printf("%s-%s\n", event.getPath(), event.getKind());
    }
}
```

onChange 方法由 @Subscribe 标记，但没有指定 topic，当有事件发送到了默认的 topic

上之后，该方法将被调用执行，接下来我们将 FileChangeListener 的实例注册给 Event Bus 并且启动 Monitor 程序，代码如下：

```
public static void main(String[] args) throws Exception
{
    ThreadPoolExecutor executor = (ThreadPoolExecutor) Executors.newFixedThreadPool(
            Runtime.getRuntime().availableProcessors() * 2);
    final EventBus eventBus = new AsyncEventBus(executor);
    // 注册
    eventBus.register(new FileChangeListener());
    DirectoryTargetMonitor monitor = new DirectoryTargetMonitor(eventBus, "G:\\monitor");
    monitor.startMonitor();
}
```

运行上面的程序，然后在子目录 G:\\monitor 下不断地创建、删除、修改文件，这些事件都将被收集并且提交给 EventBus，程序输出如下：

```
The directory [G:\monitor] is monitoring...
G:\monitor\新建文本文档.txt-ENTRY_CREATE
G:\monitor\新建文本文档 (2).txt-ENTRY_CREATE
G:\monitor\新建文本文档 (2).txt-ENTRY_DELETE
G:\monitor\a.txt-ENTRY_CREATE
G:\monitor\a.txt-ENTRY_MODIFY
G:\monitor\新建文件夹-ENTRY_CREATE
...
```

在 Apache 的 Flume 框架中提供了 Spooling source 功能，其内部使用的就是 WatchService。

28.3 本章总结

在本章中，我们实现了一个 EventBus 模式的小框架，EventBus 有点类似于 GOF 设计模式中的监听者模式，但是 EventBus 提供的功能更加强大，使用起来也更加灵活，EventBus 中的 Subscriber 不需要继承任何类或者实现任何接口，在使用 EventBus 时只需要持有 Bus 的引用即可。

在 EventBus 的设计中有三个非常重要的角色（Bus、Registry 和 Dispatcher），Bus 主要提供给外部使用的操作方法，Registry 注册表用来整理记录所有注册在 EventBus 上的 Subscriber，Dispatcher 主要负责对 Subscriber 消息进行推送（用反射的方式执行方法），但是考虑到程序的灵活性，Dispatcher 方法中又提供了 Executor 的多态方式。

CHAPTER29 · 第 29 章

Event Driven 设计模式

29.1 Event-Driven Architecture 基础

EDA（Event-Driven Architecture）是一种实现组件之间松耦合、易扩展的架构方式，在本节中，我们先介绍 EDA 的基础组件，让读者对 EDA 设计架构方式有一个基本的认识，一个最简单的 EDA 设计需要包含如下几个组件。

- Events：需要被处理的数据。
- Event Handlers：处理 Events 的方式方法。
- Event Loop：维护 Events 和 Event Handlers 之间的交互流程。

如图 29-1 所示，Event A 将被 Handler A 处理，而 Event B 将被 Handler B 处理，这一切的分配都是由 Event Loop 所控制的。

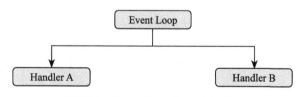

图 29-1　Event-Event Loop-Handler

29.1.1　Events

Events 是 EDA 中的重要角色，一个 Event 至少需要包含两个属性：类型和数据，Event 的类型决定了它会被哪个 Handler 处理，数据是在 Handler 中代加工的材料，下面写一个简单的程序，代码如清单 29-1 所示。

代码清单 29-1　Event.java

```java
package com.wangwenjun.concurrent.chapter29;
/**
*Event 只包含了该 Event 所属的类型和所包含的数据
*/
public class Event
{
    private final String type;
    private final String data;

    public Event(String type, String data)
    {
        this.type = type;
        this.data = data;
    }

    public String getType()
    {
        return type;
    }

    public String getData()
    {
        return data;
    }
}
```

29.1.2　Event Handlers

Event Handlers 主要用于处理 Event，比如一些 filtering 或者 transforming 数据的操作等，下面我们写两个比较简单的方法，代码如下：

```java
public static void handleEventA(Event e)
{
    System.out.println(e.getData().toLowerCase());
}
```

handleEventA 方法只是简单地将 Event 中的 data 进行了 lowerCase 之后的输出，代码如下：

```java
public static void handleEventB(Event e)
{
    System.out.println(e.getData().toUpperCase());
}
```

同样，handleEventB 方法也是足够的简单，直接将 Event 中的字符串数据变成大写进行了控制台输出。

29.1.3　Event Loop

Event Loop 处理接收到的所有 Event，并且将它们分配给合适的 Handler 去处理，代码如下：

```
Event e;
while (!events.isEmpty())
{
    // 从消息队列中不断移除 Event，根据不同的类型进行处理
    e = events.remove();
    switch (e.getType())
    {
        case "A":
            handleEventA(e);
            break;
        case "B":
            handleEventB(e);
            break;
    }
}
```

在 EventLoop 中，每一个 Event 都将从 Queue 中移除出去，通过类型匹配交给合适的 Handler 去处理，完整的例子如清单 29-2 所示。

代码清单 29-2　完整的 FooEventDrivenExample.java

```
package com.wangwenjun.concurrent.chapter29;

import java.util.LinkedList;
import java.util.Queue;
public class FooEventDrivenExample
{
    // 用于处理 A 类型的 Event
    public static void handleEventA(Event e)
    {
        System.out.println(e.getData().toLowerCase());
    }
    // 用于处理 B 类型的 Event
    public static void handleEventB(Event e)
    {
        System.out.println(e.getData().toUpperCase());
    }

    public static void main(String[] args)
    {
        Queue<Event> events = new LinkedList<>();
        events.add(new Event("A", "Hello"));
        events.add(new Event("A", "I am Event A"));
        events.add(new Event("B", "I am Event B"));
        events.add(new Event("B", "World"));
```

```
        Event e;
        while (!events.isEmpty())
        {
// 从消息队列中不断移除，根据不同的类型进行处理
            e = events.remove();
            switch (e.getType())
            {
                case "A":
                    handleEventA(e);
                    break;
                case "B":
                    handleEventB(e);
                    break;
            }
        }
    }
}
```

虽然这个 EDA 的设计足够简单，但是通过它我们可以感受到 EDA 中三个重要组件之间的交互关系，其对接下来的内容学习也会有一定的帮助，运行上面的程序，输出结果如下：

```
hello
i am event a
I AM EVENT B
WORLD
```

29.2　开发一个 Event-Driven 框架

在 29.1 节中，我们通过非常简陋的例子大致介绍了 EDA 程序设计中几个关键的组件，在本节中，我们将分别它们进行抽象，设计一个迷你的 EDA 小框架，在 29.3 节中，该框架将会被用来实现一个简单的聊天小程序。

通过 29.1 节的基础知识介绍，我们大致可以知道，一个基于事件驱动的架构设计，总体来讲会涉及如下几个重要组件：事件消息（Event）、针对该事件的具体处理器（Handler）、接受事件消息的通道（29.1.3 节中的 queue），以及对事件消息如何进行分配（Event Loop）。

29.2.1　同步 EDA 框架设计

我们先设计开发一个高度抽象的同步 EDA 框架，然后在 29.2.2 节中增加异步功能。

（1）Message

回顾 29.1 节基础部分的介绍，在基于 Message 的系统中，每一个 Event 也可以被称为 Message，Message 是对 Event 更高一个层级的抽象，每一个 Message 都有一个特定的 Type 用于与对应的 Handler 做关联，清单 29-3 是 Message 接口的定义。

代码清单 29-3　Message.java

```java
package com.wangwenjun.concurrent.chapter29;

public interface Message
{
    /**
     * 返回 Message 的类型
     */
    Class<? extends Message> getType();
}
```

（2）Channel

第二个比较重要的概念就是 Channels，Channel 主要用于接受来自 Event Loop 分配的消息，每一个 Channel 负责处理一种类型的消息（当然这取决于你对消息如何进行分配），清单 29-4 是 Channel 接口的定义：

代码清单 29-4　Channel.java

```java
package com.wangwenjun.concurrent.chapter29;

public interface Channel<E extends Message>
{
    /**
     * dispatch 方法用于负责 Message 的调度
     */
    void dispatch(E message);
}
```

（3）Dynamic Router

Router 的作用类似于 29.1 节中的 Event Loop，其主要是帮助 Event 找到合适的 Channel 并且传送给它，Dynamic Routers 代码定义如清单 29-5 所示。

代码清单 29-5　DynamicRouter.java

```java
package com.wangwenjun.concurrent.chapter29;

public interface DynamicRouter<E extends Message>
{
    /**
     * 针对每一种 Message 类型注册相关的 Channel，只有找到合适的 Channel 该 Message 才会被处理
     */
    void registerChannel(Class<? extends E> messageType,
                         Channel<? extends E> channel);

    /**
     * 为相应的 Channel 分配 Message
     */
    void dispatch(E message);
}
```

Router 如何知道要将 Message 分配给哪个 Channel 呢？换句话说，Router 需要了解到 Channel 的存在，因此 registerChannel() 方法的作用就是将相应的 Channel 注册给 Router，dispatch 方法则是根据 Message 的类型进行路由匹配。

（4）Event

Event 是对 Message 的一个最简单的实现，在以后的使用中，将 Event 直接作为其他 Message 的基类即可（这种做法有点类似于适配器模式），Event 接口的定义如清单 29-6 所示。

代码清单 29-6　Event.java

```java
package com.wangwenjun.concurrent.chapter29;
public class Event implements Message
{
    @Override
    public Class<? extends Message> getType()
    {
        return getClass();
    }
}
```

（5）EventDispatcher

EventDispatcher 是对 DynamicRouter 的一个最基本的实现，适合在单线程的情况下进行使用，因此不需要考虑线程安全的问题。EventDispatcher 接口的定义如清单 29-7 所示。

代码清单 29-7　EventDispatcher.java

```java
package com.wangwenjun.concurrent.chapter29;

import java.util.HashMap;
import java.util.Map;

/**
 *EventDispatcher 不是一个线程安全的类
 */
public class EventDispatcher implements DynamicRouter<Message>
{
    // 用于保存 Channel 和 Message 之间的关系
    private final Map<Class<? extends Message>, Channel> routerTable;

    public EventDispatcher()
    {
        // 初始化 RouterTable，但是在该实现中，我们使用 HashMap 作为路由表
        this.routerTable = new HashMap<>();
    }

    @Override
    public void dispatch(Message message)
    {
```

```java
            if (routerTable.containsKey(message.getType()))
            {
                // 直接获取对应的 Channel 处理 Message
                routerTable.get(message.getType()).dispatch(message);
            } else
                throw new MessageMatcherException("Can't match the channel for [" +
message.getType() + "] type");
        }

        @Override
        public void registerChannel(Class<? extends Message> messageType,
                                    Channel<? extends Message> channel)
        {
            this.routerTable.put(messageType, channel);
        }
}
```

在 EventDispatcher 中有一个注册表 routerTable，主要用于存放不同类型 Message 对应的 Channel，如果没有与 Message 相对应的 Channel，则会抛出无法匹配的异常，示例代码如清单 29-8 所示。

代码清单 29-8　MessageMatcherException.java

```java
package com.wangwenjun.concurrent.chapter29;

public class MessageMatcherException extends RuntimeException
{
    public MessageMatcherException(String message)
    {
        super(message);
    }
}
```

下面写一个简单的测试来看看如何使用我们之前实现的 EDA 框架，代码如清单 29-9 所示。

代码清单 29-9　EventDispatcherExample

```java
package com.wangwenjun.concurrent.chapter29;

public class EventDispatcherExample
{
    /**
     * InputEvent 中定义了两个属性 X 和 Y，主要用于在其他 Channel 中的运算
     */
    static class InputEvent extends Event
    {
        private final int x;
        private final int y;
```

```java
        public InputEvent(int x, int y)
        {
            this.x = x;
            this.y = y;
        }

        public int getX()
        {
            return x;
        }

        public int getY()
        {
            return y;
        }
    }
    /**
     * 用于存放结果的Event
     */
    static class ResultEvent extends Event
    {
        private final int result;

        public ResultEvent(int result)
        {
            this.result = result;
        }

        public int getResult()
        {
            return result;
        }
    }
    /**
     * 处理ResultEvent的Handler（Channel），只是简单地将计算结果输出到控制台
     */
    static class ResultEventHandler implements Channel<ResultEvent>
    {
        @Override
        public void dispatch(ResultEvent message)
        {
            System.out.println("The result is:" + message.getResult());
        }
    }

    /**
     * InputEventHandler需要向Router发送Event，因此在构造的时候需要传入Dispatcher
```

```java
	 */
	static class InputEventHandler implements Channel<InputEvent>
	{
		private final EventDispatcher dispatcher;

		public InputEventHandler(EventDispatcher dispatcher)
		{
			this.dispatcher = dispatcher;
		}

		/**
		 * 将计算的结果构造成新的 Event 提交给 Router
		 */
		@Override
		public void dispatch(InputEvent message)
		{
			System.out.printf("X:%d,Y:%d\n", message.getX(), message.getY());
			int result = message.getX() + message.getY();
			dispatcher.dispatch(new ResultEvent(result));
		}
	}

	public static void main(String[] args)
	{
		// 构造 Router
		EventDispatcher dispatcher = new EventDispatcher();
		// 将 Event 和 Handler（Channel）的绑定关系注册到 Dispatcher
		dispatcher.registerChannel(InputEvent.class, new 
InputEventHandler(dispatcher));
		dispatcher.registerChannel(ResultEvent.class, new ResultEventHandler());
		dispatcher.dispatch(new InputEvent(1, 2));
	}
}
```

由于所有的类都存放于一个文件中，因此看起来测试代码比较多，其实结构还是非常清晰的，InputEvent 是一个 Message，它包含了两个 Int 类型的属性，而 InputEventHandler 是对 InputEvent 消息的处理，接收到了 InputEvent 消息之后，分别对 X 和 Y 进行相加操作，然后将结果封装成 ResultEvent 提交给 EventDispatcher，ResultEvent 相对比较简单，只包含了计算结果的属性，ResultEventHandler 则将计算结果输出到控制台上。

通过上面这个例子的运行你会发现，不同数据的处理过程之间根本无须知道彼此的存在，一切都由 EventDispatcher 这个 Router 来控制，它会给你想要的一切，这是一种稀疏耦合（松耦合）的设计。图 29-2 所示的为同步 EDA 架构类图。

EDA 的设计除了松耦合特性之外，扩展性也是非常强的，比如 Channel 非常容易扩展和替换，另外由于 Dispatcher 统一负责 Event 的调配，因此在消息通过 Channel 之前可以进行很多过滤、数据验证、权限控制、数据增强（Enhance）等工作。

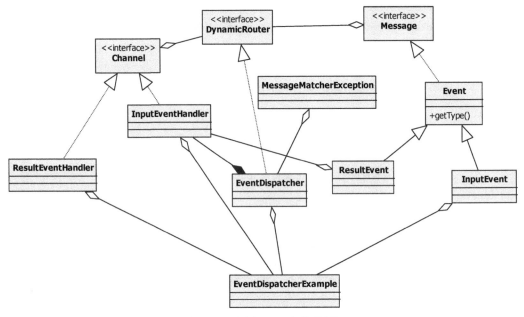

图 29-2　同步 EDA 架构类图

29.2.2　异步 EDA 框架设计

在 29.2.1 节中，我们实现了一个基本的 EDA 框架，但是这个框架在应对高并发的情况下还是存在一些问题的，具体如下。

- EventDispatcher 不是线程安全的类，在多线程的情况下，registerChannel 方法会引起数据不一致的问题。
- 就目前而言，我们实现的所有 Channel 都无法并发消费 Message，比如 InputEventHandler 只能逐个处理 Message，低延迟的消息处理还会导致 Dispatcher 出现积压。

在本节中，我们将对 29.2.1 节中的 EDA 框架进行扩充，使其可支持并发任务的执行，下面定义了一个新的 AsyncChannel 作为基类，该类中提供了 Message 的并发处理能力，代码如清单 29-10 所示。

代码清单 29-10　AsyncChannel.java

```
package com.wangwenjun.concurrent.chapter29;

import java.util.concurrent.ExecutorService;
public abstract class AsyncChannel implements Channel<Event>
{
    // 在 AsyncChannel 中将使用 ExecutorService 多线程的方式提交给 Message
    private final ExecutorService executorService;
```

```java
// 默认构造函数，提供了CPU的核数×2的线程数量
public AsyncChannel()
{
    this(Executors.newFixedThreadPool(Runtime.getRuntime()
        .availableProcessors() * 2));
}

// 用户自定义的ExecutorService
public AsyncChannel(ExecutorService executorService)
{
    this.executorService = executorService;
}

// 重写dispatch方法，并且用final修饰，避免子类重写
@Override
public final void dispatch(Event message)
{
    executorService.submit(() -> this.handle(message));
}
// 提供抽象方法，供子类实现具体的Message处理
protected abstract void handle(Event message);

// 提供关闭ExecutorService的方法
void stop()
{
    if (null != executorService && !executorService.isShutdown())
        executorService.shutdown();
}
}
```

为了防止子类在继承AsyncChannel基类的时候重写dispatch方法，用final关键字对其进行修饰（Template Method Design Pattern），handle方法用于子类对Message进行具体的处理，stop方法则用来停止ExecutorService。

其次，还需要提供新的EventDispatcher类AsyncEventDispatcher负责以并发的方式dispatch Message，其中Event对应的Channel只能是AsyncChannel类型，并且也对外暴露了shutdown方法，代码如清单29-11所示。

代码清单29-11　AsyncChannel.java

```java
package com.wangwenjun.concurrent.chapter29;

import java.util.Map;
import java.util.concurrent.ConcurrentHashMap;

public class AsyncEventDispatcher implements DynamicRouter<Event>
{
```

```java
        // 使用线程安全的 ConcurrentHashMap 替换 HashMap
        private final Map<Class<? extends Event>, AsyncChannel> routerTable;

        public AsyncEventDispatcher()
        {
            this.routerTable = new ConcurrentHashMap<>();
        }

        @Override
        public void registerChannel(Class<? extends Event> messageType,
                            Channel<? extends Event> channel)
{
    // 在 AsyncEventDispatcher 中, Channel 必须是 AsyncChannel 类型
        if (!(channel instanceof AsyncChannel))
            {
                throw new IllegalArgumentException("The channel must be AsyncChannel
                    Type.");
            }
            this.routerTable.put(messageType, (AsyncChannel) channel);
        }

        @Override
        public void dispatch(Event message)
        {
            if (routerTable.containsKey(message.getType()))
            {
                routerTable.get(message.getType()).dispatch(message);
            } else
                throw new MessageMatcherException("Can't match the channel for ["
                    + message.getType() + "] type");
        }

        public void shutdown()
        {
            // 关闭所有的 Channel 以释放资源
            routerTable.values().forEach(AsyncChannel::stop);
        }
    }
```

在 AsyncEventDispatcher 中,routerTable 使用线程安全的 Map 定义,在注册 Channel 的时候,如果其不是 AsyncChannel 的类型,则会抛出异常。

图 29-3 所示的为异步 EDA 架构类图。

好了,下面我们写一个针对异步 EDA 的测试,用法和同步的非常类似,只不过几个关键的类需要使用异步的来代替,代码如清单 29-12 所示。

第29章 Event Driven设计模式

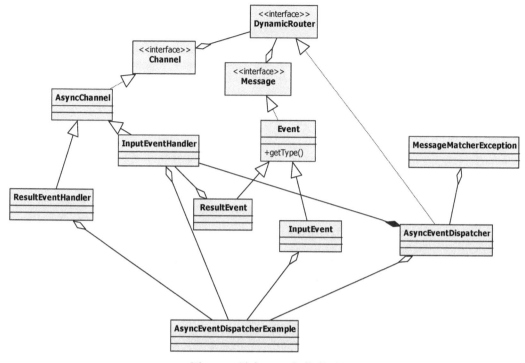

图 29-3　异步 EDA 架构类图

代码清单 29-12　AsyncEventDispatcherExample.java

```
package com.wangwenjun.concurrent.chapter29;

import java.util.concurrent.TimeUnit;

public class AsyncEventDispatcherExample
{
    // 主要用于处理 InputEvent，但是需要继承 AsyncChannel
    static class AsyncInputEventHandler extends AsyncChannel
    {
        private final AsyncEventDispatcher dispatcher;
        AsyncInputEventHandler(AsyncEventDispatcher dispatcher)
        {
            this.dispatcher = dispatcher;
        }
        // 不同于以同步的方式实现 dispatch，异步的方式需要实现 handle
        @Override
        protected void handle(Event message)
        {
            EventDispatcherExample.InputEvent inputEvent =
                    (EventDispatcherExample.InputEvent) message;
            System.out.printf("X:%d,Y:%d\n", inputEvent.getX(), inputEvent.getY());
            try
```

```java
            {
                TimeUnit.SECONDS.sleep(5);
            } catch (InterruptedException e)
            {
                e.printStackTrace();
            }
            int result = inputEvent.getX() + inputEvent.getY();
            dispatcher.dispatch(new EventDispatcherExample.ResultEvent(result));
        }
    }
    // 主要用于处理InputEvent，但是需要继承AsyncChannel
    static class AsyncResultEventHandler extends AsyncChannel
    {
        @Override
        protected void handle(Event message)
        {
            EventDispatcherExample.ResultEvent resultEvent =
                    (EventDispatcherExample.ResultEvent) message;
            try
            {
                TimeUnit.SECONDS.sleep(5);
            } catch (InterruptedException e)
            {
                e.printStackTrace();
            }
            System.out.println("The result is:" + resultEvent.getResult());
        }
    }
    public static void main(String[] args)
    {
        // 定义AsyncEventDispatcher
        AsyncEventDispatcher dispatcher = new AsyncEventDispatcher();

        // 注册Event和Channel之间的关系
    dispatcher.registerChannel(EventDispatcherExample.InputEvent.class, new AsyncInputEventHandler(dispatcher));
    dispatcher.registerChannel(EventDispatcherExample.ResultEvent.class, new AsyncResultEventHandler());
        // 提交需要处理的Message
        dispatcher.dispatch(new EventDispatcherExample.InputEvent(1, 2));
    }
}
```

当dispatcher分配一个Event的时候，如果执行非常缓慢也不会影响下一个Event被dispatch，这主要得益于我们采用了异步的处理方式（ExecutorService本身存在的任务队列可以允许异步提交一定数量级的数据）。

29.3　Event-Driven 的使用

在本节中，我们模拟一个简单的聊天应用程序，借助于我们在 29.2 节开发的 EDA 小框架，首先我们要为聊天应用程序定义如下几个类型的 Event。
- User Online Event：当用户上线时来到聊天室的 Event。
- User Offline Event：当用户下线时退出聊天室的 Event。
- User Chat Event：用户在聊天室中发送聊天信息的 Event。

29.3.1　Chat Event

首先，我们定义一个 User 对象，代表聊天室的参与者，比较简单就是一个名字，代码如下：

```java
package com.wangwenjun.concurrent.chapter29;

public class User
{
    private final String name;

    public User(String name)
    {
        this.name = name;
    }

    public String getName()
    {
        return name;
    }
}
```

下面定义一个 UserOnlineEvent，代表用户上线的 Event，代码如下：

```java
package com.wangwenjun.concurrent.chapter29;

public class UserOnlineEvent extends Event
{
    private final User user;

    public UserOnlineEvent(User user)
    {
        this.user = user;
    }

    public User getUser()
    {
        return user;
    }
}
```

下面定义一个 UserOfflineEvent，代表用户下线的 Event，代码如下：

```
package com.wangwenjun.concurrent.chapter29;

public class UserOfflineEvent extends UserOnlineEvent
{
    public UserOfflineEvent(User user)
    {
        super(user);
    }
}
```

下面定义一个 UserChatEvent，代表用户发送了聊天信息的 Event，代码如下：

```
package com.wangwenjun.concurrent.chapter29;

public class UserChatEvent extends UserOnlineEvent
{
    //ChatEvent 需要有聊天的信息
    private final String message;

    public UserChatEvent(User user, String message)
    {
        super(user);
        this.message = message;
    }

    public String getMessage()
    {
        return message;
    }
}
```

UserChatEvent 比其他两个 Event 多了代表聊天内容的 message 属性。

29.3.2　Chat Channel（Handler）

所有的 Handler 都非常简单，只是将接收到的信息输出到控制台，由于是在多线程的环境下运行，因此我们需要继承 AsyncChannel。

下面定义一个 UserOnlineEventChannel，主要用于处理 UserOnlineEvent 事件，代码如下：

```
package com.wangwenjun.concurrent.chapter29;

//用户上线的 Event，简单输出用户上线即可
public class UserOnlineEventChannel extends AsyncChannel
{
    @Override
    protected void handle(Event message)
```

```
            {
                UserOnlineEvent event = (UserOnlineEvent) message;
                System.out.println("The User[" + event.getUser().getName() + "] is
                online.");
            }
        }
```

下面定义一个 UserOnlineEventChannel，主要用于处理 UserOfflineEvent 事件，代码如下：

```
package com.wangwenjun.concurrent.chapter29;

//用户下线的 Event，简单输出用户下线即可
public class UserOfflineEventChannel extends AsyncChannel
{
    @Override
    protected void handle(Event message)
    {
        UserOfflineEvent event = (UserOfflineEvent) message;
        System.out.println("The User[" + event.getUser().getName() + "] is
offline.");
    }
}
```

下面定义一个 UserChatEventChannel，主要用于处理 UserChatEvent 事件，代码如下：

```
package com.wangwenjun.concurrent.chapter29;
//用户聊天的 Event，直接在控制台输出即可
public class UserChatEventChannel extends AsyncChannel
{
    @Override
    protected void handle(Event message)
    {
        UserChatEvent event = (UserChatEvent) message;
        System.out.println("The User[" + event.getUser().getName() + "] say: " +
        event.getMessage());
    }
}
```

29.3.3　Chat User 线程

我们定义完 Event 和接受 Event 的 Channel 后，现在定义一个代表聊天室参与者的 User 线程，代码如下：

```
package com.wangwenjun.concurrent.chapter29;

import java.util.concurrent.TimeUnit;

import static java.util.concurrent.ThreadLocalRandom.current;

public class UserChatThread extends Thread
```

```java
{
    private final User user;
    private final AsyncEventDispatcher dispatcher;

    public UserChatThread(User user, AsyncEventDispatcher dispatcher)
    {
        super(user.getName());
        this.user = user;
        this.dispatcher = dispatcher;
    }

    @Override
    public void run()
    {
        try
        {
            //User 上线,发送 Online Event
            dispatcher.dispatch(new UserOnlineEvent(user));
            for (int i = 0; i < 5; i++)
            {
                // 发送 User 的聊天信息
                dispatcher.dispatch(new UserChatEvent(user, getName() + "-Hello-"
                + i));
                // 短暂休眠 1～10 秒
                TimeUnit.SECONDS.sleep(current().nextInt(10));
            }
        } catch (InterruptedException e)
        {
            e.printStackTrace();
        } finally
        {
            //User 下线,发送 Offline Event
            dispatcher.dispatch(new UserOfflineEvent(user));
        }
    }
}
```

当 User 线程启动的时候,首先发送 Online Event,然后发送五条聊天信息,之后下线,在下线的时候发送 Offline Event,下面写一个简单的程序测试一下:

```java
package com.wangwenjun.concurrent.chapter29;

public class UserChatApplication
{
    public static void main(String[] args)
    {
        // 定义异步的 Router
        final AsyncEventDispatcher dispatcher = new AsyncEventDispatcher();
        // 为 Router 注册 Channel 和 Event 之间的关系
        dispatcher.registerChannel(UserOnlineEvent.class, new
UserOnlineEventChannel());
```

```
            dispatcher.registerChannel(UserOfflineEvent.class, new
UserOfflineEventChannel());
            dispatcher.registerChannel(UserChatEvent.class, new
UserChatEventChannel());

            // 启动三个登录聊天室的User
            new UserChatThread(new User("Leo"), dispatcher).start();
            new UserChatThread(new User("Alex"), dispatcher).start();
            new UserChatThread(new User("Tina"), dispatcher).start();
    }
}
```

在测试程序中，我们创建了三个User线程并且启动，程序的输出如下所示：

```
The User[Tina] is online.
The User[Leo] is online.
The User[Leo] say: Leo-Hello-0
The User[Alex] is online.
The User[Alex] say: Alex-Hello-0
The User[Tina] say: Tina-Hello-0
The User[Alex] say: Alex-Hello-1
The User[Alex] say: Alex-Hello-2
The User[Tina] say: Tina-Hello-1
The User[Tina] say: Tina-Hello-2
The User[Alex] say: Alex-Hello-3
The User[Alex] say: Alex-Hello-4
The User[Leo] say: Leo-Hello-1
The User[Leo] say: Leo-Hello-2
The User[Tina] say: Tina-Hello-3
The User[Tina] say: Tina-Hello-4
The User[Tina] is offline.
The User[Alex] is offline.
The User[Leo] say: Leo-Hello-3
The User[Leo] say: Leo-Hello-4
The User[Leo] is offline.
```

29.4 本章总结

Message（Event）无论是在同步还是在异步的EDA中，我们都没有使用任何同步的方式对其进行控制，根本原因是Event被设计成了不可变对象，因为Event在经过每一个Channel（Handler）的时候都会创建一个全新的Event，多个线程之间不会出现资源竞争，因此不需要同步的保护。再联想之前我们学习过的生产者消费者模式，如果使用EDA框架对其进行修改，是不是不需要多个线程之间的交互，也不需要对队列的同步呢？有消息就直接往EDA里面提交就可以了，消费的Handler自然而然地会触发消费Message。

推荐阅读

Java核心技术 卷I：基础知识（原书第10版）

书号：978-7-111-54742-6　作者：（美）凯 S. 霍斯特曼（Cay S. Horstmann）　定价：119.00元

　　Java领域最有影响力和价值的著作之一，与《Java编程思想》齐名，10余年全球畅销不衰，广受好评

　　根据Java SE 8全面更新，系统全面讲解Java语言的核心概念、语法、重要特性和开发方法，包含大量案例，实践性强

　　本书为专业程序员解决实际问题而写，可以帮助你深入了解Java语言和库。在卷I中，Horstmann主要强调基本语言概念和现代用户界面编程基础，深入介绍了从Java面向对象编程到泛型、集合、lambda表达式、Swing UI设计以及并发和函数式编程的最新方法等内容。

推荐阅读

Effective Java中文版 第2版

作者：Joshua Bloch　ISBN：978-7-111-25583-3　定价：52.00元

本书介绍了在Java编程中78条极具实用价值的经验规则，这些经验规则涵盖了大多数开发人员每天所面临的问题的解决方案。通过对Java平台设计专家所使用的技术的全面描述，揭示了应该做什么，不应该做什么才能产生清晰、健壮和高效的代码。

本书中的每条规则都以简短、独立的小文章形式出现，并通过例子代码加以进一步说明。本书内容全面，结构清晰，讲解详细。可作为技术人员的参考用书。

Effective JavaScript：编写高质量JavaScript代码的68个有效方法

作者：David Herman　ISBN：978-7-111-44623-1　定价：49.00元

作者将在JavaScript标准化委员会工作和实践的多年经验浓缩为极具实践指导意义的68个有效方法，深刻辨析JavaScript的特性和内部运作机制，以及编码中的陷阱和最佳实践。

Effective Objective-C 2.0：编写高质量iOS与OS X代码的52个有效方法

作者：Matt Galloway　ISBN：978-7-111-45129-7　定价：69.00元

本书是世界级C++开发大师Scott Meyers亲自担当顾问编辑的"Effective Software Development Series"系列丛书中的新作，Amazon全五星评价。从语法、接口与API设计、内存管理、框架等7大方面总结和探讨了Objective-C编程 中52个鲜为人知和容易被忽视的特性与陷阱。书中包含大量实用范例代码，为编写易于理解、便于维护、易于扩展和高效的Objective-C应用提供了解决方案。

推荐阅读

Java多线程编程核心技术

作者：高洪岩 书号：978-7-111-50206-7 定价：69.00元

内容亮点：

- 线程类的核心API的使用与关键技术点，掌握概念与学习路径。
- 并发访问控制技术，即如何写出线程安全的程序。
- 线程间通信技术，以提高CPU利用率和系统间的交互，增强对线程任务的把控与监督。
- Lock对象技术，以更好实现并发访问时的同步处理。
- 定时器类中的多线程技术，移动开发中使用较多，是计划/任务执行里很重要的技术点。
- 如何安全、正确地将单例模式与多线程技术相结合，避免实际应用中可能会出现的麻烦。
- 多个查疑补漏的技术案例，尽量做到不出现技术空白点。